Understanding Sustainable Development

Understanding Sustainable Development

Editor: Kane Harlow

RC CALLISTO
REFERENCE

www.callistoreference.com

Callisto Reference,
118-35 Queens Blvd., Suite 400,
Forest Hills, NY 11375, USA

Visit us on the World Wide Web at:
www.callistoreference.com

ISBN: 978-1-63239-968-7 (Hardback)

Cataloging-in-Publication Data

Understanding sustainable development / edited by Kane Harlow.
    p. cm.
Includes bibliographical references and index.
ISBN 978-1-63239-968-7
1. Sustainable development. 2. Sustainability. I. Harlow, Kane.
HC79.E5 U53 2018
338.927--dc23

Table of Contents

Preface

Our world is developing at a fast pace. The concept of sustainable development refers to the efficient utilization of natural resources. It covers the economic, social as well as environmental aspects of development. This book unfolds the innovative aspects of sustainable development which will be crucial for the progress of this field in the future. It presents researches and studies performed by experts across the globe. The text will serve as a reference to a broad spectrum of readers, including ecologists, environmentalists, economists, engineers, researchers, and students associated with this area of study.

The researches compiled throughout the book are authentic and of high quality, combining several disciplines and from very diverse regions from around the world. Drawing on the contributions of many researchers from diverse countries, the book's objective is to provide the readers with the latest achievements in the area of research. This book will surely be a source of knowledge to all interested and researching the field.

In the end, I would like to express my deep sense of gratitude to all the authors for meeting the set deadlines in completing and submitting their research chapters. I would also like to thank the publisher for the support offered to us throughout the course of the book. Finally, I extend my sincere thanks to my family for being a constant source of inspiration and encouragement.

Editor

An Assessment of Water Accessibility in the Kuvukiland Informal Settlement of Tsumeb in Namibia

Angula Nahas Enkono[1] & Alfons W Mosimane[2]

[1] Department of Geography, History and Environmental Studies, University of Namibia, Namibia

[2] Multidisciplinary Research Centre, University of Namibia, Namibia

Correspondence: Angula Nahas Enkono, Multidisciplinary Research Center, University of Namibia, Namibia. E-mail: nahasenkono@gmail.com

Abstract

Challenges of water supply in informal settlements have been observed in different parts of the world. This study evaluates accessibility to water in the Kuvukiland informal settlement. The study employed two methods: a semi structured questionnaire and in-depth interviews. Semi-structured questionnaire was used to collect the data from 50 respondents in the Kuvukiland informal settlement, and the in-depth interviews were carried out with five key informants. The findings suggest that access to water in informal settlements is a challenge, because more than half of the population in Kuvukiland live more than a kilometre from the water points. Further findings also show that affordability is a critical issue, because the more than half of the population are unemployed, and as a result they cannot afford to pay for water. Finally the findings are that, water supply in Kuvukiland does not follow an integrated water resource management approach. In addition, there is poor community involvement, and stakeholder participation is weak.

Keywords: water access, availability, affordability, community involvement, informal settlement

1. Introduction

Within today's paradigm of integrated water resource management, the management of water resources is seen as a process that aims at coordinated development and management of water and other natural resources (Global Water Partnership [GWP], 2009). The meaning of these words is reflected in the harsh realities faced by water users in drought-prone regions throughout the world. The distribution of scarce water resources therefore needs to be revised because, conflicts among water users easily erupt. Water is a critical factor for human livelihood and supply remains a serious problem throughout Namibia, as the country is considered to be one of the most arid countries in Southern Africa (Ruppel, 2013).

Many societies face enormous deficit in water supply and sanitation challenges. Currently, 900 million people around the world suffer from drinking water shortages and 2.6 billion people around the world live without safe water (Framework Programme Research for Sustainable Development [FONA], 2011). Bouwer (2000) argues that the global renewable water supply available is about 7,000 cubic meters per person per year, which means that there is enough water for at least three times the present world population. Therefore, water shortages are due to imbalances between population and precipitation distribution. The World Health Organisation (WHO) report of 2000 estimated that one sixth of humanity (1.1 billion people) lacked access to any form of improved water supply within one kilometre of their home (WHO, 2003). According to the Southern African Development Community (SADC) report, of the SADC region's total land area, only three per cent is humid; the rest of the land is moist and sub-humid (40%), dry sub-humid (19%), and semi-arid (16%). Therefore, the distribution, occurrence and availability of water resources are uneven in the region as well as in the individual countries because availability depends on rainfall (SADC, 1996).

In Namibia, water is increasingly becoming scarce, a fact compounded by two hard realities (Mendelsohn, el Obeid, & Roberts, 2002). The first is the general scarcity of water due to low, sparse and variable rainfall, coupled with high evaporation rates. The second is that a large number of people are concentrated in areas far from the major sources of water. Similarly, Schachtschneider (2000) states that Namibia is the driest country in sub-Saharan Africa, yet the concept of water demand management is not well established. The 2001 population and housing census revealed that more than half of the households have piped water within their compounds, while 35% get their water from public pipes and boreholes, and that urban households are relatively better off compared to rural households with regard to availability of piped water within their compounds (Republic of Namibia, 2003). Recent census, 2011 population and housing census reveal that, almost all urban households in Namibia have access to safe water (98%) in the form of piped water inside or outside their dwellings, or from public pipes or boreholes. 59% of rural households share the same privilege (Namibia Statistics Agency [NSA], 2011). In 2002, the Namibian Cabinet approved the National Water Policy White Paper that formed the basis for the Water Resource Management Act of 2004. The policy provides a framework for equitable, efficient and sustainable water resource management. This policy clearly states that water is an essential resource to life and that an adequate supply of safe drinking water is a basic human need (Ruppel, 2013).

This study evaluates accessibility to water in the Kuvukiland informal settlement. The study intended to inform, motivate and empower people in the Kuvukiland informal settlement at different levels of decision-making. Without providing information on water management activities and awareness of the need to conserve water to communities without water in the Kuvukiland informal settlement, this precious resource will not be sustainable. The Kuvukiland informal settlement in Tsumeb is continuously experiencing water challenges of access and availability.

The informal settlement is located on the outskirts of the town of Tsumeb, and the only access to water is through a pre-paid water meter which is shared by the community. It is on this basis that the study identified the need to assess the accessibility of water in the Kuvukiland informal settlement and evaluate the integrated water resource management in Tsumeb.

2. Integrated Water Resource Management

This study utilises the definition of integrated water resource management by the Global Water Partnership (GWP) Technical Advisory Committee (2000), which defines integrated water resource management as "a process that promotes the coordinated development and management of water, land, and related resources, in order to maximise the resultant economic and social welfare in an equitable manner, without compromising the sustainability of vital ecosystems". Viessman (1997) argues that integrated water resource management means putting all of the pieces together. Social, environmental and technical aspects must be considered. Issues of concern include: providing forums; reshaping planning process; coordinating land and water resources management; recognising water sources and water quality linkages; establishing protocol for watershed management; addressing institutional challenges; protecting and restoring natural systems; capturing society's views, educating and communicating; forming partnerships; and emphasising preventive measures.

Integrated water resource management is crucial in meeting and managing the increasing water demand in Namibia. Recent studies have shown that as part of that process, both water demand management measures and non-conventional water supply augmentation schemes are considerably cheaper than developing more traditional pipeline schemes (Biggs & Williams 2001). Grafton and Hussey (2011) argue that the driving force for the development of integrated water resource management comes from an awareness of the distinctive nature of the resource and its ubiquitous influence on human well-being and environmental sustainability. Grafton and Hussey further stress that our lives depend not only on how much water of what quality is available at any point in time and space, but crucially on what we choose to do with the resource. Therefore, integrated water resource management might be seen as a necessary measure to foil the increasing complexity of managing political economies and malfunctions as they attempt to manage challenges such as water supply and demand.

Viessman (1997) argues that there is a need for governance structures, from national to local and state agencies as well as other institutions, to strengthen or establish partnerships among themselves and with relevant publics. Such cooperative arrangements aid in conflict resolution, enhance efficiency in commitment of resources, and facilitate the identification of paths that complement or supplement each other's goals. This approach fosters learning, rather than opposing one another, because partnership is one way to bring about needed institutional reforms.

On the other hand, Soussan, Pollard, de Mendiguren, and Butterworth, (2004) argue that any discussion of water issues in contemporary South Africa must be set within the context of the existing dynamic to water laws, policies and institutional responsibilities. They further state that key aspects of the reform process are defining mechanisms to improve existing services and to allocate water to different stakeholders based on assessment of their minimum needs. For the domestic sphere, this reflects a definition of basic needs that assumes domestic water to be only about health and hygiene; this implies only water for cooking, washing, sanitation and drinking.

3. Water Supply in Informal Settlements

Muli (2013:10) defines informal settlement as: First, "Areas where groups of housing have been constructed on land that occupants have no legal claim on, or occupy illegally". Second, "Unplanned settlement and areas where housing is not in compliance with current planning and building regulations". Both definitions are emphasising the illegal character of informal settlements. From a more inclusive point of view, this paper adopts the definition by Mason and Fraser (1998, p. 313) who take the environmental, socio-economic and living conditions into account as they define informal settlement as: "….dense settlement comprising communities housed in self-constructed shelters under conditions of informal or traditional land tenure. They are common features of developing countries and are typically the product of an urgent need for shelter by the urban poor. As a result, they are characterised by a dense proliferation of small makeshift shelter built from diverse materials, by degradation of the local ecosystem, such as erosion and poor water quality and sanitation".

Challenges of water supply in informal settlements have been observed in different parts of the world. For example, municipal water supply in Karachi informal settlements in Pakistan has become grossly inadequate with regard to users' needs and expectations. Residential communities suffer from poor levels of service, and peri-urban locations, especially low-income settlements, have very limited access to municipal water supplies (Ahmed & Sohail, 2003). Informal settlements such as the one in Karachi's Orangi Township in Pakistan's inner city areas have increased in density, giving rise to acute water shortages. Squatters have sprung up in peri-urban areas, thus increasing the cost of piped water supply because of their distance from the existing water mains. Therefore, settlements located at the end of the network receive a very low level of services, since a large amount of water allocation has already been removed, legally or illegally. It is mostly the urban poor who reside on the peripheries who suffer from the water supply situation (Ahmed & Sohail, 2003).

In Namibia, Becker and Bergdolt (2001) argue that, as for potable water supply, the city of Windhoek aims to fully recover costs. Therefore, non-payment of accounts by residents results in the cutting off of the water supply. The introduction of a prepayment water management system is designed to effect substantial savings for bulk-water suppliers. Two types of payment metres are available: the yard connection for use at individual sites and the community standpoint for use at communal water supply points, mainly in informal settlements. Namibia's informal settlements, like many other international examples, are faced with challenges of water access, availability, affordability and community participation. The objective of this study was to evaluate accessibility to water in the Kuvukiland informal settlement.

4. Research Methodology

4.1 Study Site

The mining town of Tsumeb is located at 19° 15' S and 17° 42' E and it lies 1,320 m above sea level. Tsumeb is the capital of Oshikoto Region, located in north central part of Namibia (Figure 1). The climate of Tsumeb is semi-arid to arid, with an average annual rainfall of 524 mm. The rainy season is normally in summer, from December to February. The time between May and July is regarded as winter, with no or little rain. The mean annual temperature for Tsumeb is 25°C and the monthly temperature ranges between a mean maximum temperature of 26°C and a mean minimum temperature of 16°C throughout the year (Knesl, Konopasek, Kribek, Majer, Pasava, Kamona, Mapani, Mufenda, Ellmies, Hahn, Ettler, 2006).

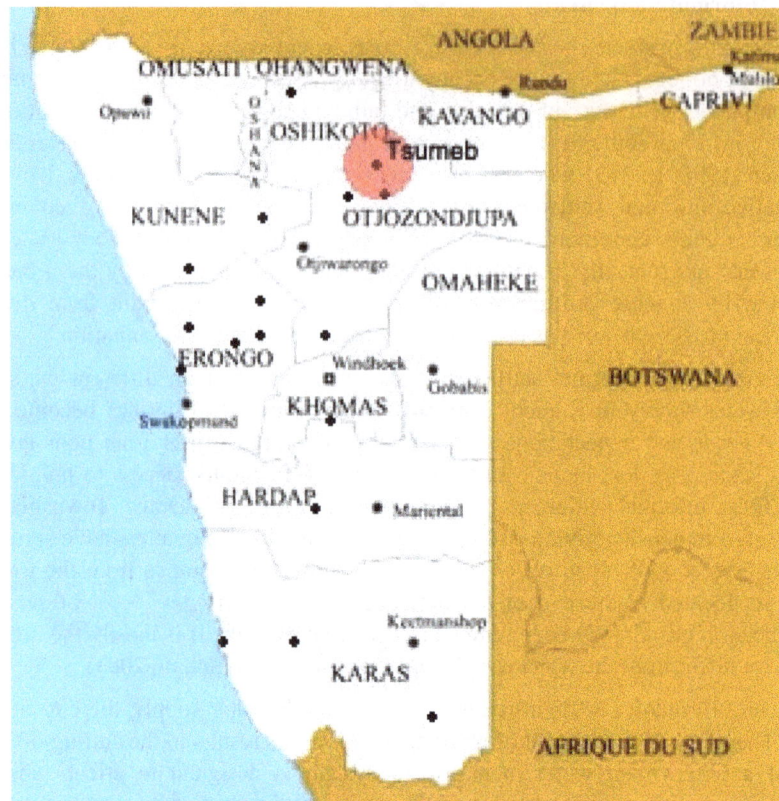

Figure 1. Map of Tsumeb (en.wikipedia.org)

According to Dierkes (2011), the Tsumeb water supply wells are drilled into Karstland with multi-aquifer systems. Tsumeb falls within the hydrological region that hosts eight major water supply schemes, of which Tsumeb's (108) waterworks are independent. Dierkes further states that until the early 1990s, the domestic water supply to the Tsumeb municipality was entirely dependent on the 2.5 cubic millimetres per year of groundwater supplied in Tsumeb and purified from the mine.

In Namibia, the need for forming institutions to manage water and other resources within the basin has been incorporated in the Water Resource Management Act of 2004 (Republic of Namibia, 2008). However, water supply remains a major challenge in Namibia, especially in the rural areas. Ruppel (2013) states that in order to organise the water supply, infrastructure has to be maintained, facilities have to be managed and fees are to be collected. The Ministry of Agriculture, Water and Forestry is the overall custodian in managing and regulating water resources in the country, with the prime objective of ensuring that water resources will be properly investigated and used on a sustainable basis to cater for the needs of the people and to sustain their environment (Republic of Namibia, 2008). Furthermore, local authorities and regional councils are responsible for the implementation of water supply and sanitation in the rural and urban settlements (proclaimed and un-proclaimed) where demand is continually increasing and a growing backlog exists. In addition, the Central Government should allocate capital resources wherever the Regional Council or local authority is unable to provide sanitation to the poor and marginalised, with strict standards that will be applied by the Directorate of Water Supply and Sanitation Coordination (Republic of Namibia, 2008).

The Kuvukiland informal settlement is located on the western part of Tsumeb, on the periphery of the town, with no services such as water, sanitation and electricity. The settlement is characterised by informal housing structures. The houses are made of makeshift corrugated iron sheets; some of the houses are built of cardboards. The informal settlement has a population of 3,300 people and over 350 households. Most houses are female-headed households, the majority of whom are unemployed and depend on pension hand-outs and wages from informal employment.

4.2 Research Methods

This study employed two methods: a semi structured questionnaire and in-depth interviews. The semi-structured questionnaire was used to collect the data from household respondents in the Kuvukiland informal settlement. 50 semi structure interviews were conducted and 5 (five) key informant questionnaires were collected. The open-ended questions allowed for the respondents to express their own opinions. Closed questions were formulated and the questions were easy to select. It was challenging to gather information from respondents who could not speak Oshiwambo, Afrikaans or English. A community member was used as a translator where there was a language barrier.

In-depth interviews were carried out with five key informants who included the Tsumeb local authority members, Kuvukiland informal settlement community leaders, and a representative from the office of the constituency counsellor. The key informants were brief and understood the questions. However, the main challenge was to schedule meetings with all the community leaders. The process of indexing was followed and it was helpful in generating the necessary information that was needed to meet the research objectives. All sets of scripts were analysed and those scripts with similar themes where grouped together, and the field notes were constantly compared to observe the significance of the notes to the objectives of the study. A Microsoft Excel spreadsheet was used to analyse data from the completed questionnaires through this process, graphs and tables were generated, only graphs that answer the questions of the study and those tables and graphs that help the research meets objectives were interpreted.

5. Results

5.1 Water Management and Supply in Tsumeb

The Tsumeb municipality, unlike other towns in Namibia that receive water from the Namibia Water Corporation Ltd. (NamWater), uses water from its own underground water aquifers. Water is pumped into the reservoirs for treatment and use for the town. The main water supplies in Tsumeb are from the Tupperware Dam, with a water holding capacity of 15,500 cubic meters. This reservoir supplies water directly to the town.

> "We are part of the sub-basin water management committee and this helps us to prolong the water aquifers" (Interviewee 2, 2013, July 9).

> "In an attempt to manage water, the town council encourages the community to use water sustainably" (Interviewee 2, 2013, July 9).

The Kuvukiland informal settlement does not have reliable and regular water supply like the formal settlement in Tsumeb. Water in Kuvukiland is supplied through a pre-paid water system; five standpipes were erected by the municipality. However, this does not address the issue of supply and access, since only two of the standpipes are working. The community members dig trenches and the municipality supplies them with water pipes to channel the water to different parts of the settlement. However the community feels that the standpipes should be improved for them to have reliable water supply.

> "We supply water to Kuvukiland informal settlement through a bulk water supply, which works like a pre-paid system" (Interviewee 5, 2013, July 9).

> "We as the community dig the trenches ourselves; what the municipality does is supply the water pipes and erect the standpipes for us. However, the standpipes are not reliable because they do not work in line with the card system; they overcharge us" (Interviewee 3, 2013, July 7).

Community leaders' claim that water supply in the Kuvukiland informal settlement is very poor; the residents sometimes harvest water from the roof during rainy seasons; long queues at the water points demonstrate slow release of water by the only operational standpipe; it can take up to 10 minutes to fill a 20-litre container of water. Further, water is sometimes supplied by sprinkler trucks, and such water is always brownish in colour.

> "During rainy seasons, residents harvest water from the roof of their iron zinc houses" (Interviewee 3, 2013, July 7).

> "Sometimes when the water situation reaches a critical point, they send sprinkler trucks filled with water, which is always brownish in colour. Water supply is very poor here and the municipality does not really look after the water in Kuvukiland" (Interviewee 3, 2013, July 7).

5.2 Community Perceptions of Water Situation in Kuvukiland

There are only five water points in Kuvukiland, and only two of the five water points were working during the interviews. The main point of water collection is the community standpipe (Figure 2).

Figure 1. Community water collection point

(Picture by AN Enkono, 2013)

The lack of water availability at the two water points has led the community members to collect water from the nearby surroundings, particularly from the Soweto public toilet. Most of the time, there is no water in the community and residents stand in long queues at water collection points.

> *"Water is rarely available here; we sleep standing in queues at the water point just to collect water"* (Interviewee 5, 2013, July 3).

> *"Water is scarce in this community, most of the time we collect water from the public toilet in Soweto"* (Interviewee 6, 2013, July, 5).

> *"Water is scarce here; we sleep at the water point just so we can get tap water. Most of the time, we collect our water from public toilets in Soweto."* (Interviewee 3, 2013, July 7).

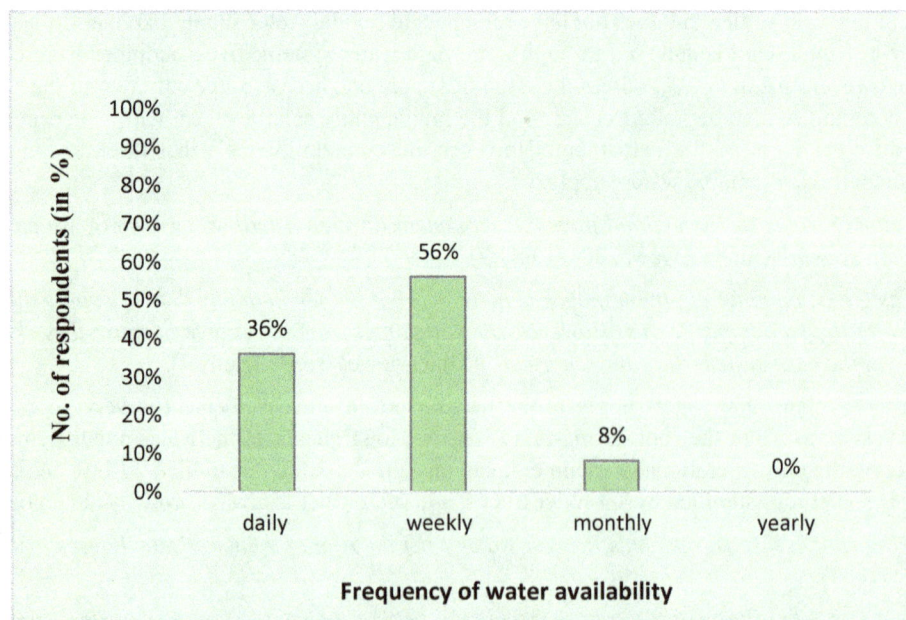

Figure 2. Access to water

As shown in Figure 3 above, most respondents claimed that water was only available once a week. However, a few respondents claimed that they had access to water on a daily basis. The divided opinions could be attributed to the fact that the community standpipes were broken most of the time, and only worked on some days. On the other hand, the majority of the respondents said that they lived further away from the water point and only collected water once a week.

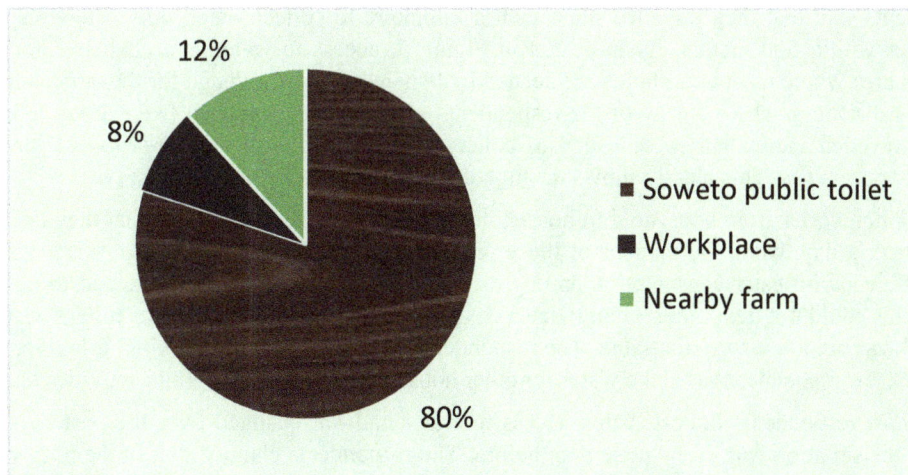

Figure 3. Alternative sources of water

Most respondents indicated that they collected water from public toilets in Soweto as an alternative source. Soweto is a semi-formalised location, which means that some parts of Soweto are formalised while a small part is informal. In Soweto, however, there are public toilets fitted with showers that are accessible to everyone. The residents do not have to pay for the water from the public toilets. However, the small part of Soweto that is not formalised also makes use of the toilets to collect water when they experience water supply problems. Some respondents said that they collected water from their workplaces, and those respondents that lived close to a private farm collected water from the nearby farm.

5.3 Distance from Water Collection Points

Despite the water collection points in the community, a larger portion of the community live far from the water source. Therefore, community members travel long distances to get water. The majority of the respondents said that they lived a long distance from Soweto, which is an alternative source of water collection, where water is collected from public toilets.

> *"The standpipes are far from our houses, they are not spread out into the community, we want them to extend the standpipes, because the distance is too much for us"* (Interviewee 6, 2013, July 5).

> *"Soweto location is very far for us and sometimes we just go without water, because we cannot afford to travel the long distance every day, especially for us women"* (Interviewee 5, 2013, July 3).

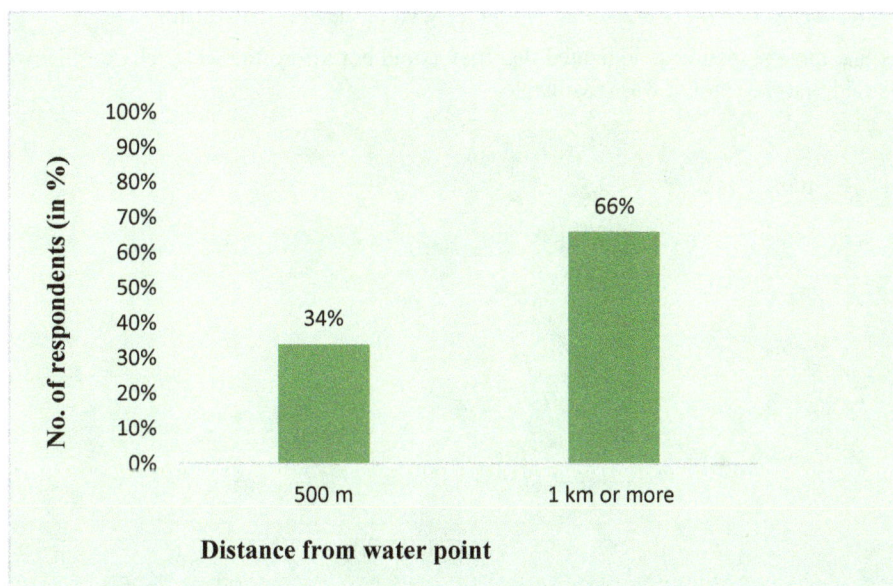

Figure 4. Distance travelled by respondents to access water

Most respondents said that they travelled more than a kilometre to collect water, only a few respondents had access to water within 500 metres. As indicated in Figure 5, access to water is a challenge for most of the residents in an area where most households are headed by females. It is a challenge for the women to carry water over such long distances. The majority of the respondents said that they travelled twice a day to collect water, while a few travelled more than twice a day to collect water, and only a small portion of the respondents collected water once a day. This clearly shows the high demand for water in the community.

Water consumption varies from household to household. Most of the respondents said that they used as much as 75 litres of water a day, while the number of those who used 50 litres and 100 litres of water was almost the same. The difference in water consumption among different households could be attributed to distance, where some households could not travel more than twice a day for more than one kilometre to collect water. Water use per household was predominantly the same. The respondents used the water for cooking, drinking, washing and bathing. Only a few respondents used the water for other household and non-household activities not listed.

Meanwhile, most respondents indicated that access to water had not changed over the past two years; some indicated that the situation was even worse than before. The respondents claimed that in the past, water was for free, but now the community had to pay for water. Therefore, they claimed that the improvement made was not significant. Even though efforts were made to increase the community standpipes, the standpipes are not evenly distributed in the community and of the five standpipes erected only two standpipes are working.

5.4 Water Quality and Affordability

The general view of the respondents was that when there was water shortage, the municipality sent a sprinkler truck filled with water to the community. However, the water was always brownish in colour and it looked dirty. Sometimes the children got sick from drinking the water.

> *"We cannot say the water is clean, because the municipality brings us dirty water, carried by trucks"* (Interviewee 5, 2013, July 3).

> *"Our kids get infected by the water they bring by trucks, and they usually suffer from diarrhoea"* (Interviewee 3, 2013, July 7).

To access water, one must buy a pre-paid card which costs US $18.00. To collect water using the card system, the individual community members load money on the card. The amount of money loaded on the card depends on what an individual can afford. The majority of the residents are not employed. Therefore, water affordability is a major challenge.

> *"Buying the pre-paid [water] is a huge challenge; we cannot afford the N$180.00 paid for the card"* (Interviewee 7, 2013, July 5).

> *"The prepaid water system is not working for us, the standpipes are ever broken and they charge us a lot. The prepaid card does not work well for us. Most of us have no jobs. How can we afford the variation charges that these machines charge us?"* (Interviewee 5, 2013, July 7).

Figure 6 shows that most respondents indicated that they could not afford the water charges. However, a few of the respondents said that the amount was reasonable.

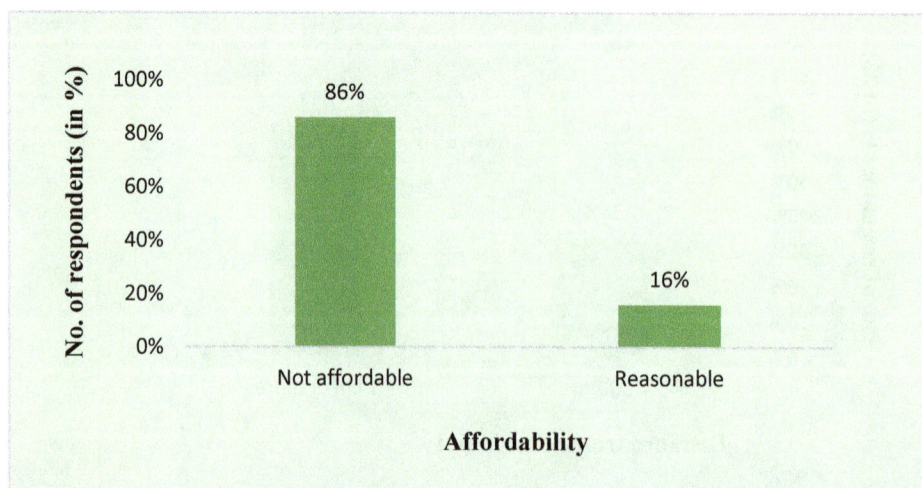

Figure 5. Perceptions of the respondents regarding water charges

5.5 Community Involvement and Stakeholders' Participation

There seems to be a lack of community involvement in making decisions that affect the community. The perception of the community is that, even though community meetings are held and they are requested to express their views, their views hardly make any difference and water supply and access challenges are hardly addressed.

> "Even if we, the community leaders, attend meetings on behalf of the community to air our views, people only listen but they do nothing about it"(Interviewee 5, 2013, July 7).

> "We do not see anything positive happening here. We are never involved in the final decisions that are made, and we want to see our views in the implementation of policies" (Interviewee 6, 2013, July 5).

Table 1. Community participation

Type of participation	No. of households respondents	Percent
Through public debate	18	36%
At organised committee meetings	32	64%
Total	**50**	**100%**

Table 1 shows that the respondents did participate in platforms that addressed water crisis issues. Most of the respondents participated at organised community meetings; however, some preferred to participate at public debates. Most respondents indicated that they were involved in decision making in most cases; however, a fair number of respondents claimed that they were less involved.

There is a reasonable degree of residents' participation in addressing water supply problems. The study found that the community members are willing to participate through community projects. The most tangible project involving the community is digging trenches to channel water to the community.

> "The residents here are very active; they always attend community meetings and they are keen to address the water issue" (Interviewee 3, 2013, July 7).

> "We dig the trenches ourselves and the municipality provides us with the water pipes" (Interviewee 6, 2013, July 5).

The views of the community leaders show that decision-making regarding addressing water problems is very poor. Therefore, the issue of water problems in the Kuvukiland informal settlement is a result of poor planning and decisions. There are quarterly meetings where residents are represented by their community leaders, and complaints are registered at stakeholders' meetings. However, community leaders feel that despite the quarterly meetings, the final decisions are not reflected in the planning process.

> "Local leaders believe that incorporated decision-making will help to solve the problem of the Kuvukiland informal settlement" (Interviewee 3, 2013, July 7).

> "When it comes to the issue of water, decisions with regard to water are very poor and they are less integrated, because we hardly see our views visible in the planning process. We need a co-ordinated effort; one that fits all parties" (Interviewee 3, 2013, July 7).

Regular stakeholder consultations have yielded some positive results. The poor people are allowed to build their low cost housing and engage in informal trading. Previously, the Council had decided against informal settlement, but with stakeholders' participation, informal development is now recognised and is made part of the economic setup of the town. The Kuvukiland informal settlement is of an economic value to the town, because they play a major part in the town's economy.

> "We have community meetings organised through the office of the Mayor, and the residents have committee members who represent them at these meetings. We also have quarterly meetings with the residents so that we listen to their complaints" (Interviewee 2, 2013, July 5).

> "We have approved a mixed development, to allow the poor to build their shacks. However, the challenges we have with the community is that the people are resistant to change" (Interviewee 2, 2013, July 5).

The study found that there are policy related challenges. As a result, the management of resources is not clearly addressed. There are no policies that govern the maintenance of the facilities. Therefore, lack of awareness has resulted in resistance from the community to change. One of the challenges that the town is faced with is to formalise Kuvukiland, because if the informal settlement is not formalised, the municipality is not supposed to supply water to the settlement. However, the municipality works in accordance with the Local Authority Amendment Act of 2002, which means that they are supposed to supply water to every residence. The most notable attempt to address policy related challenges is through committee meetings, but the challenge is that most members of the community are resistant to change, as they find it hard to adopt to new ways, unlike in the past where community used to get water for free, however now that water has become a commodity, most members of the community cannot afford to pay for the water.

> *"The policy related challenges that the town faces are from the town's residents. People are resistant to change. This is because there is lack of awareness* (Interviewee 2, 2013, July 9).

> *"We need to alert the community on the management of resources. There is still a lot of awareness to be done in order to enforce policies"* (Interviewee 5, 2013, July 9).

> *"We have challenges of water supply to Kuvukiland because the town is not formalised, so we are not supposed to supply water to the informal settlement. We also work in accordance with the Local Authority Act and, according to the Act, we are supposed to supply water to every residence"* (Interviewee 2, 2013, July 5).

> *"There are no rules on water supply; the only rule is pay for water and get water"* (Interviewee 5, 2013, July 9).

There seems to be lack of infrastructural development and social support from the local authority. The community expressed interest in working together with the local authorities to address the issues related to water. To a certain degree, the perception of the community is that the municipality has not shown interest in the community. Water is clearly a major concern for the community, and they find the municipality to be too relaxed on the issue of water supply in the area.

> *"The local community feels that there is a lack of interest from the local authorities with regard to addressing water issues and incorporating decisions"* (Interviewee 3, 2013, July 7).

> *"The municipality does not look after the water in Kuvukiland, and most of the time we have no water to sustain ourselves"* (Interviewee 3, 2013, July 7).

6. Discussion

6.1 Water Supply in Tsumeb

Integrated water resource management works best when all aspects of governance are accorded recognition, which addresses issues of socio-economic characteristics of the people, technical as well as environmental aspects, thereby leading to sound decisions that include all members of the society. Viessman (1997) similarly argues that integrated water resource management should include social, environmental and technical aspects for it to work. This view is further supported by Grafton and Hussey (2011), who content that the driving force for the development of integrated water resource management comes from an awareness of the distinctive nature of the resource and its ubiquitous influence on human well-being and environmental sustainability. However, guided by the definitions of integrated water resources management provided above, this research found the concept of integrated water resource management in Tsumeb to be weak. The implementation of the integrated water resource management approach in Tsumeb lacks the recognition of the community's views, hence the challenges of addressing water supply in Kuvukiland. Communication flow in Kuvukiland is poor due to poor education, which results in lack of awareness by the community. Therefore, water supply in Kuvukiland does not follow the integrated water resource management approach.

6.2 Water Supply in Kuvukiland

Water supply in Kuvukiland is through bulk water supply, where the pre-paid water system is used. Similar pre-paid water systems are used in other urban areas in Namibia, such as Windhoek, Swakopmund and Walvis Bay. Standpipes are erected in the community which serve as water collection points, but the standpipes cannot supply enough water for the community because they are broken most of the time. In addition, the water collection points are not evenly distributed across the informal settlement where houses are sparsely located. However, the Government of Namibia's position statement is that water supply in urban areas (all non-farming areas) where people reside on a permanent basis, such as in Kuvukiland, should be approached in the same way as water supply in the formal municipal areas (Republic of Namibia, 2008).

Water supply to informal settlement is a global challenge. Ahmed and Sohail (2003) found that municipal water supply in informal settlements has become grossly inadequate with regard to users' needs and expectations. The findings in this research suggest that there is a serious water supply problem in the Kuvukiland informal settlement. Ahmed and Sohail's findings (2003) give guidance to this research to conclude that issues of water supply to informal settlements are very common elsewhere in the world and the water supply problem in Kuvukiland is one of those issues that need a holistic approach to address. Generally, informal settlements experience poor services because they are not formalised and most of the residence do not legally occupy the land. In addition, these are areas where the poor people live and they cannot afford to pay for the services. Indongo, Angombe and Nickanor (2013) point out that in most developing countries, over 30% of the urban population are living in slums or informal settlements, where vacant state-owned land or private land is occupied illegally and is used for illegal informal housing.

6.3 Water Availability and Access

Mendelsohn (2002) found that lack of access to water is attributed to the fact that a large number of people in Namibia are concentrated in areas far from the major sources of water. In Kuvukiland, most community members do not have access to water, and when water is available, it is a long distance away from their homes. There is an observed difference in water consumption per household. This difference is determined by the distance that the residents walk to reach the access water access points. Residents closer to the water source are able to collect more water for consumption, compared to those that live a distance from the water source.

In South Africa, the Government has pledged to try and fulfil the millennium development goals by ensuring that all households have access to clean and safe running water by 2014. This will address the issues related to water shortages (Farrar, 1994). Similar attempts have been made in Namibia, where the Cabinet has endorsed a national water policy that includes decentralisation, with the main emphasis being to decentralise the resource to the lowest level, while at the same time managing the resource at basin level (Ruppel, 2013). The national water policy therefore needs to be more effective in addressing access to water supply in informal settlements across Namibia and, in particular, the policy must be applied in practice to improve the issues of water availability for the community, including the Kuvukiland informal settlement.

6.4 Water Affordability

Water affordability in Kuvukiland is hampered by the Tsumeb municipality's focus on trying to recover the cost of water supply while the majority of the informal settlement residents cannot afford to pay for the services. Becker and Bergdolt (2000) made similar findings in the city of Windhoek where non-payment of accounts by residents resulted in the cutting off of the water supply. These trends can be observed in informal settlements across the towns in Namibia. This is a policy related challenge that hinders the provision of safe affordable and continuous water supply to the Kuvukiland informal settlement, where more than half the population are unemployed, and a situation similar across informal settlements in Namibia. According to Soussan, Pollard and de Mendiguren (2002), these policy related challenges happen when institutions such as the Tsumeb municipality, which are responsible for implementation, are ineffective.

Two reasons why water in Kuvukiland is not affordable are: Firstly, more than half of community are unemployed and they have no source of income, making it impossible to afford the pre-paid card of N$180.00 from the municipality. Secondly, there is no standard fixed price for water per collection. The charges for water by the standpipes are not consistent and charges differ per individual collecting water at the water point. Therefore, the community cannot afford the different amounts that the standpipes charge them. The findings have shown that the standpipes are not well maintained and, as a result, they malfunction and draw inappropriate fees from the card. Therefore, issues of water supply should be addressed both from a social as well as from an economic point of view to ensure that those who cannot afford have equal access to water supply to meet their basic needs.

Soussan, Pollard and de Mendiguren (2002) contend that the discussion of water issues must be set within the context of the existing dynamic changes to water laws, policies and institutional responsibilities. This study found that there are policy related challenges, where water laws and policies in Tsumeb need to be more inclusive of informal settlement residents so that water supply is accessible to the community. The fact that Kuvukiland is not formalised is a policy related challenge, which hinders water supply to the informal settlements across Namibia.

6.5 Community Involvement

Results indicate that partnership between the local authorities and the local community of Kuvukiland is weak. This is contrary to the findings of Desert Research Foundation of Namibia [DRFN], (2005) that generally, communities are interested in being involved in the management of their natural resources.

The findings are that community involvement is very strong. The community, through their leaders, hold regular meetings and amongst the issues they discuss are water shortages. In addition, it is also observed that the community participates actively in discussions. The community leaders represent the community at local authority meetings but, despite the regular meetings, implementation and planning by the Tsumeb municipality seldom includes the views of the community in decision making. However, the results show that the community is allowed to participate in activities such as digging trenches and the municipality provides them with pipes and with the erecting of the standpipes.

There was an assumption by the national government, that basin management committees will provide an opportunity for the Government and communities to work together to ensure that integrated water basin management is achieved (Republic of Namibia, 2000). On the contrary, this research found that decisions with regard to addressing water supply and shortages are very poor. The main principle of integrated water resource management is good governance, and the characteristics of good governance include good communication flow, consultation, stakeholder participation in decision-making and most importantly the political will (Viessman, 1997). The Tsumeb local authority governance structure does not incorporate these characteristics on purpose in its attempt to address the water supply problems in Kuvukiland. Thus, issues relating to water supply and shortage in the Kuvukiland informal settlement reflect a result of poor planning. The study therefore found that the views of the stakeholders are not reflected in the planning process and this underlines the fact that there is poor community involvement.

7. Conclusion

The study concludes that water supply in informal settlements is characterised by challenges of access to water, distance to water points, and availability and affordability of water to the resident communities. To improve good water governance, the Tsumeb municipality needs to improve the distribution of water points in the community and develop a water pricing policy for the poor and pensioners in the Kuvukiland informal settlement. Water supply to informal settlements in Namibia needs a holistic approach supported by policies that address issues of access, availability, affordability, and community involvement. Improvement of water supply in informal settlements should be a development priority in developing countries. Future research should focus on understanding social, and economic challenges in informal settlements that affect residents' quality of life. This study provides a foundation for similar studies in Namibia in order to contribute to the body of knowledge as well as provide evidence-based information for national development.

References

Ahmed, N., & Sohail, M. (2003). Alternate water supply arrangements in peri-urban localities: awami (people's) tanks in Orangi Township, Karachi. *Environment and Urbanization, 15*(2), 33-42.

Becker, F., & Bergdolt, A. (2001). *"Managing urban sprawl in Windhoek: Problem, formations of potable water supply and housing development."* In Proceedings of the International Symposium on government, governance and urban territories in Southern Africa, edited by Mulenga MC. Pages 135-54. Lusaka, Zambia: University of Zambia.

Biggs, D., & Williams, R. (2001). A case study of integrated water resource management in Windhoek, Namibia. *Frontiers in urban water management: Deadlock or hope?* Technical Documents in Hydrology. No 45. Paris: UNESCO.

Bouwer, H. (2000). Integrated water management: emerging issues and challenges. *Agricultural water management, 45*(3), 217-228. http://dx.doi.org/10.1016/S0378-3774(00)00092-5

Desert Research Foundation of Namibia. (2005). *Basin Management Approach: A Guidebook by DRFN.* Windhoek: Desert Research Foundation of Namibia.

Dierkes, K. (2011). *Water supply of the Etosha National Park.* Regensburg: University of Regensburg.

Farrar, L. (1994). *Is South Africa's free basic water policy working? A quantitative analysis of the effectiveness of the policy implementation.* Cape Town: University of Cape Town.

FONA. (2011). *Integrated Water Resources Management: From Research to Implementation.* Leipzig: Helmholtz Centre for Environmental Research.

Global Water Partnership. (2009). *the GWP Strategy 2009–2013.* London: Scriptoria.

Grafton, Q. R., & Hussey, K. (2011). *Water resource planning and management: Urban water supply and management.* Cambridge, UK: Cambridge University Press. http://dx.doi.org/10.1017/CBO9780511974304

Indongo, N., Angombe, S., & Nickanor, N. (2013). *Urbanisation in Namibia: Views from semi-formal and informal settlements.* Windhoek: University of Namibia.

Knesl, I., Konopasek, J., Kribek, B., Majer, V., Pasava, J., Kamona, F., ... Ettler, V. (2006). Assessment of the mining and processing of ores on the environment in mining districts of Namibia: Kombat, Berg Aukas and Kaokoland. Czech Geological Survey.

Kundell, J. E. (2007). University water resources research and state legislature. *Journal of the American Water Resources Association, 22*(5), 785–789. http://dx.doi.org/10.1111/j.1752-1688.1986.tb00752.x

Mason, S. O., & Fraser, C. S. (1998). Image sources for informal settlement management. *Photogrammetric Record, 16*(92), 313-330. http://dx.doi.org/10.1111/0031-868X.00128

Mendelsohn, J. M., el Obeid, S., & Roberts, C. S. (2002). *A profile of north-central Namibia.* Windhoek: Gamsberg Macmillan.

Muli, E. M. (2013). *Change detection of informal settlements using remote sensing and Geographic Information Systems: case of Kawangware, Nairobi* (Doctoral dissertation, University of Nairobi).

Namibia Statistics Agency (NSA). (2011). Namibia 2011 Population and Housing Census Main Report.

Republic of Namibia. (2000). *National water policy white paper.* Windhoek: Ministry of Agriculture, Water and Forestry.

Republic of Namibia. (2003). *2001 Population and Housing Census.* Windhoek: Central Bureau of Statistics (CBS) - National Planning Commission, Government of Republic of Namibia.

Republic of Namibia. (2008). *Water supply and sanitation.* Windhoek: Ministry of Agriculture, Water and Forestry.

Ruppel, C. O. (2013). *Water management: Environmental law and policy in Namibia.* Windhoek: Orumbonde Press.

Schachtschneider, K. (2000). *Water demand management study of Namibian tourist facilities.* Windhoek: Ministry of Agriculture, Water and Rural Development.

Soussan, J., Pollard, S., de Mendiguren, J. C. P., & Butterworth, J. (2004). Allocating water for home-based productive activities in Bushbuckridge, South Africa. In: Asia Development Bank (2004). *Water and Poverty: The realities experiences from the field* (pp. 139-151). Asia Development Bank. Water for all Publication Series.

Southern African Development Community (SADC). (1996). *Water in Southern Africa.* Windhoek: SADC.

Viessman, W. (1997). *Integrated water resource management.* Florida, USA: University of Florida.

World Health Organisation (WHO). (2003). *Domestic water quantity, services, levels and health.* Geneva, Switzerland: WHO.

2

Effects of Fertilization Rate and Water Availability on Peanut Growth and Yield in Senegal (West Africa)

Babacar Faye[1,3], Heidi Webber[2], Thomas Gaiser[2], Mbaye Diop[3], Joshua D. Owusu-Sekyere[4] & Jesse B. Naab[5]

[1] West African Science Service Center on Climate Change and Adapted Land Use (WASCAL), School of Agriculture, University of Cape Coast, Ghana

[2] Crop Science Group, Institute of Crop Science and Resource Conservation (INRES), University of Bonn, Katzenburgweg 5, D-53115 Bonn, Germany

[3] Institut Sénégalais de Recherches Agricoles (ISRA), Bambey BP: 211, Senegal

[4] Department of Agricultural Engineering, School of Agriculture, University of Cape Coast, Cape Coast, Ghana

[5] West African Science Service Center on Climate Change and Adapted Land Use (WASCAL) Competence Center, Ouagadougou, Burkina Faso

Correspondence: Babacar Faye, School of Agriculture, University of Cape Coast, P. O. Box 5007, Ghana; Institut Sénégalais de Recherches Agricoles (ISRA), Bambey BP: 211, Senegal.
E-mail: bbcrfy@yahoo.fr

Abstract

The effects of fertilization rate and water availability on peanut growth and yield of two cultivars were investigated in a series of field experiments at Bambey, Nioro and Sinthiou Malem in Senegal. Both rainy and dry season experiments were conducted over two years between 2014 and 2015, for a total of seven experiments. The first set of four experiments were to evaluate fertilizer application rate on peanut production. One experiment was conducted in the dry season 2014 in Nioro with four levels of fertilizer and one experiment in the rainy season 2014 in each of Bambey, Nioro and Sinthiou Malem with six levels of fertilizer in a RCBD with four replications both. The second set of experiments were to evaluate the effect of different water regimes on peanut production. Experiments were conducted in the dry season of 2014 and 2015 in Bambey and in Nioro 2015. The experimental design was a split plot design with four replications and three levels of water, namely, E, S1 and S2. The effects of fertilization rate on peanut in three different sites were not significantly different between fertilizer levels. However, irrigation treatments were significantly different in all sites during the two years. Under water stressed conditions, the seed yield was more affected than the biomass yield. Seed yield decreased by 33% when stress occurred at flowering period and by 50% when stress occurred during seed filling. The most sensitive period for yield declined was observed during the period of maturation followed to the flowering stage. The interaction between irrigation and fertilizer was not signification in both Bambey and Nioro sites of field experiments. Such experiments should be conducted in field based conditions where occur limited soil nutrients to test higher dose of NPK.

Keywords: water stress, irrigation, fertilizer, peanut, Senegal

1. Introduction

Peanut (*Arachis hypogaea L.*) is one of the world's most important legumes, grown primarily for its high quality edible oil and protein (Kambiranda et al., 2011). It is cultivated in over 100 nations around the world with the main producers being China and India, with more than 60% of total production while Africa has 25% of the production (Noba et al., 2014). Most of the production is domestically used and only small proportion of the world production is devoted to imports and exports, therefore, the world trade market can be considered as a residual market (Revoredo and Fletcher, 2002). The proportion of peanut used for food purposes increases compare to the proportion used to produce vegetable oil. Africa is the more affected for this changes due to the lower quality of the production which contained an important level of aflatoxin (Bankole and Adebanjo, 2004; Martin et al., 1999). The contamination of peanut to aflatoxin in Africa is also due to increase drought (Kambiranda et al., 2011). West African countries are the main producer of peanut in Africa where Nigeria and

Senegal occupied the first place followed by Mali and Niger (Singh et al., 2013).

Peanut is the country's primary industrial crop and constitutes the principal source of agricultural incomes for the majority of farmers in Senegal (Pélissier, 1966).

However, since the 1990s, peanut value chains has entered a deep crisis and various agricultural policies have yet to succeed in boosting the sector (Freud. et al., 1997).

A reduction of 25.8% of the production is observed when we compare the mean from 1972 to 1975 with the mean from 1996 to 2000 (Revoredo and Fletcher, 2002).

The decline in peanut production Senegal is mainly due to climate variability and lack of input supply and specially low quality of seeds (Foncéka, 2010). Other factors that have led to the decline of production are soil degradation, reduction of cultivated area, bad agriculture practices, lack of and poor maintenance of agricultural machinery and difficult access to credit (Gaye, 2013; Montfort, 2005).

The effect of drought stress on peanut has been an important subject investigation and was used to support development of adapted varieties (Gautreau, 1982). It depends of the duration of the drought stress, on the stage of peanut growth and the intensity of the stress. While the application of fertilizer on peanut result in enhanced normally the yield up to certain level but can have negative effect on yield if the application is not reasoned mainly for over applications and the time of applications.

The effects of drought stress is known to be more drastic on peanut when occurred during the reproductive phases than the vegetative phases (Hemalatha et al., 2013; Jongrungklang et al., 2013). Drought stress in pre-flowering has no effect in peanut yield. While the greatest reduction in kernel yield due to drought stress occurred during the seed filling phase whereas an increase of pod yield by at least 13% occurred when drought stress is imposed during the early phase (Rao et al., 1985; Stirling et al., 1989). Above ground biomass is affected negatively whenever the drought stress imposed, however, the leaf area of index of peanut stressed during the reproductive phase was lower than the leaf area index of peanut during the vegetative phase (Nautiyal et al., 2002). Drought at the end of the season affected more pod yield than number of mature pods which is most sensitive to mid season drought (Kenchanagoudar et al., 2002).

The mechanisms of physiological adaptation to drought were also investigated on peanut in Senegal which aim to select adapted peanut varieties (Annerose, 1988; Clavel et al., 2007; Clavel et al., 2005) and to identify the sensitivity of peanut growth to drought (Annerose, 1985; Annerose, 1990).

A part drought and high temperature stress, nutrient deficiencies mainly N, P and K caused significant yield losses in semiarid regions but they are lower than for most other crops with a general requirement of 20kg N/ha, 50-80kg P/ha, and 30-40kg K/ha (Prasad et al., 2010). Phosphorus is the most important nutrient for peanut in semiarid zone (Naab et al., 2015). In most case in Africa, peanut is grown in soil with phosphorus deficiency which is a limited yield factors under on-farm conditions (Naab et al., 2009; Ogeh and Oyibo, 2015).

No studies were identified that were conducted in Senegal to assess the effect of drought stress in the different stage of peanut growth associated to the response of different fertilizers application.

The aim of this study is to assess the effects of fertilizer response and water stress on peanut development, growth and yield in Senegal. To achieve this aim, field experiments were conducted in three different part in Senegal to (i) to evaluate the most sensitive period of peanut to water stress on yield reduction and (ii) to determine the effect of mineral fertilizer rate on peanut in Senegal.

2. Materials and Methods

2.1 Study Sites

The study was conducted in Senegal in three different agro-climatic zones. All trials are conducted at the Senegalese Institute of Agricultural Research (ISRA) sites (Figure 1). The first site was Bambey research station located at, 14°42' N and 16°29' W. The second site was Nioro research station located at 13°45' N and 15°46' N. The third site was the Sinthiou Malem research station located at 13°49' N and 13°54' W. Bambey and Nioro belong to the Senegal's Peanut Basin zone which constitutes the country's most important peanut production area.

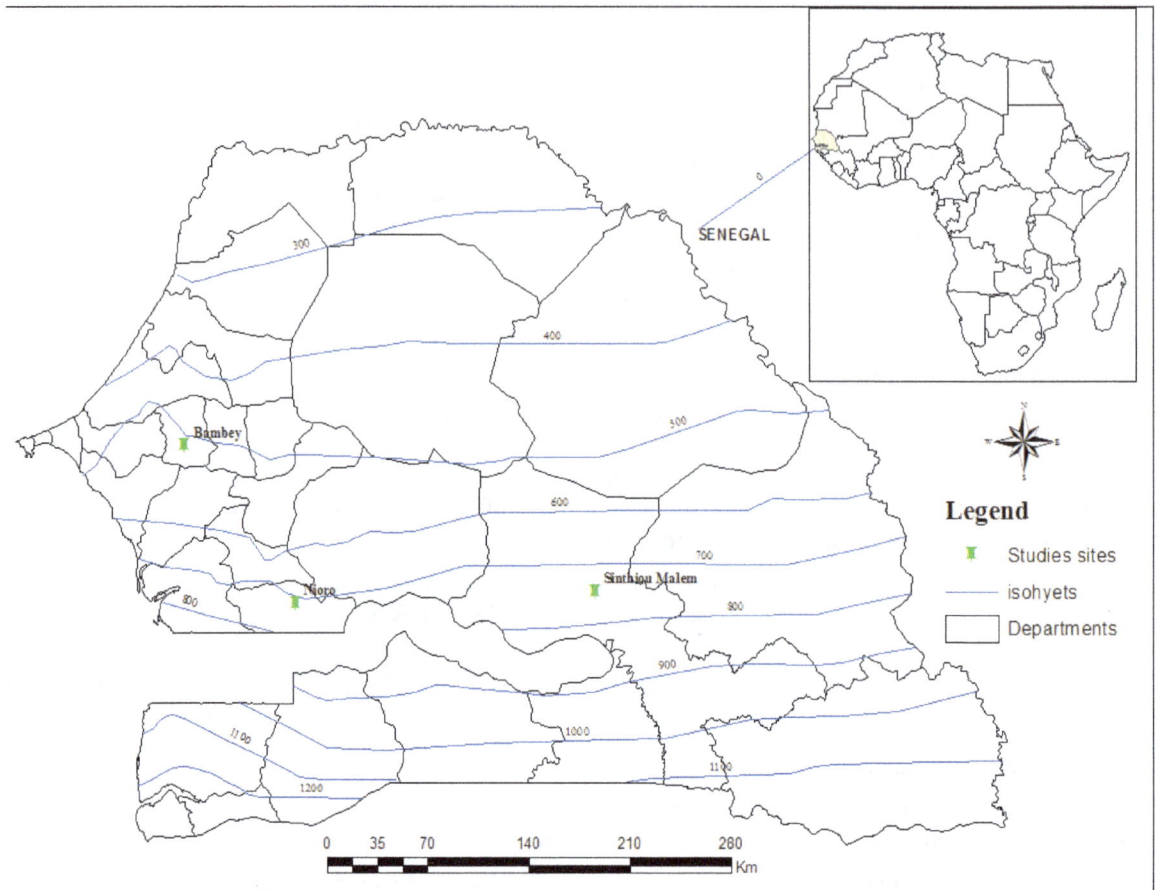

Figure 1. Study Area, normal rainfall: period 1961-1990 mm

The zone is characterized by a Sudano-Sahelian climate (Bambey) to a Sudanian climate (Nioro and Sinthiou Malem) where the variation in rainfall ranges between 400 to 800mm per year (Ganry and Gueye, 1992).

The rainy season is uni-modal and the rain is mostly concentrated in three to four months between June to September. The highest amount of rainfall is recorded in August (Figure 2). The minimum temperature varies between 18 to 20°C in December to January, while the maximum temperature varies between 40 to 42°C and occurs in April to May with a maximum value of 42.6°C in Sinthiou Malem on April, the hottest month in all sites. Maximum solar radiation occurs in April (25 MJ m^{-2} day^{-1}) and the minimum in December (15 MJ m^{-2} day^{-1}). The lowest mean relative humidity is observed during the dry season and varies between 25 to 30%, in contrast during the rainy season it varies between 70 to 80 %.

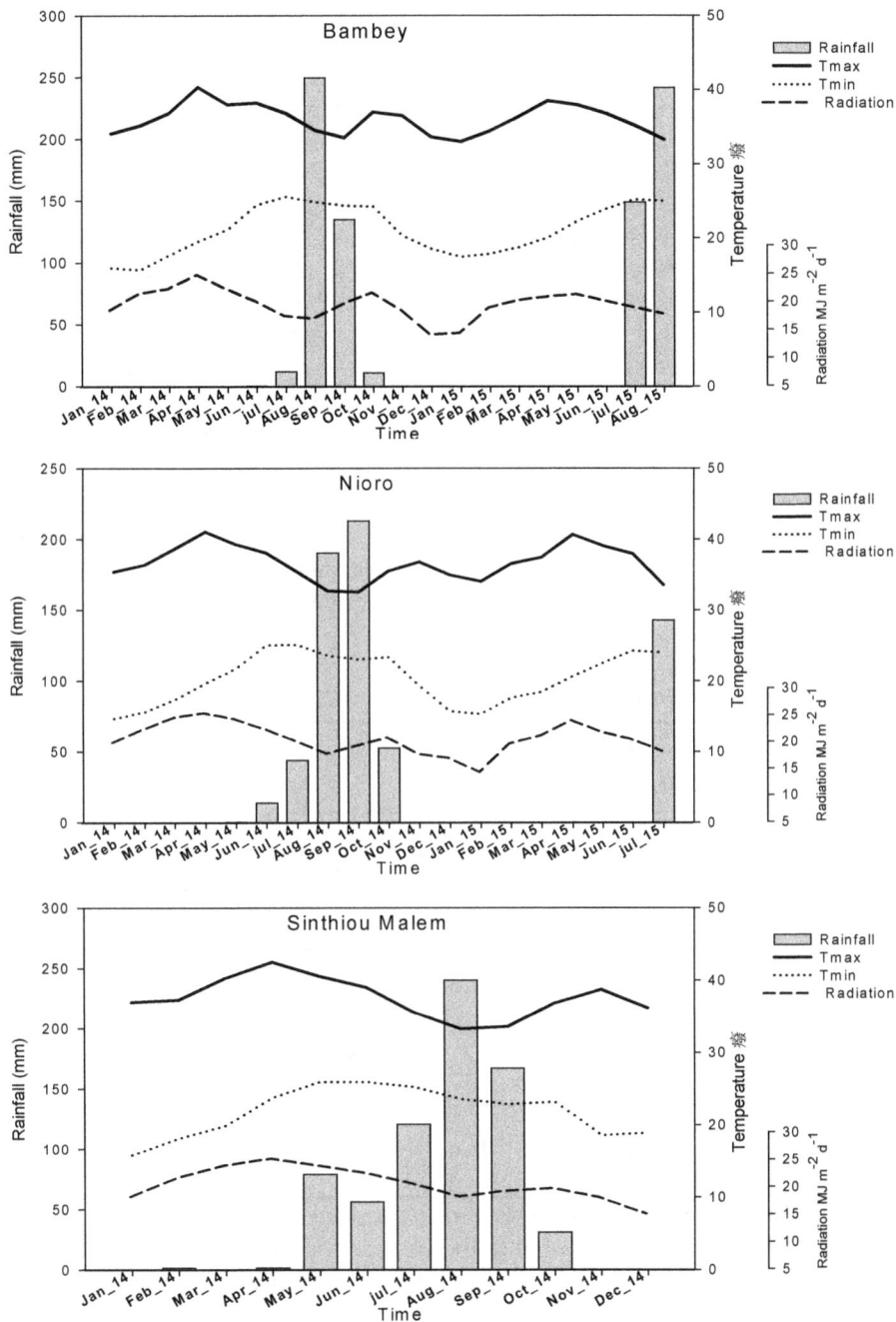

Figure 2. Seasonal cycle for climate variables during the growing season 2014 and 2015 Bambey (January 2014 to August 2015), Nioro (January 2014 to July 2015) and Sinthiou Malem (January to December 2014), with monthly rainfall (bars), monthly maximum (solid lines) and minimum (dashed lines) temperatures and monthly radiation (dotted lines)

Source: CNRA Bambey and ANACIM

There are three different types of soil in the study area depending on the average percentage of silt and clay in the top 40 cm layers. Tropical ferruginous soils commonly called "sol Dior" which are found mainly in Bambey (Clay + Silt <12%), tropical ferruginous leached soils commonly called "sol Deck-Dior" at all sites (12% < Clay + Silt < 15%), and tropical ferruginous hydromorphic soils commonly called "sol Deck" found mainly in Nioro and Sinthiou Malem (Clay + Silt >15%). These soils are low in nitrogen content, which ranges between 0.2 and 0.3‰ and generally low in available phosphorus which is less than 30 ppm (AGETIP, 1995).

2.2 Description of the Experimental Data

Field experiments were conducted in Bambey, in Nioro and in Sinthiou Malem research stations of ISRA during the dry seasons 2014 and 2015 and the rainy season 2014.

A total of seven field experiments were carried out in these three sites. Two dry season field experiments in Bambey and in Nioro and one rainy season field experiment in each site (Table1). The peanut cultivars selected were Fleur 11 (V1) Spanish type and 73-33 (V2) Virginia type which are known to be early (90 days) and medium (110 days) maturity cultivars respectively. Composite fertilizer 6-20-10 was applied just after sowing as recommended by the National Agricultural Research Institute of Senegal (ISRA) for a dose of 150kg/ha. N as total nitrogen, P as single superphosphate tripe (P_2O_5) and K as potassium oxide (K_2O) during de dry season. However, during the rainy season single doses of N (urea 46%), P (DSP 24% P_2O_5) and K (KCL with 60% K_2O) where applied in treatment T4 and T5 (Table 1). The three irrigation levels were: E, irrigation applied at field capacity on plants with no water stress; S1, irrigation applied on plants with water stress during the flowering period (25 days after sowing); and S2, irrigation applied on plants with water stress during seed filling (70 days after sowing). Experimental units measured 16m^2 (4mx4m). Row spacing was 50 cm between rows (inter-row) and 15 cm in the rows (intra-row).

Before sowing each year, the field was disc-plowed to a depth of 12 cm, harrowed and levelled. The seeding was done by hand at a depth of 4 cm with two seeds per seed hole. Seeds were treated with Saxal fungicide to protect them from insects and diseases. Thinning to one plant per seed hole was done after emergence at 11 days after sowing (DAS). Weed control was done by hoe and pest was controlled by chemical pesticide where attacks occurred.

Table 1. Summary of the treatments in the seven field experiments

Sites	Seasons	Irrigation Levels	Fertilizer Levels	Variety Levels	Repetition	Design	Planting Month
Bambey Nioro Sinthiou Malem	Rainy season 2014	No irrigation	Six (T0, T1, T2, T3, T4, T5)	Two (V1,V2)	Four	RCBD	August July July
Bambey	Dry Season 2014	Three (E,S1,S2)	Two (T0, T3)	Two (V1,V2)	Four	Split plot	March
Nioro		One (E)	Four (T0, T1, T2, T3)	Two (V1,V2)	Four	RCBD	March
Bambey Nioro	Dry Season 2015	Three (E,S1,S2)	Two (T0, T3)	Two (V1,V2)	Four	Split plot	February February

Footnote: T0 = without fertilizer, T1 = with 33% of recommended dose, T2 = 66% with recommended dose and T3 = recommended dose that is 150kgha^{-1} of 6-20-10, T4 = same level of T3 without Phosphorus and T5 = same level of T4 with 50% of phosphorus. RCBD = Randomized Complete Block Design

2.3 Field Observations and Measurements

Phenology observations were taken at interval of 7 days to determine parameters such as, day of emergence, day of flowering, beginning of peg, beginning of pod formation, beginning of seed and physiological maturity as described in (Boote, 1982; Meier, 2001). A given stage was considered achieved when 50% of the plants sampled had achieved the specified node number or have one or more flowers, pegs, pods, or seeds exhibiting the specified trait. A total of five plants were selected and tagged in the yield square and followed during the growing season for each plot. The observed emergence occurred on average in all the sites at 6 days after sowing . However, the date of emergence was faster during the rainy season. The early emergence occurred at five days after sowing where the quantity of rain received before sowing was more than 20mm, case of Sinthiou Malem and Bambey. During the dry season, emergence could be delayed up to eight days after sowing when the soil was not humid enough,. The appearance of the first leaf occurred on average two to three days after emergence. The two varieties had the same number of leaves in the main stem (Figure 3). Vegetative stage is defined as one developed node with one tetrafoliate leaf unfolded and its leaflets flat (Boote, 1982). The observed flowering date occurred on average in all sites and for both years, for Fleur 11 from 24 DAS during the rainy season and 30 DAS during the dry season and for 73-33 from 32 DAS during the rainy season to 36 DAS during the dry season.

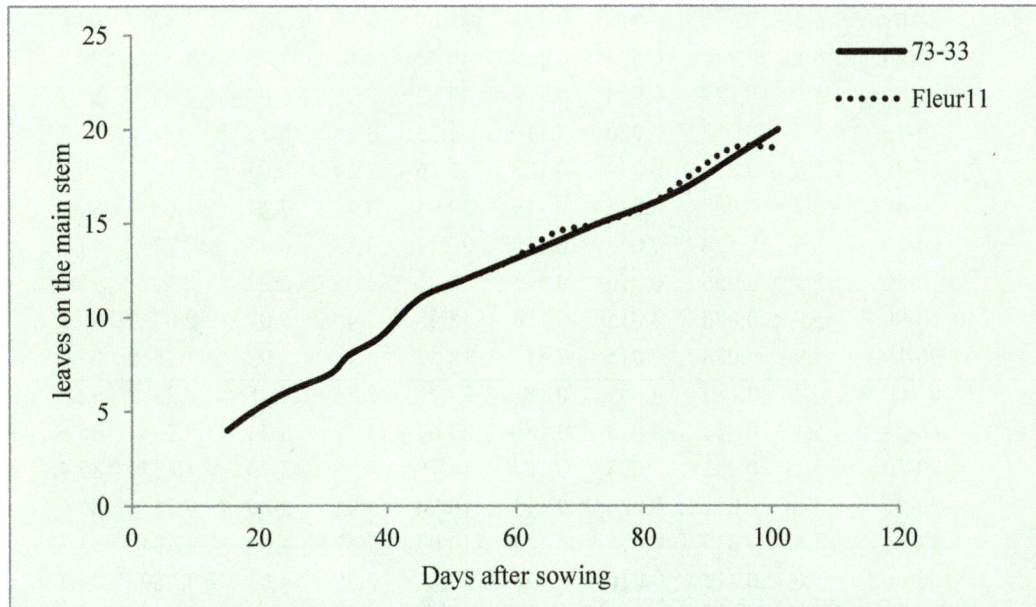

Figure 3. Number of leaves on the main stem of Fleur 11 and 73-33

Total dry matter was determined in leaves, stems and pods at different stages of the growing season on weekly basis. Final harvest was determined at maturity in each plot in an area of $3.9m^2$ (2m x 1.95m). Time series of leaf area index (LAI-2000, 1992; Webb et al., 2008) were measured before each biomass sampling in all sites with LAI 2000 early in the morning or late in the afternoon at Bambey and Nioro or SunScan at noon at Sinthiou Malem.

Composite soil samples were collected in 10 cm intervals from 0 to 100cm depth using an auger in all sites. Samples were air-dried, kept in polythene bags and brought to the laboratory for physical and chemical analysis in 2014 in ISRA's plant-soil laboratory and in 2015 in the National Institute of Pedology (INP).

Data analysis showed a low contain of organic carbon in Bambey both in 2014 and 2015 in the top 50cm of the soil with an average of 0.2% nitrogen content (0.014% to 0.028%). However, phosphorus content was higher in 2014 (28ppm) compared to 2015 (2.62ppm) in which the phosphorus content was lower (Tables 2 and Table 3). The clay content was less than 5% with a depth of 1 metre which ensured a low cation exchange capacity. The percentage of sand reached 90% or more with low water holding capacity. The soil moisture content at wilting point (LL) was 14mm/m and the soil moisture content at field capacity (DUL) was 134.5mm/m that gives a maximum Agricultural Water Reserve (AWR) of 120.5mm/m (Sarr et al., 1999).

In Nioro, the soil was higher in organic carbon in the top soil (0.7%) which allowed a higher cation exchange capacity, but the pH was more acid than Bambey. Soil clay content was less than 10%. The phosphorus content was 12.91ppm in 2014 and 4.97ppm in 2015 because of the phosphorus uptake by the previous maize crop.

Table 2. Analysis of soil physical and chemical properties at the experimental sites in 2014

location	horizon	pH	C	N	K	P	CEC	A	L	S	bd
BAMBEY	0-10	6.3	0.160	0.015	0.139	45.08	3.39	2.82	1.53	94.95	1.48
	10-20	5.8	0.144	0.024	0.119	40.71	2.67	3.77	1.7	92.7	1.48
	20-30	5.8	0.128	0.024	0.112	39.83	2.90	5.07	1.63	91.8	1.47
	30-40	5.8	0.152	0.020	0.119	36.33	3.36	6.15	1.72	91.15	1.57
	40-50	5.8	0.256	0.039	0.125	26.26	3.59	6.95	1.77	89.85	1.44
	50-60	5.7	0.088	0.015	0.119	24.95	3.36	7.17	1.65	89.7	1.44
	60-70	5.6	0.128	0.015	0.134	20.13	4.33	6.45	2.22	91.5	1.35
	70-80	5.7	0.056	0.010	0.115	18.82	4.79	6.22	2.63	91.05	1.38
	80-90	5.8	0.080	0.015	0.108	18.38	4.45	6.07	2.03	90.35	1.31
	90-100	5.9	0.088	0.015	0.112	18.82	4.41	5.92	1.25	87.6	1.44
NIORO	0-10	6.2	0.487	0.049	0.183	37.2	3.51	4.45	4.32	92.8	1.40
	10-20	5.1	0.447	0.029	0.139	23.64	5.30	3.24	9.73	88.65	1.40
	20-30	5.1	0.383	0.024	0.104	16.19	4.76	3.43	10.32	86.45	1.47
	30-40	5.6	0.359	0.020	0.113	10.94	3.31	3.06	9.21	86.6	1.43
	40-50	5.6	0.335	0.015	0.114	10.07	2.67	4.01	12.04	84.25	1.41
	50-60	5.6	0.211	0.010	0.125	7.44	3.59	4.63	13.89	80.7	1.35
	60-70	5.7	0.208	0.015	0.125	6.13	4.05	18.42	3.58	77.55	1.24
	70-80	5.8	0.192	0.015	0.147	5.25	4.51	21.4	4.12	74	1.20
	80-90	5.9	0.207	0.010	0.135	7	4.55	14.82	11.03	73.05	1.17
	90-100	6	0.168	0.029	0.154	5.25	4.88	23.9	4.25	72.5	1.15

Table 3. Analysis of soil physical and chemical properties at the experimental sites in 2015

localities	horizon	pH	C	MO	N	K	P	CEC	A	L	S
BAMBEY	0-10	7.9	0.156	0.269	0.014	0.012	0.04	8	4	17.135	78.87
	10-20	7.8	0.293	0.504	0.028	0.02	0.04	9	5.75	15.065	79.19
	20-30	7.7	0.215	0.370	0.014	0.016	0.13	9	7	15	78.00
	30-40	6.9	0.312	0.538	0.028	0.02	7.68	10	6.25	19.74	74.01
	40-50	6.6	0.273	0.471	0.028	0.012	6.53	18	9.25	20.72	70.03
	50-60	6.6	0.351	0.605	0.028	0.008	3.33	19	10.25	48.26	41.49
	60-70	6.3	0.234	0.403	0.028	0.04	2.99	20	11	32.425	56.58
	70-80	6.1	0.176	0.303	0.014	0.02	1.79	22	10.5	34.555	54.95
	80-90	6.3	0.195	0.336	0.014	0.016	1.75	20	14.5	27.11	58.39
	90-100	6.2	0.195	0.336	0.014	0.012	1.92	21	15.75	26.205	58.05
NIORO	0-10	5.8	0.488	0.840	0.042	0.016	15	24	5.5	25.545	68.96
	10-20	5.0	0.371	0.639	0.028	0.02	10	33	8.5	28.58	62.92
	20-30	4.8	0.234	0.403	0.014	0.02	11	34	10.25	31.705	58.05
	30-40	4.9	0.429	0.740	0.042	0.016	5	34	11	24.915	64.09
	40-50	5.1	0.215	0.370	0.014	0.012	2	31	9.5	29.585	60.92
	50-60	5.0	0.234	0.403	0.028	0.008	2	34	9.25	26.51	64.24
	60-70	5.4	0.244	0.420	0.028	0.04	2	30	12	29.905	58.10
	70-80	5.5	0.254	0.437	0.028	0.02	1	29	19.25	14.11	66.64
	80-90	5.8	0.215	0.370	0.014	0.016	1	27	15	28.36	56.64
	90-100	6.0	0.205	0.353	0.014	0.012	1	26	16.5	27.12	56.38

pH=pH (1:2.5 H2O), C=Organic carbon (%), N=Total nitrogen (%), P=Available Bray P (mg/kg), K=Available Bray K (meq/100g), MO=organic matter (%) CEC=cation exchange capacity (meq/100g), S=Sand (%), L=Silt (%), A=Clay (%), bd= bulk density

Soil moisture content was measured only in Bambey during the dry seasons of 2014 and 2015. It was measured to a depth of 160 cm at 10 cm intervals twice a week by a Diviner 2000. Readings were taken through the wall of a PVC access tube. Data was collected from a network of 24 access tubes installed in site.

Weather stations were located at each site at less than 1kilometre distance from the field experiments. Rainfall, maximum and minimum air temperature, sunshine hours, maximum and minimum relative humidity and wind speed were measured. The rainy season was cooler than the dry season due to a higher relative humidity. The high amount of water during the rainy season was recorded in Sinthiou Sinthiou Malem (696mm). Plants during the dry season 2014 suffered more from heat stress than plants during the dry season 2015 although they received nearly the same amount of water due to difference in sowing date. However, plants received less water in rainy season, therefore they suffered more for water stress than heat stress.

2.4 Data Analysis

The data were processed using Sigma plot 12.0 for figures and R 3.2.2 (https://www.r-project.org/) in RStudio, which is an Integrated Development Environment (IDE) for the statistical analysis (Team, 2014). The analysis of variance was used to analyze the differences between treatments. The Tukey's HSD test was performed to determine the significant differences of means between treatments at 5% after proceeding the analysis of variance (ANOVA).

3. Results

3.1 Field Experiments Analysis

To assess the effects of fertilizer response and water stress on peanut development, growth and yield in Senegal, we conducted field experiments in three different parts in Senegal.

3.1.1 Above Ground Biomass (AGB)

Peanut is an indeterminate plant (Cattan, 1996). The evolution of the total AGB was mostly linear from emergence to maturity during the dry season for all the experiments due to the irrigation effect (Figure 4.a,b,c,d). However, during the rainy season, leaf defoliation at maturity have effect to decline the total AGB (Figure 4.e,f,g). The evolution of the total AGB is father for the early maturity variety (Fleur11) than the medium maturity variety (73-33) for all experiments and for all seasons. However, greater values were recorded for 73-33 at maturity. During the dry season, the percentage of leaf defoliation was small due to the effect of irrigation. Therefore, there was no reduction of total AGB at harvest except in Nioro in dry season 2014 and 2015 for the variety Fleur11. During the rainy season, a period of rapid growth was observed, followed by a decline towards the harvest because of leaf defoliation which decreased by 30% of the total AGB. During the dry season the total AGB produced was higher in Nioro than Bambey. Furthermore, the total AGB produced during the dry season was greater (maximum value recorded in Nioro 2014 with 8695kg/ha for Fleur 11 and 10271kg/ha for 73-33) than the total AGB produced during the rainy season (minimum value recorded in Bambey 2014 with 3051kg/ha for Fleur 11 and 3264kg/ha for 73-33).

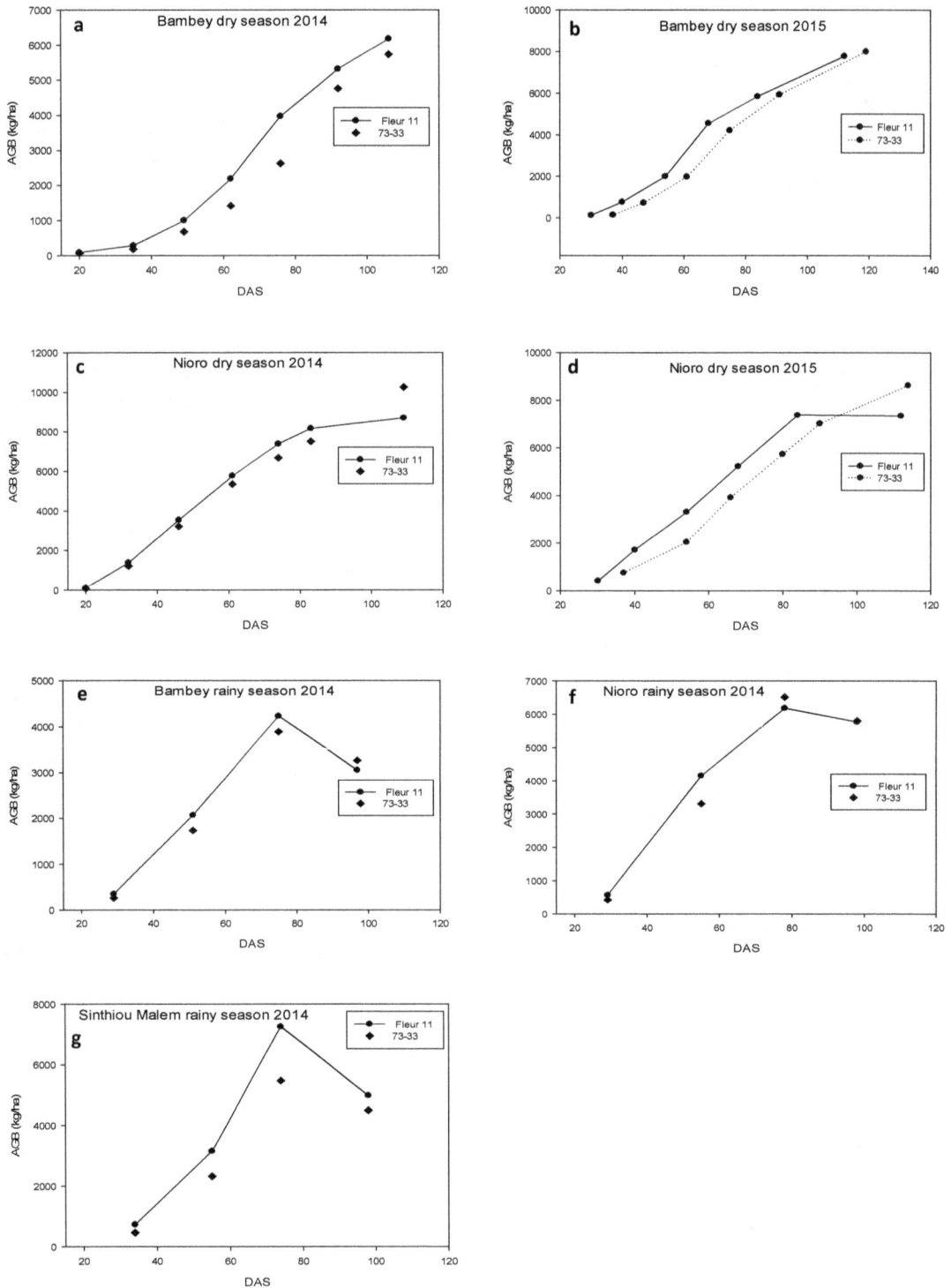

Figure 4. Evolution of above ground biomass (AGB)

3.1.2 Leaf Area Index (LAI)

The two varieties showed similar evolution of LAI. However, 73-33 tended to produce higher LAI at the end of the growing season in both dry and rainy season due to the cycle length except in Bambey during the dry season 2014 (Figure 5.a). The LAI increase was linear during the dry season with a slight decline at maturity (Figure 5,a,b,c,d). Under rainy season condition, it increased quickly and declined towards maturity.

This phenomenon of decline was also observed during the dry season when the plants were under water stress treatments. However, the effect of fertilization rate did not show any difference between treatments.

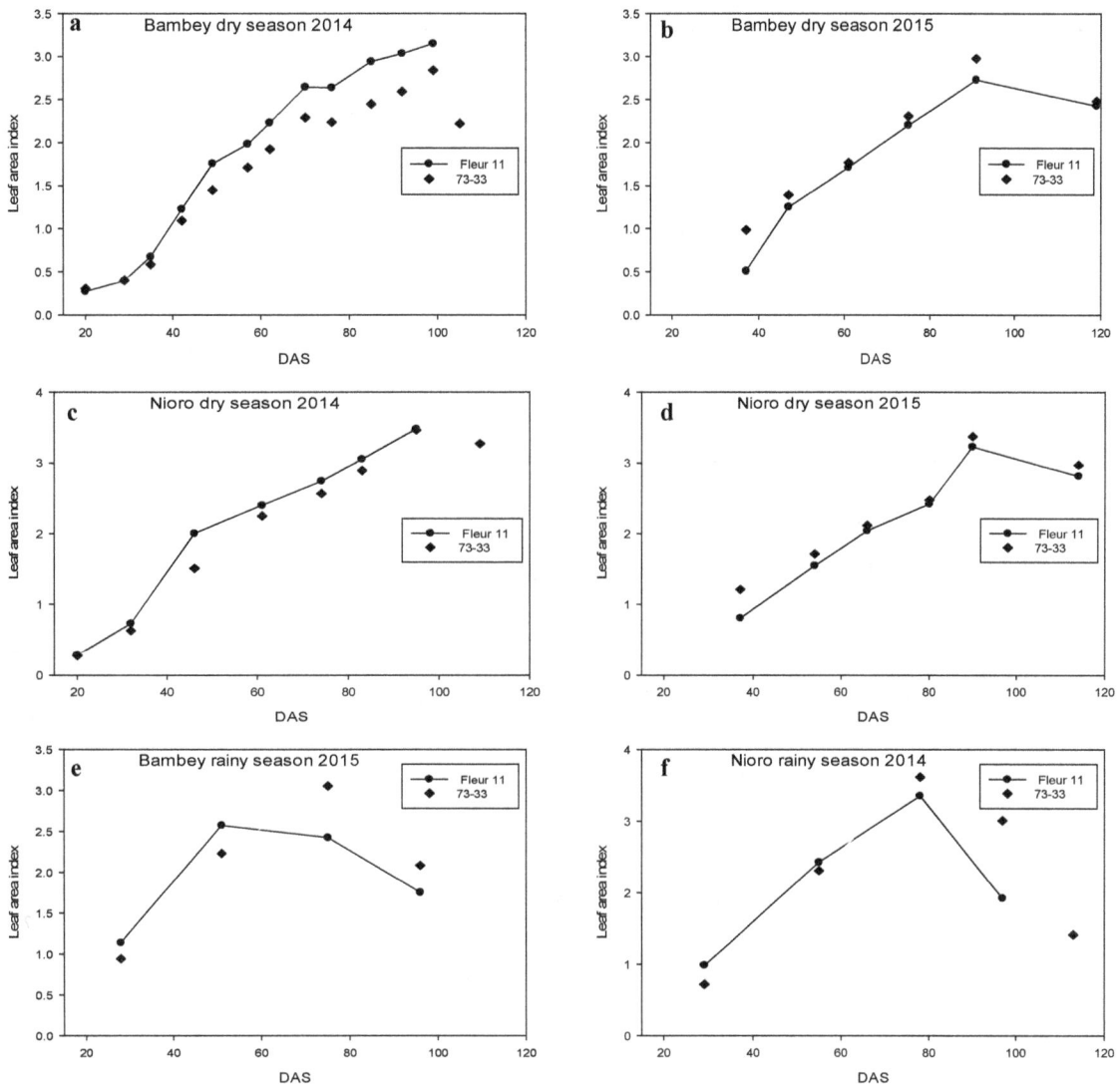

Figure 5. Leaf area index under growing season

3.1.3 Analysis of Variance (ANOVA)

The first step for presenting result was to make an ANOVA for all single experiment to determine the interaction between factors. Table 4 and table 5 present a summary of the ANOVA results for biomass and seed yield respectively.

Table 4. Analysis of variance for biomass yield

Season	Sites	Irrigation	Fertilizer	Variety	Irrigation Fertilizer	Irrigation Variety	Fertilizer Variety	Irrigation Fertilizer Variety
Dry season 2014	Bambey	S	NS	NS	NS	NS	NS	NS
	Nioro		NS	S			NS	
Rainy season 2014	Bambey		NS	S			NS	
	Nioro		NS	S			NS	
	Sinthiou		NS	NS			NS	
Dry season 2015	Bambey	S	NS	NS	NS	NS	NS	NS
	Nioro	S	NS	S	NS	NS	NS	NS

Table 5. Analysis of variance for seed yield

Season	Sites	Irrigation	Fertilizer	Variety	Irrigation Fertilizer	Irrigation Variety	Fertilizer Variety	Irrigation Fertilizer Variety
Dry season 2014	Bambey	S	NS	S	NS	S	NS	NS
	Nioro		NS	NS			NS	
Rainy season 2014	Bambey		NS	S			NS	
	Nioro		NS	S			NS	
	Sinthiou		NS	NS			NS	
Dry season 2015	Bambey	S	NS	NS	NS	NS	NS	NS
	Nioro	S	NS	NS	NS	NS	NS	NS

S = significant, NS = no significant

3.2 Effect of Water Stress on Peanut Yield

The results presented in this section regarded experiments conducted in dry season 2014 and 2015 in Bambey and in Nioro 2015. A split plot design was used for these experiments.

3.2.1 Biomass Yield

Biomass yield was higher in Bambey than Nioro during the dry season in 2015 under full irrigation treatment ET0 and ET1 (Figure 6) for both varieties except for 73-33 under ET0 treatment. In opposite, for the stress treatment S1T0, S1T1, S2T0 and S2T1, it was higher in Nioro than in Bambey due to duration where plants were exposed under drought which was longer in Bambey (25 days) than in Nioro (20). The comparison of the two seasons in Bambey showed that biomass was higher in year 2015 than in year 2014 under full irrigation and stress conditions (Figure 6.a,b). The temperatures during the growth season were lower in year 2015, where peanut was showed in February. The treatments were significantly different ($p<0.05$) for irrigation level (E, S1 and S2) under dry season 2014 and 2015 for both sites (table 4) with the greatest mean values observed in field capacity irrigation (E) followed by the first stress (S1) and then the second stress (S2). Any significant difference between fertilizer levels was observed in both Bambey and Nioro, as well as for the interaction between fertilizer levels and varieties. Difference between varieties was observed in Nioro 2015 (Table 4). However, a comparison of the analysis of variance between the two sites in 2015, showed a significant difference between sites, between varieties, between irrigation, between variety and sites and between irrigation and sites (Table 6).

Table 64. ANOVA for the biomass yield between Bambey and Nioro for the dry season 2015

	Df	Sum Sq	Mean Sq	F value	Pr (>F)	
Variety	1	9006201	9006201	9.169	0.00341	**
Fertilizer	1	1256208	1256208	1.279	0.26184	
Irrigation	2	135581676	67790838	69.019	< 2e-16	***
Site	1	8584792	8584792	8.74	4.21E-03	**
Variety:Fertilizer	1	526263	526263	0.536	0.46656	
Variety:Irrigation	2	1493782	746891	0.76	0.47119	
Fertilizer:Irrigation	2	1553950	776975	0.791	0.45727	
Variety:Site	1	5154473	5154473	5.248	0.0249	*
Fertilizer:Site	1	1869507	1869507	1.903	0.17197	
Irrigation:Site	2	25136656	12568328	12.796	1.76E-05	***
Var:Fert:Irri	2	109224	54612	0.056	0.94596	
Var:Fert:Site	1	1515597	1515597	1.543	0.21819	
Var:Irri:Site	2	834412	417206	0.425	0.65555	
Fert:Irri:Site	2	1710738	855369	0.871	0.42295	
Var:Fert:Irri:Site	2	44165	22083	0.022	0.97777	
Residuals	72	70718621	982203			

Var = Variety, Irri = Irrigation, Fert = Fertilizer

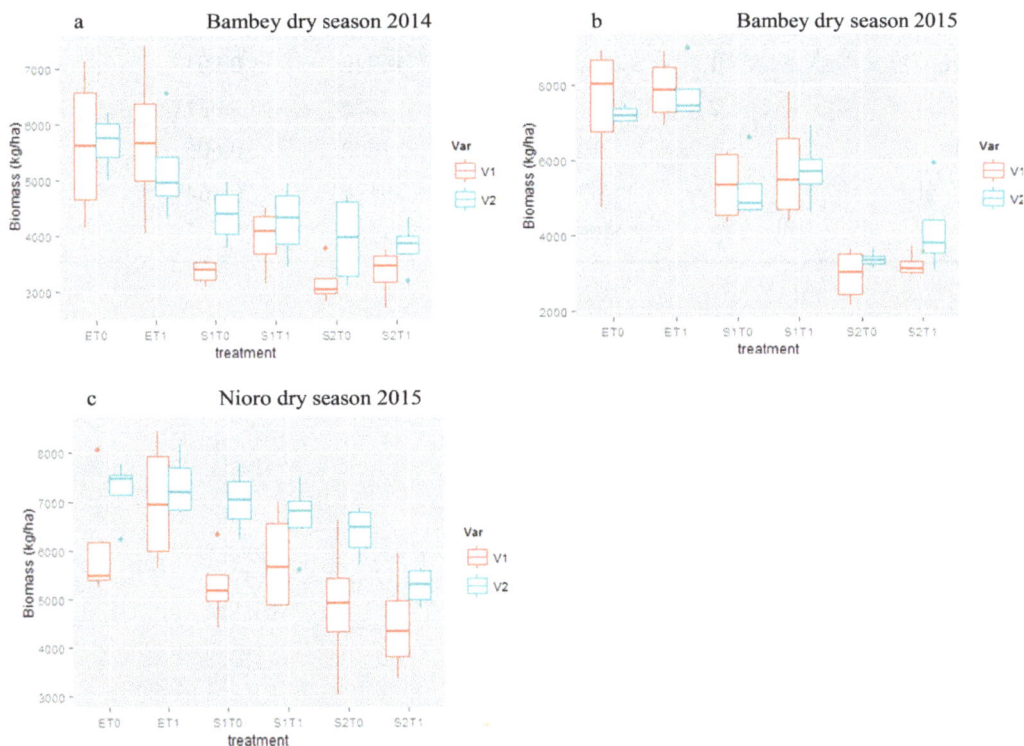

Figure 6. Biomass yield under different irrigation regimes

3.2.2 Seed Yield

Under full irrigation application seed yield varied between 2000kg/ha to 2500kg/ha for both seasons and both varieties excepted for 73-33 in Bambey in 2014 (Figure 7.a). Under stress condition higher values of seed yield were recorded in Nioro 2015 for both S1 and S2 (Figure 7.c). In both sites seed yield was higher under full irrigation followed by S1 and S2 (Figure 7). The analysis of variance in each site showed a significant difference for irrigation in both sites and in both seasons, while, any difference were observed for fertilizer. In Bambey 2014, a significant difference was observed between varieties and the interaction between variety and irrigation (Table5). The results in Bambey 2014 and 2015 showed that higher values was recorded in year 2015 than in year 2014 under full irrigation and S1 and S2 conditions.

However, a comparison between the two sites in 2015, showed a significant difference between irrigation, between sites and between site and irrigation (Table 7).

Table 7. Seed yield interaction between Bambey and Nioro under dry season 2015

	Df	Sum Sq	Mean Sq	F value	Pr(>F)
Variety	1	6095	6095	0.072	0.788734
Fertilizer	1	67501	67501	0.801	0.373759
Irrigation	2	37728097	18864049	223.866	< 2e-16 ***
Site	1	1007315	1007315	11.954	0.000919 ***
Variety:Fertilizer	1	3925	3925	0.047	0.82973
Variety:Irrigation	2	87740	43870	0.521	0.596372
Fertilizer:Irrigation	2	17375	8688	0.103	0.90217
Variety:Site	1	26424	26424	0.314	0.577228
Fertilizer:Site	1	46463	46463	0.551	0.460163
Irrigation:Site	2	562212	281106	3.336	0.041157 *
Var:Fert:Irri	2	6295	3148	0.037	0.963353
Var:Fert:Site	1	16948	16948	0.201	0.655155
Var:Irri:Site	2	3552	1776	0.021	0.97915
Fert:Irri:Site	2	102179	51090	0.606	0.548126
Var:Fert:Irri:Site	2	78252	39126	0.464	0.630431
Residuals	72	6067065	84265		

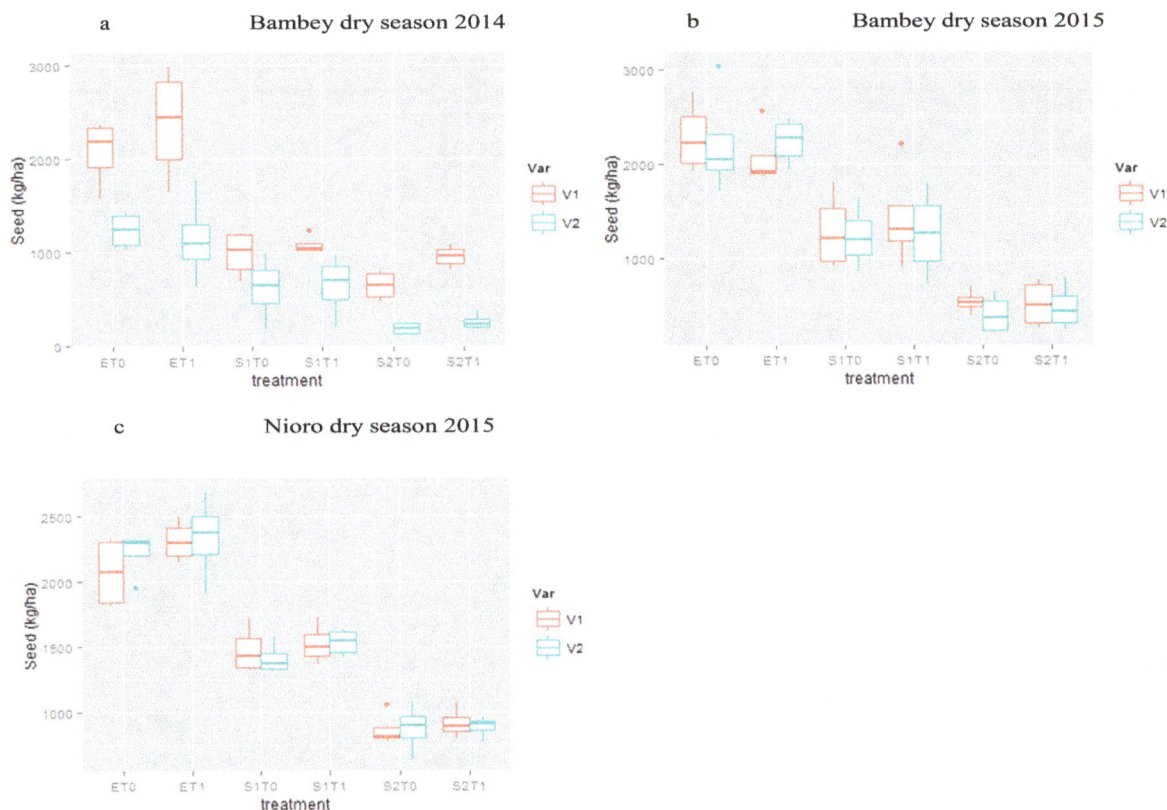

Figure 7. Seed yield under different irrigation regimes

3.3 Effect of Fertilization Rate on Peanut Yield

Datasets for field experiments conducted during the dry season 2014 in Nioro and the rainy season in Bambey, Nioro and Sinthiou Malem 2014 were used in a randomized complete block design (RCBD) with four replications.

3.3.1 Biomass Yield

Aboveground biomass was greater during the dry season (Figure 8.d) than the rainy season (Figure 8.a,b,c). Any difference were observed between the fertilizer levels in all sites and in both seasons. However greater values were recorded in treatment with fertilizer application T1 for Fleur 11 and T4 for 73-33 in Bambey (Figure 8.a), T5 for both Fleur 11 and 73-33 in Nioro (Figure 8.b), T5 for both Fleur 11 and 73-33 in Sinthiou Malem (Figure 8.c), T3 for both Fleur 11 and 73-33 in Nioro in dry season 2014 (Figure 8d). Significant difference between varieties was recorded in Nioro during the dry season 2014 and in Bambey and Nioro during the rainy season 2014, whereas in Sinthiou Malem any difference was observed. Nevertheless, higher values of biomass were recorded in Sinthiou Malem during the rainy season 2014 with a maximum value for Fleur11, of 4500kg/ha and 5200kg/ha for 73-33 compared to Bambey and Nioro (Figure 8. a, b, c). Table 8 presents an analysis of the interaction between sites under rainy season condition. It showed a significant site effect on biomass and a significant difference between varieties and sites.

Table 8. Biomass yield interaction in three different sites in rainy season

	Df	Sum Sq	Mean Sq	F value	Pr(>F)
Variety	1	764491	764491	1.634	0.20389
Fertilizer	5	4316427	863285	1.845	0.1101
Site	2	134453126	67226563	143.68	<2e-16***
Variety: Fertilizer	5	2590121	518024	1.107	0.36091
Variety:Site	2	5464224	2732112	5.839	0.00391**
Fertilizer:Site	10	6681037	668104	1.428	0.17767
Variety:Fertilizer:Site	10	3699469	369947	0.791	0.63768
Residuals	108	50530028	467871		

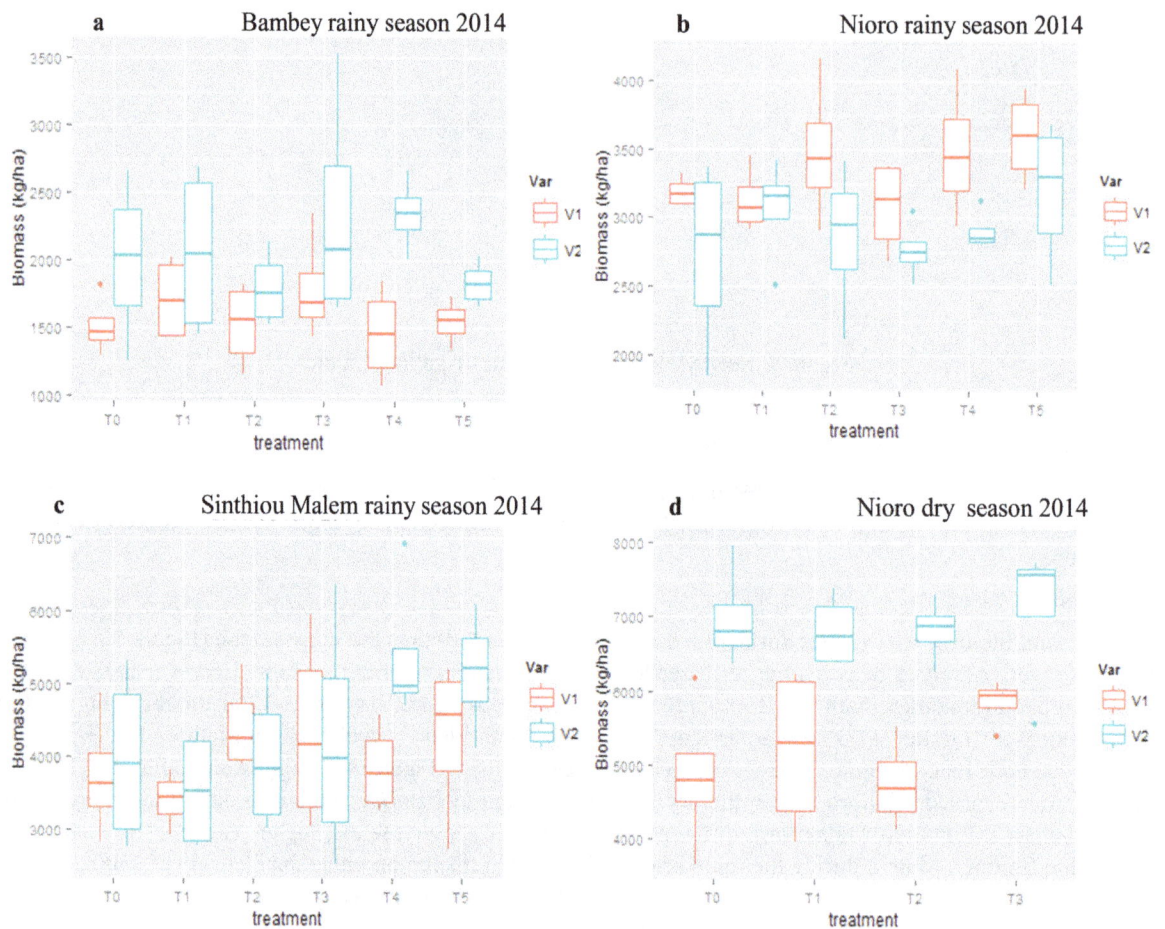

Figure 8. Biomass yield for fertilization rate evaluation

3.3.2 Seed Yield

Seed yield was lower in Bambey followed by Sinthiou Malem and than in Nioro under rainfed condition (Figure 9.a,b,c). The ANOVA showed a significant difference between varieties for Bambey and Nioro, whereas there was no significant difference between varieties in Sinthiou Malem and Nioro during the dry season 2014 (Table 5). During the rainy season, the variety Fleur 11 gave higher yield than 73-33 in all fertilizer levels in Bambey (Figure 9.a), in opposite in Nioro where the variety 73-33 gave higher yield than Fleur 11 (Figure 9.b) for all fertilizer levels. While in Sinthiou Malem these difference between variety were not significant (Figure 9.c). In Nioro during the dry season 2014 no significant difference was showed between varieties but the variety Fleur11 gave slightly high yield for all the fertilizer level (Figure 9.d) with a maximum mean value of 2615kg/ha for T3 level.

The fertilizer response was not significant at any sites, as well as the interaction between variety and fertilizer level. A significant difference between sites and between varieties and sites was observed for seed yield while there were no differences for the interaction between variety, fertilizer and site (Table 9).

Table 9. Seed yield interaction in three different sites in rainy season 2014

	Df	Sum Sq	Mean Sq	F value	Pr(>F)	
Variety	1	39419	39419	0.404		0.527
Fertilizer	5	216479	43296	0.443		0.817
Site	2	27107482	13553741	138.823	< 2e-16 ***	
Variety: Fertilizer	5	526490	105298	1.079		0.376
Variety: Site	2	2686364	1343182	13.757	4.76e-06 ***	
Fertilizer: Site	10	1431817	143182	1.467		0.162
Var: Ferti: Site	10	999769	99977	1.024		0.428
Residuals	108	10544388	97633			

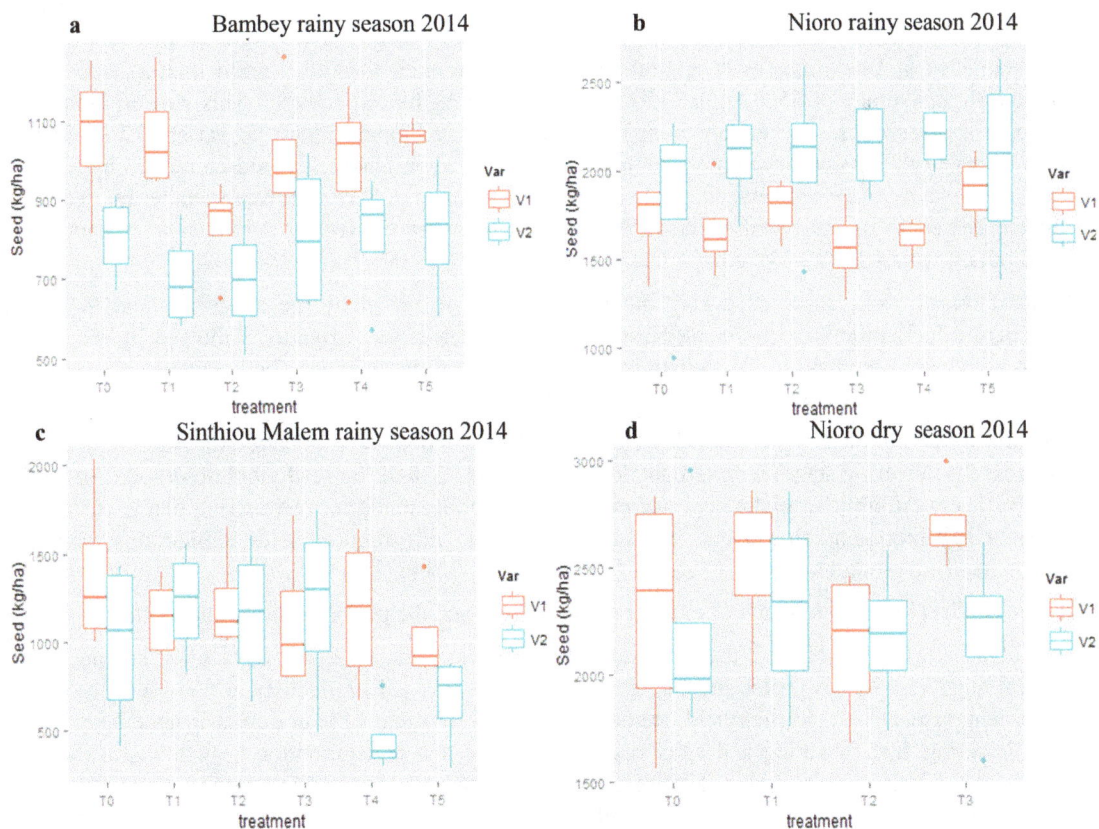

Figure 9. Seed yield for fertilization rate evaluation

4. Discussion

A low content of organic carbon in Bambey in the top 50cm could be attributed to the low contain of soil residues and the high percentage of soil degradation which was already shown by (Diouf, 2000). While in Nioro the higher value of organic carbon could be explained by the effect of soil residues because of the fallow in 2014.

The smaller quantity of organic carbon, in Bambey is associated also to a smaller clay contain (5.65 %) which explains the low cation-exchange capacity (CEC). The higher percentage of sand in Bambey explains the low capacity of soil to retain water. In contrast, in Nioro the medium clay contain (10.1%) in the top soil explains the higher values of CEC.

The low pH value recorded in both sites explains the low amount of base cations due to leaching. The low values of pH were reported by (Sarr et al., 1999) for the soil in Bambey. The smaller quantity of total soil nitrogen content (0.02%) in Bambey was as a result of the smaller quantity of the soil organic matter (SOC), which was due to crop residue deficiencies contrary in Nioro. The phosphorus content was higher in both sites. It could be attributed in Bambey 2014 to the phosphorus uptake by the previous millet grew in rainy season 2013 and previous maize grew in rainy season 2014 in Nioro 2015 which values were above the threshold of phosphorus deficiencies (Diouf, 2000).

The decline of peanut production through recent decades (DAPSA., 2014) was mostly explained by lack of input, soil degradation and limited water availability (Montfort, 2005).

These two varieties grown during the field experiments are drought tolerant and they are part of the most varieties with high yield most grown in the country.

Drought stress was known to substantially reduce peanut yield when it occurred at reproductive stage (Annerose, 1990; Cecilia et al., 2013; Pandey et al., 2001; Reddy et al., 1996; Wright et al., 1991; Wright and Rao, 1994; Yao et al., 1982).

The substantial reduction of biomass yield, pod yield and kernel yield were recorded in all the water stress treatment in two different reproductive stages of the growth of peanut. These are in the agreement with results obtained by (Annerose, 1985) which showed that biomass and pod yield is linear according to the quantity of water received.

The yield obtained at field capacity irrigation for the two varieties showed a good biomass, pod and seed production with an average of 6500 kg/ha, 3500kg/ha and 2000kg/ha respectively. This was different from the production in stress phases. The variety Fleur11 produced more biomass than the variety 73-33 in Bambey whereas the varieties 73-33 produced more biomass than Fleur 11 in Nioro. It could be related by the fact that the Fleur 11, known by early maturity, adapted more in Bambey than in Nioro. However, under rainfed condition, these differences could be attributed to the higher amount of water received in Nioro than in Bambey which allow medium cycle variety (73-33) to complete its growth.

The higher biomass yield observed during the full irrigation followed by the stress application during the flowering period in Bambey. In Nioro, biomass was also higher in full irrigation followed by the first stress application with a high biomass in Nioro than in Bambey in stress application. The difference was attributed to the duration of stress on plants which was longer in Bambey (25days) than in Nioro (20days).

Pod yield and seed yield observed are also lowest during the second stress application (at maturation stage) than the first stress (at flowering stage) with a reduction of 50% and 33% of the seed yield observed (Figure 6.a.b.c) respectively. These differences might have occurred due to the fact that after re-watering plants during the first stress, peanut still producing flower and was able to recover and produce yield. Similar observations were reported for Bambara groundnut in Botswana by (Vurayai et al., 2011).

The amount of yield losses is function of number of days stress and the period of stress application.

Vaghasia et al. (2010) established that the stress at flowering stage (25-47 days after sowing) and pod development stage (50-72 days after sowing) gave 18.45% and 30.63% reduction in pod yield than no water stress treatment, respectively. Kernel yield decreased by 28%, 36% and 41% in deficit-irrigated late vegetative and early flowering, late flowering and early pegging, pegging and pod formation growth stages, respectively compared with full irrigation treatment (Kheira, 2009). The reduction of yield for peanut due to water stress is known to occur during the pod filling stage (Rao et al., 1974; Vaghasia et al., 2010; Vurayai et al., 2011) which was confirmed in this study at both sites and both varieties. However, the water stress management could depend on the variety cycle. In this study the reduction of the pod and seed yield is less affected by the Fleur 11 variety than the 73-33 variety. It could be explained by the fact that the Fleur11 variety achieved the physiological maturity faster than the 73-33 variety. Therefore, water deficit reduced pod and seed yield by causing smaller and younger pods to terminate the growth and eventually by reducing the growth rate of old pod. These results were corroborated by (Rao et al., 1985) in India.

The stress application on the flowering stage and the pod filling stage was justified by the fact that peanut was most sensitive for these periods to water stress (Billaz and Ochs, 1961; Black et al., 1985; Nageswara Rao et al.,

1985; Patel and Golakiya, 1991; Reddy et al., 2003; Stirling et al., 1989) where the greatest reduction in kernel yield occurred when stress was imposed during the seed filling phase.

Most of the flowers did not form pegs during the stress at the flowering period (30-45 days after sowing) but flowers produced after re-watering compensated for this loss (Gowda and Hegde, 1986). It confirms the greater yield obtained during the flowering period compared to the yield obtained during the pod and seed filling in the experimental sites.

It is observed a slight increase of biomass, pod and seed yield when fertilizer was applied but no significant difference was observed between fertilizer level applications. There was also no significant difference between the interaction fertilizer and irrigation. The lowest difference may be explained by the lowest quantity of NPK applied which contained 9kg/ha of nitrogen 14 kg/ha of phosphorus and 9kg/ha of potassium.

Peanut response to NPK fertilizer was studied in Nioro during the dry season 2014 and in Bambey, Nioro and Sinthiou Malem during the rainy season 2014. In Sinthiou Malem, low harvest yield could be attributed to the dry spell that occurred earlier in October when plants were at maturity stage (seed filling) while in Bambey the low harvest could be explained by the low quantity of water received during the rainy season 2014 (407mm).

There were any significant responses of fertilizer in all sites. However, the plot with fertilizer application resulted in the greatest biomass yield, pod yield and seed yield. These results showed that the quantity of fertilizer used was not in the optimum dose to determine a difference between fertilizer level on one hand and on other hand, the initial soil properties contained considerable quantity of nutrients. This may be because the experiments were conducted at field research stations which received ample fertilization. However, the quantity of nitrogen used was sufficient (9kg/ha) because of the capacity of peanut to fix nitrogen. Campbell et al. (1980) showed that there was no significant difference in kernel yield from plot received 10, 50 and 100kg/ha of nitrogen. Therefore, it is an economic disadvantage for increasing nitrogen levels beyond 10kg/h. In addition, organic amendments increased microbial population and resulted in a positive correlation between population of symbiotic bacteria and nitrogen fixation. Similar observations were reported by (Lee et al., 2004; Limtong and Piriyaprin, 2006). This could also be explained by the higher nitrogen contain in Bambey and Nioro with received 5750 kg/ha of cow manure incorporated in the soil before sowing in the year 2014.

The quantity of phosphorus used was too low to give a response on peanut growth where the appropriate application rate for peanut is 60 kg P_2O_5/ha for poor alluvial soils and 90 kg P_2O_5/ha for sandy soils (Ha, 2003; Mirvat et al., 2006; Ogeh and Oyibo, 2015). Naab et al. (2009) demonstrated positive effects of phosphorus application from 30kg/ha on biomass and seed yield but no difference was found between 30kgP/ha and 60kgP/ha application and between 60kgP/ha and 90kgP/ha application on farm experiments in Ghana.

5. Conclusion

Field experiments were carried out in three different agro-climatic zones in Senegal. Results of assessing the effect of water stress on peanut and the fertilizer rate response indicated that the reduction of peanut yield was highly associated to water stress than to fertilizer application for the range of conditions considered. Therefore, addressing the issues of water stress is pivotal in increasing the yield of peanuts as opposed to fertilizer application. Furthermore, the quantity of fertilizer applied was insufficient to show a difference between the fertilizer levels and the interaction between irrigation and fertilizer during the two growth seasons. Thus, adequate fertilizer application at the right dose and time could influence the performance of the peanut. From this point of view, the application of fertilizer should be split in different phases of the growing season but not only applied just after sowing.

Further investigations have to be made to test higher single dose of nitrogen or phosphorus or potassium in split applications. After successful results are obtained, in an ideal field based situation they can be evaluated under limited water conditions. This investigation could further be conducted in field based conditions where limited soil nutrients occur.

Acknowledgments

This work was funded by German Federal Ministry of Education and Research (BMBF) through the West Africa Science Service Center on Climate Change and Adapted Land Use (WASCAL).

References

AGETIP. (1995). *Programme de réhabilitation des stations ISRA. Informations pédologiques et étude cartographique des sols de la station de Nioro.*

Annerose, J. M. D. (1985). *Réponses d'une variété d'arachide soumise à des sécheresses d'intensité croissante durant différentes phases de son cycle.* Dot. ISRA/CNRA Bambey. 41.

Annerose, J. M. D. (1988). *Critères physiologiques pour l'amélioration de l'adaptation a la sécheresse de l'arachide, Oléagineux,* Vol. 43, numéro 5 6p.

Annerose, J. M. D. (1990). *Recherches sur les mécanismes physiologiques d'adaptation à la sécheresse. Application au cas de l'arachide (Arachis hypogea L.) cultivée au Sénégal.* Paris VII University

Bankole, S., & Adebanjo, A. (2004). *Mycotoxins in food in West Africa: current situation and possibilities of controlling it. African Journal of Biotechnology, 2,* 254-263.

Billaz, R., & Ochs, R. (1961). *Stages of susceptibility of groundnut to drought oleaginous.* 605-611.

Black, C., Tang, D. Y., Ong, C., Solon, A., & Simmonds, L. (1985). *Effects of soil moisture stress on the water relations and water use of groundnut stand*s. *New phytologist, 100,* 313-328. https://doi.org/10.1111/j.1469-8137.1985.tb02781.x

Boote, K. J. (1982). *Growth stages of peanut (Arachis hypogaea L.).* Peanut science 35-49.

Campbell, V. A., Wahab, A. H., & Murray, H. (1980). *Response of peanut (Arachis hypogaea L.) to nitrogen, minor elements and phosphorus fertilization on a newly terraced ultisol in Jamaica. In* "Legume Seminar.,(Jamaica)., 28 Feb 1980..".

Cattan, P. (1996). T*he components of groundnut yield, Agriculture et Development,*33-38

Cecilia, M. T. S., Ayman, S., Jakarat, A., Ian, F., & Gerrit, H. (2013). *Scheduling irrigation with a dynamic crop growth model determining the relation between simulated drought stress and yield for peanut. Irri Sci* 889-901.

Clavel, D., Baradat, P., Khalfaoui, J. L., Drame, N. K., Diop, N. D., Diouf, O., & Zuily-Fodil, Y. (2007). *Adaptation à la sécheresse et création variétale : le cas de l'arachide en zone sahélienne.* Deuxième partie : une approche pluridisciplinaire pour la création variétale 16.

Clavel, D., Drame, N. K., Diop, N., & Zuily-Fodil, Y. (2005). *Adaptation à la sécheresse et création variétale : le cas de l'arachide en zone sahélienne.* Première partie : revue bibliographique. *OCL 2005,* 249-260. https://doi.org/10.1051/ocl.2005.0248

DAPSA. (2014). *Direction de l'Analyse, de la Prévision et des Statistiques Agricoles. Résultats définitifs de la campagne agricole.* 16.

Diouf, O. (2000). *Réponses agrophysiologiques du mil (Pennisetum glaucum (L.) R. Br.) à la sécheresse: influence de la nutrition azotée.* Thèse de Doctorat, Université libre de Bruxelles, 160p.

Foncéka, D. (2010). *Elargissement de la base génétique de l'arachide cultivée (Arachis hypogaea): Applications pour la construction de populations, l'identification de QTL et l'amélioration de l'espèce cultivée,* Montpellier University.

Ganry, F., & Gueye, F. (1992). *La mise en valeur des bas-fonds de la zone soudano-sahélienne par Sesbania rostrata est-elle possible?* L'agronomie Tropicale 46, 155-159.

Gautreau, J. (1982). *Améliorations agronomiques par le développement de variétés d'arachide adaptées aux contraintes pluviométriques.* Oléagineux. 469-475.

Gaye, M. (2013). *Le Sénégal pourra-t-il revivre sa belle économie arachidière, Communication personnelle. 18e* Mardi du BAME, Thème : la filière arachidière au Sénégal, Communication orale. 27 Août 2013, ISRA, Dakar.

Gowda, A., & Hegde, B. (1986). *Moisture stress and hormonal influence on the flowering behavior and yield of groundnut (Arachis hypogaea L.).* Madras Agric. J 73, 82-86.

Ha, T. T. T. (2003). *Effect of phosphorus fertilizer on groundnut yield in poor alluvial and sandy soils of Thua Thien Hue.* Better Crops International 17, 16.

Hemalatha, S., Rao, V. P., Padmaja, J., & Suresh, K. (2013). *An overview on role of Phosphorus and Water Deficits on growth, yield and quality of Groundnut (arachis hypogaea L.).* International Journal of Applied Biology and P...

Jongrungklang, N., Toomsan, B., Vorasoot, N., Jogloy, S., Boote, K., Hoogenboom, G., & Patanothai, A. (2013). *Drought tolerance mechanisms for yield responses to pre-flowering drought stress of peanut genotypes with different drought tolerant levels.* Field Crops Research 144, 34-42. https://doi.org/10.1016/j.fcr.2012.12.017

Kambiranda, D. M., Vasanthaiah, H. K., Katam, R., Ananga, A., Basha, S. M., & Naik, K. (2011). *Impact of drought stress on peanut (Arachis hypogaea L.) productivity and food safety.* Plants and Environment. In Tech Publisher.(http://www. intechopen. com/books/plants-and-environment/impact-of-drought-stress-onpeanut-arachis-hypogaea-l-productivity-and-food-safety), 249-272.

Kenchanagoudar, G., Nigam, S., & Chennabyregowda, M. (2002). *Effect of drought on yield and yield attributes of groundnut.* Karnataka Journal of Agricultural Sciences 15, 364-366.

Kheira, A. A. A. (2009). *Macromanagement of deficit-irrigated peanut with sprinkler irrigation.* Agricultural water management 96, 1409-1420. https://doi.org/10.1016/j.agwat.2009.05.002

LAI-2000. (1992). *Plant Canopy Analyser, Operating Manuel.* LI-COR, Inc, USA, 179p.

Lee, J., Park, R., Kim, Y., Shim, J., Chae, D., Rim, Y., & Kyoon, B. (2004). *Effect of food waste compost on microbial population, soil enzyme activity and lettuce growth.* Bioresource Technol 93, 21-28. https://doi.org/10.1016/j.biortech.2003.10.009

Limtong, P., & Piriyaprin, S. (2006). *Effect of compost with chemical fertilizer on biological properties in different fertility soil.* 10p.

Martin, J., Bâ, A., Dimanche, P., & Schilling, R. (1999). *Comment lutter contre la contamination de l'arachide par les aflatoxines? Expériences conduites au Sénégal.* Agriculture et développement, 58-67.

Meier, U. (2001). *Stades phénologiques des mono-et dicotylédones cultivées BBCH Monographie,* 2e éd., Uwe Meier (réd.). Centre fédéral de recherches biologiques pour l'agriculture et les forêts.

Mirvat, E. G., Magda, H. M., & Tawfik, M. M. (2006). *Effect of Phosphorus Fertilizer and Foliar Spraying with Zinc on Growth, Yield and Quality of Groundnut under Reclaimed Sandy Soils.* Journal of Applied Science Research, 491-496.

Montfort, M. A. (2005). *Filières oléagineuses africaines. Notes et Études Économiques* n°23, Septembre. 55-85.

Naab, J., Boote, K., Jones, J., & Porter, C. H. (2015). *Adapting and evaluating the CROPGRO-peanut model for response to phosphorus on a sandy-loam soil under semi-arid tropical conditions.* Field Crops Research 176, 71-86. https://doi.org/10.1016/j.fcr.2015.02.016

Naab, J., Prasad, P., Boote, K., & Jones, J. (2009). *Response of Peanut to Fungicide and Phosphorus in On-station and On-farm Tests in Ghana.* Peanut Science 36, 157-164. https://doi.org/10.3146/PS08-017.1

Nageswara Rao, R., Singh, S., Sivakumar, V., Srivastava, K., & Williams, J. (1985). *Effect of Water Deficit at Different Growth Phases of Peanut. I.* Yield Responses. 5p.

Nautiyal, P., Joshi, Y., & Dayal, D. (2002). *Response of groundnut to deficit irrigation during vegetative growth.* Food and Agricultural Organization of the United Nations (FAO)(ed.), Deficit Irrigation Practices. Rome, Italy, 39-46.

Noba, K., Ngom, A., Guèye, M., Bassène, C., Kane, M., Diop, I., Ndoye, F., Mbaye, M. S., Kane, A., & Ba, A. T. (2014). *L'arachide au Sénégal : état des lieux, contraintes et perspectives pour la relance de la filière.* OCL 2014, 21(2) D205), 5p. https://doi.org/10.1051/ocl/2013039

Ogeh, J. S., & Oyibo, R. O. (2015). *Phosphorus Fertilization Under Different Land Preparation Methods and Performance of Groundnut (Arachis hypogea L.) in Rainforest Zone of Southern Nigeria.* Jurnal TANAH TROPIKA (Journal of Tropical Soils) 19, 1-7.

Pandey, V., Shekh, A. M., Vadodaria, R. P., & Bhat, B. K. (2001). *Evaluation of CROPGRO-Peanut model for two genotypes under different environments. Paper presented at the National seminar on Agro Meteorological* Research for Sustainable Agricultural Production at GAU Anand.

Patel, M. S., & Golakiya, B. A. (1991). *Effect of water stress on yield and yield attributes of groundnut (Arachis hypogaea L.).* Madras Agricultural Journal, 178-181.

Pélissier, P. (1966). "*Les paysans du Sénégal,*" *Imprimerie Fabrègue Saint Yrieux.*

Prasad, P. V., Kakani, V. G., & Upadhyaya, H. D. (2010). *Growth and production of groundnut.*UNESCO Encyclopedia, 1-26.

Rao, I. V. S., L., N. R., & R., a. U. (1974). *Effect of moisture stress at different growth stages on yield and oil content of groundnut.* The Andhra Agricultural Journal 21, 111-116.

Rao, R., Singh, S., Sivakumar, M., Srivastava, K., & Williams, J. (1985). *Effect of water deficit at different growth phases of peanut. I. Yield responses.* Agronomy Journal 77, 782-786. https://doi.org/10.2134/agronj1985.00021962007700050026x

Reddy, P., Reddy, B., Praveen Rao, V., & Sarma, P. S. (1996). *Effect of water deficits at different crop growth periods on growth and yield of groundnut.* Journal of Research, Acharya N. G. Ranga Agricultural University 24(2), 147-149.

Reddy, T., Reddy, V., & Anbumozhi, V. (2003). *Physiological responses of groundnut (Arachis hypogea L.) to drought stress and its amelioration: a critical review, Plant growth regulation.*41, 75-88. https://doi.org/10.1023/A:1027353430164

Revoredo, C. L., & Fletcher, S. (2002). "*World peanut market: an overview of the past 30 years,*" Georgia Agricultural Experiment Stations, College of Agricultural and Environmental Sciences, the University of Georgia.

Sarr, B., Ndiendole, S., Diouf, O., Diouf, M., & Roy-Macauley, H. (1999). Suivi de l'état hydrique du sol et de la température du couvert du maïs (Zea mays L.) au Sénégal. *Sécheresse 1999* 10 (2), 129-135.

Singh, P., Nedumaran, S., Ntare, B., Boote, K., Singh, N., Srinivas, K., & Bantilan, M. (2013). *Potential benefits of drought and heat tolerance in groundnut for adaptation to climate change in India and West Africa.* Mitigation and Adaptation Strategies for Global Change.

Stirling, C., Black, C., & Ong, C. (1989). *The Response of Groundnut (Arachis hypogaea L.) to Timing of Irrigation II. 14C-PARTITIONING AND PLANT WATER STATUS.* Journal of experimental botany 40, 1363-1373. https://doi.org/10.1093/jxb/40.12.1363

Team, R. (2014). *RStudio: integrated development for R.* RStudio, Inc., Boston, MA. URL http://www. RStudio. com/ide.

Vaghasia, P. M., Jadav, K. V., Jivani, L. L., & Kachhadiya, V. H. (2010). *Impact of water stress at different growth phases of summer groundnut (Arachis hypogaea L.)on growth and yield.* 693-696.

Vurayai, R., Emongor, V., & Moseki, B. (2011). *Effect of water stress imposed at different growth and development stages on morphological traits and yield of Bambara Groundnuts (Vigna subterranea L. Verdc).* American journal of plant physiology 6, 17-27. https://doi.org/10.3923/ajpp.2011.17.27

Webb, N., Nichol, C., Wood, J., & Potter, E. (2008). *User Manual for the SunScan Canopy Analysis System,(2.0 Version).* Delta-T Devices Ltd.: Cambridge, UK, 83.

Wright, G., Hubick, K., & Farquhar, G. (1991). *Physiological analysis of peanut cultivar response to timing and duration of drought stress.* Crop and Pasture Science 42, 453-470. https://doi.org/10.1071/AR9910453

Wright, G., & Rao, R. N. (1994). *Groundnut water relations. In "The Groundnut Crop",* pp. 281-335. Springer. https://doi.org/10.1007/978-94-011-0733-4_9

Yao, J. P., Luo, Y. N., & Yang, X. D. (1982). *Preliminary report on the effect of drought and seed development and quality of early groundnut.* Chinese Oil Scops 3, 50-52.

3

Passing the Scepter, not the Buck
Long Arms in EU Climate Politics

Doris Fuchs[1] & Berenike Feldhoff[1]

[1] University of Münster, Germany

Correspondence: Doris Fuchs, University of Münster, Institute of Political Sciences, Scharnhorststraße 100, 48151 Münster, Germany. E-mail: sustainability@uni-muenster.de

Abstract

This paper investigates power dynamics between and within the core institutions of the European Union (EU) in the development of the 2030 EU climate & energy framework. Starting from the widely studied change in the EU's climate policy from "leadership by example" to "slow motion", it integrates a specific focus on power dynamics between the core EU institutions and non-state actors. Interestingly, perspectives on power relations between and within EU institutions and the power of non-state actors in EU governance are rarely integrated explicitly in analyses of the EU's shift in its stance on climate policy. In this article, we aim to draw together the analyses of institutional power dynamics in EU climate policy with analyses of the exercise of power by non-state actors to explain the dynamics leading to the change in EU climate politics. Using data from document research, secondary literature and interviews conducted with members of EU institutions as well as non-state actors, we show that an absence of clear leadership from the European Council in the end furthered an extended influence of business actors via the strengthening of the Commission as well as particular DGs. The results show how inadequate guidance from the Council and divisions within the Commission facilitated the informal passing of the scepter of EU climate policy to business, even as EU institutions retained formal responsibility, thus not passing the buck in a corresponding manner. The analysis further reveals strong reasons for doubt towards future EU ambitions in global climate change politics.

Keywords: business, climate change, European Union, power

1. Introduction

The 2009 climate summit in Copenhagen marks a perceptible shift in the EU's policy efforts on climate. Before 2009, the EU seemed to pursue a global leadership role in climate policy, pledging carbon emission reductions beyond what other countries were promising. With its "2020 Climate & Energy Package", the European Union set the 20-20-20 targets, referring to a 20% cut in greenhouse gas emissions from 1990 levels, 20% of EU energy being produced via renewables, and an improvement in energy efficiency of 20%. Moreover, the EU member states agreed on binding annual targets for the reduction of emissions that were not yet included in the Emissions Trading System (ETS). (Note 1) The dominant mindset in Europe seemed to be to "lead by example," thereby aiming to motivate other countries to follow suit (Oberthür & Dupont, 2011). The literature has categorized such intentions under the term of "normative power Europe" (Diez, 2005; Manners, 2002).

After 2009, such commitment seemed largely absent. The disappointment from the failure of the Copenhagen summit alongside the rise of competing issues drawing the full attention of member state governments contributed to a tangible shift in European climate policy. As a result, the new climate objectives the EU formulated for 2030 and the COP in Paris did not go far beyond what would be reached with existing policies and, most importantly, were (partly) defined as "non binding". The targets were set at 40% cuts in greenhouse gas emissions from 1990 levels, a 27% share for renewable energy and 27% improvement in energy efficiency. (Note 2)

This change in the EU's climate change policy from "leadership by example" to "slow motion" has received considerable attention in the scientific literature. Beyond the enabling conditions of disappointment with Copenhagen and competition with other issues, particularly the financial crisis, scholars have attributed this shift to an internal division within the European member states, notably between northern/western and eastern member countries on concerns like costs, competition, and the EU's competencies (Unbehaun, 2016). In an excellent analysis, Bürgin (2015) added attention to the power dynamics within the Commission to this picture. He argued that the ebbing attention of the Council made room for a strengthened role of the Commission and that divergent interests within the Commission enabled actors opposed to more ambitious goals to demand a compromise.

Behind many of the analyses of the dynamics leading to the change in EU climate politics lies an – at times more, at times less observable – argument about the political power of business interests. It is visible, for instance, in Bürgin's (2015) reference to Commissioner Oettinger's close ties to the car industry. Similarly, recent studies emphasize the impact that the lobbying by trade associations has on EU climate policy (Fagan-Watson, Elliott &Watson, 2015) as well as the general role that the European industry plays as a central stakeholder, target group, lobbyist, and implementer of policies (Boasson & Wettestad, 2013). Interestingly, though, perspectives on power relations between and within EU institutions and the power of non-state actors in EU governance are rarely explicitly and comprehensively integrated into analyses of the EU's shift in its stance on climate policy. In this article, we therefore aim to draw together the analyses of institutional power dynamics in EU climate policy with analyses of the exercise of power by non-state actors. Specifically, we argue that the Council's lack of attention strengthened the role of the Commission vis-à-vis the Council and the Parliament. In turn, divisions within the Commission fortified certain DGs and Commissioners. These changes, finally, were crucial for enabling business interests to play a pivotal role in EU climate politics. We demonstrate that only an understanding of institutional change in terms of its impact on stakeholder influence will reveal core drivers of the EU's climate policy, specifically the long arm of business interests.

Our argument supports and complements the work by Bürgin (2015) on the institutional dynamics behind the change in the EU's stance on climate policy and adds further pieces to that picture. Similarly, our findings correspond to and expand upon research on lobbying in the EU. Most importantly, the integration of these aspects contributes to our understanding of the political and scientific implications of the EU's climate policy shift. More specifically, it allows us to gain a clearer idea of what the future of EU climate policy may look like. Moreover, it highlights reasons for continued concerns about the asymmetric influence of stakeholders on EU politics and policy, and thus, about the democratic deficit in the EU (Fuchs, Gumbert, Schlipphak forthcoming). Both climate change as a continued challenge to the future of human kind and the rise of anti-EU sentiments within Europe, most notable in the Brexit vote, should draw our attention to these concerns.

Power is a core concept in political science. Yet, it is extremely complex. Scholars work on relative or productive, potential or actual, de facto or de jure, agenda-setting or decisional power (Baldwin, 2002; Barnett & Duvall, 2005; Fuchs & Graf 2015; Partzsch & Fuchs 2012; Shepsle, 1979). When it comes to (inter)governmental organizations, one can attribute power to their legislative competencies, their ability to sanction "misbehavior", their procedural role, or their resources in terms of finances, manpower and expertise, among others (Barnett & Finnemore, 1999; Karns & Mingst, 2004). In terms of the power of non-state actors, scholars tend to work with frameworks differentiating between dimensions of power, such as instrumental, structural, and discursive power (Arts, 2003; Fuchs 2007; Levy & Newell, 2005). In such conceptualizations, instrumental power refers to the influence on policy output that non-state actors gain via lobbying, sponsoring or campaign finance. Structural power points to the influence non-state actors gain via "exit" threats (agenda-setting power) or private standards (rule-setting power). Discursive power indicates efforts to exercise influence on the public discourse via norms, frames and ideas (Fuchs 2013).

Given the broad focus of our analysis in terms of the actors involved, we cannot carry out a differentiated review of all the specifics of the types, sources, and dimensions of power involved for all of the actors. Therefore, we restrict our consideration of power to the following aspects. First and foremost, we focus on relative power, as we study the relationship between the EU institutions, parts of the EU institutions, as well as between the EU institutions and external stakeholders in the classical political sense of competition and contest. As we are looking at specific developments, we also focus on the actual power exercised rather than the potential power of the actors, i.e. take aspects such as attention and salience into account. In terms of the EU institutions, we consider their procedural role and their resources highly relevant which is in line with much of EU politics scholarship and which translates into questions of agenda-setting and decisional power. Finally, we consider the instrumental, structural (agenda-setting), and discursive dimensions of power when looking at the role non-state actors have played in the processes analyzed. We base our analysis on a triangulation of analyses of official EU documents and stakeholder position papers, media reports, secondary literature, and structured interviews on the EU's climate and energy policy. We conducted semi-structured interviews with members and staff of the European Parliament, of the Commission, as well as with non-state actors.

The article proceeds as follows. In the next section, we consider the first layer of the picture, focusing on the role of power dynamics between and within the EU institutions in shaping climate policy outcomes. Section 3 then examines the layer beneath this picture, describing the long arm of business influence on the EU's 2030 targets. Finally, section 4 summarizes our analysis and lays out implications for science and politics.

2. The 2030 Targets as a Result of Power Dynamics between and within the EU's Core Institutions

2.1 Power between and within the EU's Institutions

The EU politics and policy literature has long emphasized the different roles that the EU's core institutions play in the policy-making process (see Tsebelis & Garrett, 1996; Crombez, 1997; Thomson & Hosli, 2006; König, Lindberg, Lechner & Pohlmeier, 2007). The central actors in the traditional policy-making process are the Commission, the Council of Ministers, and the European Parliament. (Note 3) They form the so-called institutional triangle and are interdependently linked (Delreux & Happaerts, 2016). Although the European Council is not part of this triangle, it is equally important for understanding EU policy-making. In providing political direction and determining the broad strategies for future EU policies, it has a decisive role in the proceedings prior to the actual policy-making process that takes place within the institutional triangle, a role that - as studies show - has actually increased in recent years (Costello & Thomson, 2013; Thomson, 2015; Bocquillon & Dobbels, 2014), and that was particularly important in EU climate politics before and after Copenhagen. Hereafter, we will focus on the Commission, the European Council, and the Parliament, and discuss their relative power in the EU policy-making process leading up to the decision about the 2030 targets.

The supranational European Commission is the institution which, at least formally, has exclusive agenda-setting power, i.e. the power to propose new EU legislation, which makes it a pivotal actor in EU environmental governance (Selin & VanDeveer, 2015). However, its formal agenda-setting power may be challenged by internal dynamics that originate from the informal distribution of power between the EU institutions. Since environmental issues, such as the decision on long-term goals for energy and climate policy, are highly politicized topics, relevant debates have increasingly shifted to the level of the heads of state and government. Therefore, the agenda-setting power in environmental governance is "in practice often the result of a ping-pong game between the Commission and the European Council" (Delreux & Happarts, 2016: 64), going hand in hand with an increase in the European Council's powers as *de facto* agenda-setter (Bailer, 2014; Bocquillon & Dobbels, 2014). Some studies argue that the Commission's structural agenda-setting power has, in fact, been eroded by the European Council (Wessels & Höing, 2013).

Research has shown that the European Council is the most powerful agenda-setter when the state leaders in the Council are united (Bürgin, 2015; Bocquillon & Dobbels, 2014; Skovgaard, 2014). Conversely, "the discretion level of the Commission President, but also the individual commissioners, increases in absence of strong signals from the Council in favour of a certain policy" (Bürgin 2015: 704). The implication is that the interest constellations in the Council are pivotal. Whereas the interests were quite homogenous during the drafting of the 2020 climate and energy package with the strong personal engagement of the political leaders of core member states, the Council sent much more ambivalent signals during negotiations for the 2030 framework (op.cit.). Since the failure of the COP 15 in Copenhagen in 2009, more and more EU member state governments have voiced skepticism regarding the Commission's ambitious targets (Phillips, 2010; Fischer & Geden, 2015). Scholars document a "trend towards polarization on most energy and climate related topics in the EU with two camps facing off against one other" (Fischer, 2014:2). On the one hand, there are the more proactive member states with Sweden and Denmark at the front. On the other hand, there are the Visegrád countries (Poland, Slovakia, the Czech Republic, Hungary) as well as Lithuania, Bulgaria and Romania, who tend to be very reluctant and are often described as hampering the process (Interviews 1, 2 and 3; Fischer & Geden, 2015; Teffer, 2014). Scholars attribute their reluctance regarding ambitious climate targets to "their still weak overall economic performance, their energy-inefficient industries and their political sensitivities vis-à-vis energy dependence on Russia" (Delreux & Happaerts, 2016: 224). Poland, which draws its electricity primarily from coal, emerged as the most vocal representative of the interests of the Central and Eastern European member states in the Council (Interview 3; see also Fischer, 2014: 2; Evans, 2014). Moreover, in the case of the negotiations for the 2030 framework, even core member states like France and Germany, once considered to be strong supporters of ambitious climate targets, failed to provide the leadership necessary for a proactive agreement (Interviews 1, 2 and 3; Fischer & Geden, 2015; van Renssen, 2014). Between the proactive and the reluctant member states were countries such as the UK and the Netherlands who, "while advocating for an ambitious climate target, … wanted to prevent the EU from expanding its powers in the fields of energy efficiency and initiating another round of binding targets for renewable energies" (Fischer, 2014: 2). One interviewee observed that the Eastern European countries may publicly oppose proposals, but that states like the

Netherlands like to hide behind such opposition (Interview 1). Not surprisingly, then, several interviewees perceived getting all the member countries of the EU 'on board' as one of the main challenges for future EU climate governance (most strongly Interviews 1 and 3). The heterogeneity of interests in the Council, then, encouraged member states opposing strict targets to voice their opposition even more vehemently (Bürgin 2015: 699f.). With the European Council split over the desirable level of ambition for the 2030 energy targets, it exerted little restraint on the agenda-setting role of the Commission (op.cit.).

As climate politics requires a high degree of technical expertise, it is important to also consider the Commission's cognitive powers. Through the development of technical and political expertise, the Commission has become a "key institutional actor, operating at the heart of EU environmental policy" (Schön-Quinlivan, 2013: 96). The Commission's competencies are difficult to compare with those of the European Council, due to the latter's ability to rely on national bureaucracies (Thomson 2015: 203). There is a clear difference to the resources of the EP, however. The MEPs are only modestly resourced in comparison to the Commission and the European Council. Taking into account that "institutions require resources to give meaning to formal powers" (Thomson, 2015: 207), the cognitive powers of the Commission provide it with more power in practice than formal rules suggest (op.cit.: 201). In contrast, the "EP oftentimes does not or cannot make full use of the power granted to it in the rules of procedures" (op.cit.: 199). Indeed, the EP perceives itself as rather dependent on adequate proposals by the Commission (see below).

Importantly, the Commission applies its cognitive powers not only in the agenda-setting process, where it has a formal role. The introduction of the co-decision procedure and the increased heterogeneity and workload between and within the EU's core institutions led to an intensification of direct informal relations between the Council, the EP and the Commission. Today, so called informal trialogues, typically between the rapporteur, the relevant committee chair, delegation chairs from the Council and the EP, and a representative of the Commission tend to be used when seeking a compromise (Burns, 2013: 140f). (Note 4) Due to its participation in these trialogues, the Commission etches closer to exercising decisional power, even though formal rules do not provide for that (Thomson, 2015: 199).

In contrast to the Commission, the EP has gained formal decisional power over an increasing range of policy areas over the last decades. The consequences of this development, however, are not as straightforward as one may think. The EP used to be considered the green voice in European politics that "forced its way into the traditional bipartite relationship between the Commission and Council" (Burns, 2013: 147). At least partly, the EP still perceives itself as THE environmental actor (Interviews 1, 2, and 3). In particular, the "Committee of Environment, Public Health and Food Safety" (ENVI), the main committee dealing with environmental and climate change policies, had developed a reputation of often favoring environmental over industry concerns (Rasmussen, 2012: 249).

Now, however, there is also the perception of a certain weakness, of the EP alone being powerless and dependent, as pointed out above (Interview 2). An interviewee specifically stated that the MEPs have to wait for a proposal of the Commission, as well as then adjust their positions to the ones of the Council (ibid.). The traditionally most environmentally progressive actor is forced into a waiting position (ibid.). Interviewees also noted that there is a certain pressure within the EP to produce results (Interview 4). This increases the need to find compromises in order to maintain the EP's reputation as an effective institution, even if that means that a lot of MEPs will not be 100 % content with the result (Interview 1). These findings are in line with Rasmussen's (2012) argument that the EP has increasingly been departing from its traditional role and has turned into an "environmental pragmatist", due to asymmetric lobbying from the industry (see also below), a strategic use of "realistic" negotiating positions towards the Council, and increasing cooperation and coordination between EP committees. Similarly, other scholars argue that an increase in the Parliament's formal power, namely the co-decision procedure, has led to a decline in its willingness and ability to adopt radical environmental amendments and to set the wider EU environmental policy agenda (Burns & Carter, 2010; Burns, Carter & Worsfold, 2012).

2.2 The 2030 Targets

With respect to the 2030 targets, we detect a struggle in the EP between its traditional ambition in environmental leadership and the wish to adopt "pragmatic" positions. In February 2014, the EP adopted a non-legislative resolution on the 2030 framework demanding three binding targets: one for the reduction of greenhouse gas emissions by at least 40 %, one of producing at least 30 % of total final energy consumption from renewable energy sources, and an energy efficiency target of 40 % (European Parliament, 2014). The debate surrounding these targets can be illustrated in the roles of ENVI rapporteur Delvaux, who supported the decision even against her own group, and ITRE [Committee on Industry, Research and Energy] rapporteur Szymanski, who withdrew his name from the report arguing that binding targets do not sufficiently consider the different capacities of member states as well as individual sectors. In the end, the proposal from the Commission reflected these targets, but not the EP's wish for their binding nature.

As the above discussion illustrates, dynamics in the Council, resource asymmetries between the EU's core institutions, and informal procedures in the EU institutional interplay strengthened the power of the Commission in the political process leading up to the definition and adoption of the 2030 targets. Therefore, it is appropriate to pay special attention to the power dynamics within the Commission during that phase.

Delreux and Happaerts describe the Commission as a "hybrid institution insofar as it is partly political (the College of the Commissioners) and partly bureaucratic (the administrative services and Directorate-Generals)" (Delreux & Happarts, 2016: 59). The Commission's President can potentially deploy considerable structural, entrepreneurial or cognitive leadership, which, however, depends on each individual Commission President and on the support of the Council and/or EP (ibid.). The College of Commissioners, in turn, consists of twenty-eight Commissioners who each hold a special portfolio. The autonomy of individual Commissioners is restricted, however, as the College strives for consensus decision-making. (Note 5) Yet, Bürgin convincingly argues that in the absence of a clear position of the European Council, the influence of an individual Commissioner can be noteworthy (Bürgin, 2015: 703f.). The Commissioners are considered to represent the EU as a whole. Thus, they should not represent the national interests of their home countries (Selin & VanDeveer, 2015: 53). However, it is widely acknowledged that Commissioners can assume multiple roles, the 'country role' being one of them (Delreux & Happarts, 2016: 59; Egeberg, 2006).

Next to the Commission's political component, which we described above, its bureaucratic element is crucial for understanding the power dynamics within the Commission. The Commission's bureaucracy consists of horizontal services (Publication Office, Legal Service etc.) and vertical Directorates-General (DGs). The DGs support the College of Commissioners and the relevant DGs typically become the "lead service" for the drafting of proposals. When a DG is in the 'lead' for a policy proposal, it may, in theory, promote its own interests in the drafting process, which is why "the appointment of the lead service can sometimes be heavily disputed" (Delreux & Happaerts, 2016: 65). Apart from the lead, the staff size of the individual DGs influences their relative power in the intra-Commission dynamics (Delreux & Happaerts, 2016). In the case of cross-sectoral policy issues, the DGs having the lead need to consult other relevant DGs. However, the institutional fragmentation of the Commission as well as the inadequate willingness for inter-DG cooperation frequently hamper the coordination of policy areas (Smith, 2014; Knill & Liefferink, 2007). Instead, "[r]ivalries between DGs and the interests related to their policy domain often characterize the internal decision-making process in the Commission" (Delreux & Happaerts, 2016: 66). One initiative to ameliorate the intra-Commission coordination was the strengthening of the role of the Secretariat-General (SG), which Barroso converted into a personal service to the President and thereby enabled "to turn from the guardian of procedure into an interventionist body, with an involvement in policy, and able to force DGs to co-operate more effectively with each other" (Kassim et al., 2016: 8).

Against this institutional context, two factors can help explain the change in the ambitiousness of EU climate policy from pre- to post-Copenhagen: a reduction in climate leadership by the President and divisions within the Commission creating a window of opportunity for Commissioners opposed to stringent targets. Turning to the President's role first, we find that Barroso reduced his efforts at EU climate leadership in his second term. As Barnes (2011) notes, however, "effective and/or radical climate policy proposals cannot come from the Commission without some measure of commitment from the President" (43). The global economic and financial crises and its aftershocks may be cited as a legitimate stumbling block for the Commission's attention to environmental issues (Čavoški, 2015). However, scholars also argue that the strengthened role of the Commission Presidency contributed to these changes in environmental activism (Steinebach & Knill, 2016). "[S]ince 2004, Commission Presidents have centralized decision-making authority and used the power of the office to control policy activism in the organization" (Kassim et al. 2016: 2). Kassim et al. claim that Barroso, first, and then Juncker used this control to reduce the Commission's production of legislative proposals in all policy areas (op.cit.: 13f). (Note 6) Interviewees from different institutional backgrounds also made note of this strengthened role of the presidency and the simultaneous reduction in attention to environmental objectives when discussing the prospects for environmental policy ambitions in the EU (Interviews 1 and 2). With regards to the apparent ineffectiveness of the EU's leadership norm up to Copenhagen, however, a strong Presidential agenda would have been more necessary than ever to internally and externally promote ambitious EU climate and energy targets (Bürgin, 2015; Steinebach & Knill, 2016).

The weakening of presidential leadership, in turn, allowed divisions in the Commission to become more prominent and influential. While a close collaboration between the Environment Commissioner and the Energy Commissioner developed during the first Barroso Commission and led to the further integration of energy and climate change issues (Barnes, 2011: 49f.), the post-Copenhagen situation looked much different. In particular, the lead DGs, DG Energy and DG Climate Action, were divided – even internally – on questions like the number of climate targets (one or three) as well as the level and the bindingness of the targets (Bürgin, 2015: 700f.). In addition, the inclusion of many representatives from other DGs in the drafting process diluted the agenda-setting power of the lead DGs. In this context, energy Commissioner Günther Oettinger took a particular active role, promoting a 35% GHG reduction target and nationally non-binding renewables targets (ibid.). The negotiations on the 2030 framework in the Commission ended with a compromise agreement put forward by Barroso that retained the 40% GHG reduction goal in exchange for the abandonment of the national binding renewable energy targets (Bürgin, 2015: 703). Commissioner Oettinger's role in this drafting process was especially interesting because of his "more pronounced business orientation compared to his precursor, Piebalgs" (op.cit.: 702; Hall, 2012), which we will discuss in more detail below.

3. Non-State Actors and the 2030 Targets

3.1 Non-state Actors in EU Politics

Interest groups are omnipresent in European policy processes. The EU, in particular, is notorious for both huge numbers of lobbyists in general as well as a particularly large presence of business lobbies (Beyers, 2004; Bouwen, 2002; 2004; Eising, 2007; Greenwood, 2011; Knodt, Quittkat & Greenwood, 2012). Moreover, critical observers have noted that regulations trying to reign in lobbying activities have been mostly ineffective (Greenwood, 1998).

Formal and informal lobbying (including the sponsorship of events) is the primary form of the exercise of instrumental power in the EU, reflecting attempts to influence political output via communication with politicians and bureaucrats (Fuchs 2007). (Note 7) As early as the 1990s, scholars and activists pointed out the sheer numbers of lobbyists in Brussels. While these lobbyists and interest groups represent business as well as civil society interests, lobbying research on the EU has generally found that there is a stark imbalance in the numbers in the representation of business interests versus civil society interests (Nollert, 1997; Ronit & Schneider, 2000): "business groups are the predominant category of European interest group" (Grant, Matthews & Newell, 2000: 46).

This is partly because business actors lobby via European and national associations, as well as – in the case of large business actors - individually or in small groups (see below). Moreover, business associations and large business actors tend to have staffed offices in Brussels but also hire professional lobbyists there. They also bring in personnel from headquarters if necessary. Business actors, particularly large business actors, are advantaged because the higher costs of interest representation at the supranational level work to the benefit of resource rich actors relative to those with lesser financial resources, such as smaller businesses or NGOs. Given that most business actors lobby on rather specific issues affecting their sector or company, while most environmental organizations cover a multitude of issues and legislative proposals at any one time, the effect of this imbalance tends to be felt even more strongly. Official efforts to create a more balanced playing field by strengthening civil society representation in the EU, however, have mostly been failures (Geiger, 2005, Hüller, 2010).

One should not assume that this preponderance in "heads" and money always translates into a similar preponderance in the successful exercise of influence, of course. As some scholars have pointed out, citizen groups can at times successfully attain their desired policy outcomes despite strong opposition from business actors (Dür, Bernhagen & Marshall, 2015: 24). One condition that increases the chances of such outcomes is a lack of cohesion among business interests, as Falkner (2008) suggests.

The asymmetric relationship between civil society and business is paralleled by an asymmetry in the representation of the interests of small versus large and especially corporate business actors. Next to their individual presence in terms of offices, hired lobbyists and support from headquarters, many corporate actors lobby through "several different types of organization at the European level through which [they] can represent [their] views" (Grant et al., 2000: 47), ranging from large associations to informal dining clubs. They generally draw on the club strategy because of the lower organizational costs and greater flexibility of this organizational form, and in particular, in cases where corporate interests diverge from the interests of smaller business actors (Coen, 2005, Grande, 2001, 2003). (Note 8) In this context, scholars have drawn particular attention to the role of the European Roundtable of Industrialists (ERT), founded in 1983 as an informal group of around 50 CEOs of the largest European multinational companies. While the public may know little about this organization, it has been identified as the most influential lobby organization in the EU (Brand et al., 2000: 147; see also Nollert,

2016). Another type of exclusive club, which constitutes a grey area between private lobbying associations and intergovernmental organizations, is the 'Transatlantic Business Dialogue (TABD)', created in 1995 by the EU Commission, the US Ministry of Commerce, and the ERT. It consists of chief executive officers and executives from leading American and European companies. The TABD was established to facilitate intergovernmental negotiations between the EU and the US and, thus, is granted favoured access to the State Department and the European Commission (Coen & Grant, 2001). Because of the TABD's ability to provide quick and focused policy responses, it has evolved "into a business-government policy-making body" (op.cit: 37).

Importantly, the exercise of instrumental power in European politics is a multi-level game, which again underlines the advantages of resource rich actors:

> Europe-wide business groupings such as the European Roundtable of Industrialists of the Union of Industrial and Employers' Confederations of Europe (UNICE) assist the European Commission in agenda setting. However, the ratification process in Europe is conducted by national governments, so lobbies must focus at this stage on national capitals. The Europe-wide industry groupings noted above tend to flexibly dissolve into domestic constituents when necessary, and to reconstitute at the European level when pressure needs to be applied on the European Commission or the European Council (Walter, 2001: 55).

Lobbying plays a peculiar role in European politics also due to the European Commission's need for expertise as well as for democratic legitimacy and political support (Broscheid & Coen, 2003; Delreux & Happaerts, 2016; Grant, 2013). (Note 9) Due to the Commission's limited resources and the complexities involved in creating regulation for 28 member countries, the Commission regularly asks for input and invites especially business representatives to contribute to working groups and committees if not the drafting of regulatory proposals (Eising, 2001), a process that contributes to the development of clientilist relationships (Nollert, 1997). (Note 10) In consequence, scholars consider the interests and views of business lobby groups to be deeply embedded within the Commission (Coen, 2007). ENGOs have been making conscious efforts to improve their situation and skills in this respect, especially through the creation of a loose but coordinated network called the Green 10, which consists of the ten largest environmental organizations and networks active on the European level (Grant, 2013). (Note 11) Through the Green 10, the ENGO community has managed to foster the development of a "network of informal contacts" and regular meetings with Commission officials (Delreux & Happaerts, 2016: 133).

Structural forms of power present a second way for non-state actors to exercise power in European politics. The form of structural power that is relevant for our analysis is agenda-setting power, i.e. it allows non-state actors to get issues on political agendas or just as importantly to keep issues off political agendas (Note 12). The source of this power is business' ability to threaten the moving of investments and jobs to other countries with more attractive policy environments (Frank, 1978). In environmental contexts, scholars developed and debated the concept of "pollution havens", allowing for a race to the bottom in environmental standards, or at least barriers to improvements due to competitiveness concerns in political debates (Braithwaite & Drahos, 2000). Today, such competitiveness concerns still frequently influence political debates and agendas, even as subnational units compete with each other for investments (Altvater & Mahnkopf, 1996; Amoore, 2006; Brand et al., 2000; Gill, 1995; Newell & Levy, 2006; Strange, 1998). Moreover, capital has become ever more flexible in its movements over the last decades, as firms frequently work with subcontracting in production rather than the ownership of production sites. Debates about the constraints that global competition imposes on policy choices are also well known in the EU (Mazey & Richardson, 1997; Marazzi, 1995) and scholars have identified the ERT, in particular, as influential in the corresponding framing of political debates (van Apeldoorn, 2002).

Finally, non-state actors exercise discursive power in attempts to influence the policy process, i.e. aim to shape policy via the framing of policy problems, solutions, political actors and contexts. To this end, they reference and strategically use norms and ideas in efforts to influence public opinion (Levy & Newell, 2005). Discursive power can intervene in the policy process at the earliest possibility, i.e. before identities and interests are even formed. Therefore, it is the most subtle form of political power, but also the most difficult to contest (Graf & Fuchs 2014). Given that discursive power depends on the perceived legitimacy of the speaker, it is considered one of the strong points of NGOs, who tend to benefit from ascriptions of non-profit interests (Auer, 2000; Holzscheiter, 2005). At the EU level, however, the lack of transparency of policy processes and cognitive distance to the population renders traditional discursive civil society instruments in terms of protest and public debate less useful (Nollert, 1997). At the same time, in the era of mediatized politics, discursive power also relies on financial resources, which further advantages business actors (Fuchs 2007). Still, scholars of interest group politics in the EU point out that NGOs have been relatively successful in framing and influencing environmental policy debates, in particular (Dür et al., 2015).

3.2 Non-state Actors and the EU's 2030 Targets

When it comes to the 2030 targets, our interviews show evidence of the exercise of instrumental, structural, and discursive power by non-state actors as well as coordinated multi-level activities (see also Boasson & Wettestad, 2013; Fagan-Watson et al., 2015; Grant, 2013). Moreover, the predominance of business becomes visible again. In the public consultation, which was launched after the Commission's adoption of a Green Paper on the 2030 framework in March 2013, 59% of submissions came from business (industry associations 41%, individual power sector companies 10%, other individual companies 8%) compared to 8% from NGOs. When asked for the ten most influential individuals in EU climate policy, several of our interviewees pointed out the extensive influence of business and their associations. While they noted the simultaneous presence of NGOs, some interviewees saw the industry lobbies as more influential, in the end, especially because of their influence on the Commission and the member states (Interview 2). According to the interviewees, the pressure created in such situations may become so strong that the Commission feels almost forced to react in proposing corresponding legislation (ibid.).

Likewise, several studies document the predominance of business in lobbying on EU environmental policy in general and EU climate policy in particular (Bunea, 2013; Coen, 2007; Greenwood, 2011; Knill & Liefferink, 2007; Rasmussen, 2012). Fagan-Watson et al. (2015) show how industry associations such as Business Europe, Fuels Europe (formerly known as EUROPIA), the International Association of Oil and Gas Producers (IOGP), the Confederation of European Paper Industries (CEPI) and the European Steel Association (EUROFER), established formal and informal relationships with policymakers, submitted position papers and policy proposals, and stayed engaged in every subsequent stage of the policy process, also functioning as a "revolving door" between the public and the private sectors. While acting as associations, they also brought in their members at important points in the policy process. Fagan-Watson et al. conclude that the extent of the associations' engagement with the EU policy process is "both impressive and (to some commentators) a source of serious concern" (op.cit.: 73). Similarly, Ydersbond (2016) finds the energy-intensive industry and Business Europe to have been very influential in lobbying the Commission in the multi-stakeholder negotiations leading up to the 2030 framework (98).

At the same time, the Commission also proved particularly open to business in the negotiation process. Given their comparatively small staff as well as the importance of technical expertise, DG ENV and DG CLIMA are considered particularly accessible to business from the onset (Coen & Katsaitis, 2013). In fact, Coen and Katsaitis consider DG Environment to be the most intensely lobbied DG in the entire Commission (op.cit.: 1113). In the negotiations for the 2030 targets, moreover, other DGs and Commissioners with close ties to industry took on a pivotal role (Bürgin, 2015). Indeed, the role of Energy Commissioner Oettinger in the drafting process deserves particular attention. Critical observers connect him to the German car industry, large energy utilities, and the oil industry, specifically BP (Hall, 2012; Nelsen, 2016). In 2012, Oettinger, himself, argued that "Europe is at a competitive disadvantage because of a reluctance to take risks on offshore oil drilling and tar sands, and a failure to fully explore its shale gas options" (Neslen, 2012). This situation supports recent lobbying research suggesting that business exercises its influence not necessarily by swaying politicians but by cultivating "industry champions" among them.

Business associations and individual companies also relied extensively on the use of structural power in this phase. The lobbies of the energy-intensive industry, in particular, all emphasized the threat of a loss in competitiveness and jobs and the risk of 'carbon leakage' in their position papers (Business Europe, 2013; CEPI, 2013; ERT, 2013; EUROFER, 2013; Fuels Europe, 2013; IOGP, 2013): "These concerns were then frequently used to support calls for a 'rebalancing' of the three headline policy objectives of security of supply, climate mitigation and cost-effectiveness" (Fagan-Watson et al., 2015: 34). Next to the position papers, letters to Barroso as well as individual Commissioners emphasized the same frames. In an open letter to Commission President Barroso in January 2014, for instance, Business Europe warned of the likelihood of carbon leakage: "After a series of wrong signals, […], we must strengthen carbon leakage protection otherwise this would have immediate and serious consequences for industrial investments in Europe" (Business Europe, 2014). Likewise, BP warned Energy Commissioner Oettinger that an exodus of the oil industry from Europe would occur if the proposed laws to regulate tar sands, cut power plant pollution and accelerate the uptake of renewable energy were passed (Nelsen, 2016).

Given that business' campaigns in the negotiation of the 2030 targets included open letters, the briefing of journalists as well as the hosting of relevant events, we can also note business' attempts to exercise discursive power in the process. Indeed, the case is particularly illustrative of how the different types of power – which we have conceptually separated for analytical purposes – tend to work jointly in practice. Thus, business associations combined the provision of "technical" information to the Commission and submission of position papers with discursive strategies aiming at lobbying the public.

Importantly, the above discussion neither aims to suggest that business should be considered as a monolithic bloc when it comes to climate policy (see also Falkner, 2008; Grant, 2013) nor that the eventual outcome was exactly what business had asked for. Business associations representing renewable energies were also present in the process and lobbying both the European as well as national actors, and had different interests than the energy intensive industries, for example (Interview 5). Yet, we find a clear predominance of business interests on the side of non-binding national targets. Many of the associations sought an overarching GHG target rather than three targets, furthermore, citing inefficiencies resulting from multiple, overlapping targets in the 2020 package (Business Europe, 2013; CEPI, 2013; Fuels Europe, 2013; IOGP, 2013). They opposed the energy efficiency and renewable energy targets, in particular. However, in the end, they were not able to achieve these goals. According to our interviews, this was, to a large extent, a result of the influence of the EP.

ENGOs, especially the Green 10, also participated in the consultation process for the 2030 targets, though to a quantitatively lesser extent than business actors. As documented by their position papers, the majority of ENGOs supported three ambitious and binding targets, but failed to agree on the specifics of them. (Note 13) Moreover, they tried to use their discursive power to oppose business structural agenda-setting power by framing climate policy in terms of job growth rather than loss and by deconstructing the supposed contradiction between environmental policy and economic growth (see also Ydersbond, 2016). WWF (2013) emphasized that ambitious 2030 renewable energy targets as well as investing in energy efficiency in buildings can create jobs. FoEE (2013) called measures of reducing energy consumption and developing renewables a "huge stimulus for European businesses and jobs" and the EEB (2013) argued that ambitious targets will "boost Europe's competitiveness".

ENGOs also tried to exercise power jointly with business actors that were closer to their positions. The 'Broad Green Community', an alliance between the Coalition for Energy Savings (CoE) (Note 14), CAN Europe and the European Renewable Energy Council (EREC) (Note 15), campaigned for three legally binding climate and energy targets and tried to coordinate the lobbying activities of its members on the European as well as national level (Ydersbond 2016). Again, they exercised discursive power, in particular, publishing scientific analyses, and arranging dinner debates and other events (op.cit.: 62).

Our interviewees found business to be more influential during the development of the proposal, however. One interviewee stated that NGOs may be good in moving and mobilizing civil society and public opinion but that they were not able to influence politicians' opinions (Interview 2). In the 'battle' with the industry lobbies, NGOs tended to lose, whereas industry lobbies successfully exercised influence on member states and the Commission (ibid.). Yet, some influence can be attributed to ENGOs in the development of the 2030 targets after all. Scholars argue that, "[t]heir actions helped to pave the way for the more radical recommendations of the Parliament and enhanced the legitimacy of a triptych approach with targets that could be adjusted upwards, and for setting targets that were higher than the 'business-as-usual' scenarios" (Ydersbond, 2016: 97).

The predominance of business interests continued beyond the proposal development stage, however. This should not come as a surprise. The European Commission may still be understood as the most important target for lobbying strategies in Brussels, but the European Parliament and its committees have increasingly entered into the limelight of interest groups as well (Adelle & Anderson, 2013: 165). Even in the EP, the actor traditionally more open to NGOs influence than the Commission (Interview 1), the powerful and intense lobbying by business, both by associations and individual companies, was noteworthy, according to almost all interviewees. One interviewee described business lobbying on the 2030 targets as sort of aggressive, with position papers approaching political declarations and fervent attempts to get these positions into the EP's papers (Interview 3). Interviewees also commented on the nature of the process, pointing out numerous requests for meetings with nine out of ten of them taking place, followed by subsequent email exchanges and frequently the submission of proposals for text passages for legislation (Interview 4). Not surprisingly, one interviewee noted a particularly strong lobby on the part of companies with a lot at stake in European environmental negotiations (Interview 1), while another considered the negative lobby from many businesses as one of the main obstacles to reaching more ambitious climate targets (Interview 3).

These indications of business' predominance in the lobbying of the EP on the 2030 targets correspond to Rasmussen's (2015) general findings on non-state actors' access to and influence on the EP. In the past, the European Parliament was considered a "lobbying sideshow" (op.cit.: 365) targeted by interest groups that were unsuccessful in lobbying the European Commission. According to Rasmussen, however, today the European Parliament is targeted by a plethora of interest groups with business interests becoming more and more influential. As with the Commission, an asymmetry in lobbying does not mean that business will get all it wants. Indeed, the chances of business winning legislative contests at the EP depend on the extent of consensus among business actors, the degree of technicality of an issue, and the handling of dossiers by mainstream committees (op.cit.: 380). Yet, the asymmetry in lobbying does have an impact according to Rasmusssen (see also Smith, 2008): "when MEPs are exposed asymmetrically to industry lobbying on a specific dossier, they are less likely to support ambitious environmental provisions than those who are more evenly exposed to lobbying from opposing interest groups" (253).

An analysis of the role of non-state actors in the EU's development of its 2030 climate targets shows a strong presence of non-state actors, but particularly the predominance of business. Given the nature of multi-level governance in the EU, non-state actors simultaneously targeted policy at the European and the national level. At the European level, both the Commission and the Parliament were the focus of attempts to exercise instrumental, and – in the case of business – also structural power, while discursive power was exercised in efforts to sway public opinion. In this context, the Commission, especially individual Commissioners, proved particularly accessible to business lobbies. Accordingly, business interests opposed to ambitious targets were able to exploit and foster the divergence in positions in the Commission and the Parliament.

4. The Future of the EU's Climate Policy?

In this article, we investigated power dynamics between and within the EU institutions as well as between the EU institutions and non-state actors engaged in the development of the 2030 climate & energy framework. Employing an analysis of official documents, stakeholder position papers, interviews with representatives of the EU institutions and non-state actors, and secondary literature, this paper highlighted the central role the Commission and, in turn, individual DGs and Commissioners were able to play, given the heterogeneity of positions in the Council and the influence that that situation awarded to business interests. Moreover, unveiling the power dynamics between and within the EU institutions in terms of their implications for the influence of non-state actors on EU climate policy serves to identify different layers of the picture. Most importantly, it reveals how inadequate guidance from the Council and divisions within the Commission facilitated the informal passing of the scepter of EU climate policy to business, even as EU institutions officially retained responsibility.

Our findings speak to the future of EU climate policy as well as future political dynamics in the EU, in general. In terms of EU climate policy, they demonstrate that we should not expect too much ambition with regards to environmental policy from the EU in the future. As long as climate action is not in the focus of and desired by a majority of the member states, the Commission will continue to play a core role. Within the Commission, in turn, power dynamics have shifted against environmental interests. Given Juncker's 'Agenda for Jobs and Growth', in particular, scholars consider it unlikely "that we will experience a reinvigoration of environmental policy in the near future" (Steinebach & Knill, 2016: 15). Interestingly, many public statements by EU officials and heads of the large member states after the Brexit vote emphasized the need for jobs and growth rather than fairness and democratic change, thereby further underscoring the priorities of the relevant political elites. Furthermore, the number of Commissioners for the core environment-related portfolios (environment, maritime affairs and fisheries, climate, and energy) has been reduced from four to two from the second Barroso Commission to the Juncker Commission. In the Juncker Commission, the Environment Commissioner is also in charge of maritime affairs and fisheries, while the Climate Action Commissioner also has energy in his portfolio (Delreux & Happarts, 2016: 59f). (Note 16) Juncker's merging of these portfolios had immediately raised "concerns among environmental NGOs and members of the European Parliament about a downgrading of environmental issues in the Juncker Commission" (op.cit.: 60), a concern the above discussion seems to justify. Interviewees, similarly, attributed changes in the Commission's environmental ambitions to such structural changes. One interviewee stated a clear hierarchy in the Juncker Commission, with the Commission President at the top, DG ENV increasingly losing influence, and DG Growth more and more taking the lead on climate related issues, even overruling what DG ENV is doing (Interview 1). Furthermore, the interviewee pointed out a strong influence of the SG arguing that the SG is increasingly doing things that should 'normally' be done by DG ENV. (Note 17) Another interviewee echoed these concerns about the relative importance of environmental objectives in the Juncker Commission, seeing the Commission as much more sensitive to economic and financial interests than to environmental policies, also pointing out its long-time failure to reference sustainability issues at all (Interview 2). (Note 18) Interviewees further noted that the current Commission's proposals on EU climate policy closely resemble the European Council's conclusions, indeed, that the Commission seemingly just 'copies and pastes' those conclusions, having given up on any ambition (Interview 1). Finally, interviewees also stated that industrial associations are especially influential in the Juncker Commission, thus underlining reasons for continued concern about the democratic deficit in the EU (Interview 2).

These developments in the Commission occur at a time, when we are also witnessing the change in the EP from an 'environmental champion' to an 'environmental pragmatist'. This change has gone hand in hand with a decline in the privileged position of diffuse interests, such as environmental ones, and provided "concentrated interests with a more favorable European parliamentary arena in which they can advance their demands" (Rasmussen, 2012: 257). Moreover, this transition is in line with a general change in the EU's environmental regulatory discourse: "Environmental legislation is increasingly viewed through a competitive lens – rather than merely being embedded in an environmental frame – with closer attention paid to the competitive implications of environmental policies" (op.cit.: 255).

Overall then, there is little reason to be optimistic for the future of EU climate policy and its ambitions. Upcoming negotiations of strategies to pursue the adopted targets including revisions of the ETS are likely to show core actors dragging their feet rather than pushing ahead. On a broader level, there is also little reason to expect serious efforts to reduce the democratic deficit in EU politics. Current developments in terms of the predominance of certain ideas and actors point in the opposite direction.

Acknowledgments

The authors are grateful for grant FU434/5-1 by the German Research Association (DFG), which funded our research on the EU's fisheries and climate policy between 2012 and 2016, and the research assistance of Julia Henn, Nina Hilgenböcker, Naemi Hüsemann, Alva Hoffmann, Jana Baldus, Carolin Bohn, Sarra Amroune, Barbara Elpers, Rebekka Stadler and Johannes Grünecker. The interviews were conducted by Antonia Graf and Stephan Engelkamp.

References

Adelle, C., & Anderson, J. (2013). Lobby groups. In Jordan, A., & C. Adelle (Eds.), *Environmental Policy in the EU*. London: Routledge.

Altvater, E., & Mahnkopf, B. (1996). *Grenzen der Globalisierung*. Münster: Westfälisches Dampfboot.

Amoore, L. (2006). Making the Modern Multinational. In May, C. (Ed.), *Global Corporate Power*. Boulder: Lynne Rienner.

Apeldoorn, B. V. (2002). The European Round Table of Industrialists. In Greenwood, J. (Ed.), *The Effectiveness of EU Business Associations*. Basingstoke: Palgrave.

Arts, B. (2003). *Non-State Actors in Global Governance*. Preprint, Max-Planck-Projektgruppe, Recht der Gemeinschaftsgüter, 003/4. Bonn: Max-Planck-Gesellschaft.

Auer, M. (2000). Who Participates in Global Environmental Governance? *Policy Sciences, 33*, 155-180.

Bailer, S. (2014). An Agent Dependent on the EU Member States? *Journal of European Integration, 36*, 37-53. http://dx.doi.org/10.1080/07036337.2013.809342

Baldwin, D. (2002). Power and International Relations. In Carlsnaes, W., T. Risse, & B. Simmons (Eds.), *Handbook of International Relations*. London: Sage.

Barnes. (2011). The role of the Commission of the European Union. In Wurzel, R., & J. Connelly (Eds.), *The European Union as a Leader in International Climate Change Politics*. London: Routledge.

Barnett, M., & Duvall, R. (Eds.). (2009). *Power in global governance*. Cambridge: CUP.

Barnett, M., & Finnemore, M. (1999). The Politics, Power, and Pathologies of International Organizations. *International Organization, 53*, 699-732. http://dx.doi.org/10.1162/002081899551048

Beyers, J. (2004). Voice and access. Political practices of European interest associations. *European Union Politics, 5*, 211-240. http://dx.doi.org/10.1177/1465116504042442

Boasson, E., & Wettestad, J. (2013). *EU Climate Policy. Industry, Policy Interaction and External Environment*. Farnham: Asghate Publishing.

Bocquillon, & Dobbels, M. (2014). An elephant on the 13th floor of the Berlaymont? *Journal of European Public Policy, 21*, 20-38. http://dx.doi.org/10.1080/13501763.2013.834548

Bouwen. (2002). Corporate lobbying in the European Union. *Journal of European Public Policy, 9*, 365-390. http://dx.doi.org/10.1080/13501760210138796

Bouwen. (2004). The logic of access to the European Parliament. *Journal of Common Market Studies, 42,* *473-495.* http://dx.doi.org/10.1111/j.0021-9886.2004.00515.x

Braithwaite, J., & Drahos, P. (2000). *Global Business Regulation.* Cambridge: CUP.

Brand, U., Brunnengräber, A., Schrader, L., Stock, C., & Wahl, P. (2000). *Global Governance. Alternative zur neoliberalen Globalisierung?* Münster: Westfälisches Dampfboot.

Broscheid, A., & Coen, D. (2003). Insider and outsider lobbying of the European Commission. *European Union Politics, 4,* 165-189. http://dx.doi.org/10.1177/1465116503004002002

Bunea, A. (2013). Issues, preferences and ties. *Journal of European Public Policy, 20,* 552-570. http://dx.doi.org/10.1080/13501763.2012.726467

Bürgin, A. (2015). National binding renewable energy targets. *Journal of European Public Policy, 22,* 690-707. http://dx.doi.org/10.1080/13501763.2014.984747

Burns, C. (2013). The European Parliament. In Jordan, Andrew & Camilla Adelle (Eds.), *Environmental Policy in the EU. Actors, Institutions and Processes.* London: Routledge.

Burns, C., & Carter, N. (2010). Is codecision good for the environment? *Political Studies, 58,* 128-142. http://dx.doi.org/10.1111/j.1467-9248.2009.00782.x

Burns, C., Carter, N., & Worsfold, N. (2012). Enlargement and the Environment: The Changing Behaviour of the European Parliament. *Journal of Common Market Studies, 50,* 54-70. http://dx.doi.org/10.1111/j.1468-5965.2011.02212.x

Business Europe. (2013). *A Competitive EU Energy and Climate Policy. Businesseurope Recommendations for a 2030 Framework for Energy and Climate Politics.* Retrieved July 1, 2016, from http://www.bdi.eu/download_content/KlimaUndUmwelt/20130618_FINAL_Brocure_2030_energy_and_climate_LOW_RESOLUTION.pdf

Business Europe. (2014). *Industrial competitiveness and the 2030 climate and energy package.* Retrieved July 7, 2016, from http://www.foe.co.uk/sites/default/files/downloads/businesseurope-letter-barroso-21976.pdf

Čavoški, A. (2015). A post-austerity European Commission. *Environmental Politics, 24,* 501-505. http://dx.doi.org/10.1080/09644016.2015.1008216

CEPI. (2013). *Europe needs more than a 2030 climate target.* Retrieved July 15, 2016, from http://ec.europa.eu/ energy/en/consultations/consultation-climate-and-energy-policies-until-2030

Coen, D. (2005). Environmental and Business Lobbying Alliances in Europe. In Levy, D., & P. Newell (Eds.), *The Business of Environmental Governance.* Cambridge: MIT.

Coen, D. (2007). Empirical and Theoretical studies in EU lobbying. *Journal of European Public Policy, 14,* 333-345. http://dx.doi.org/10.1080/13501760701243731

Coen, D., & Grant, W. (2001). Corporate political strategy and global policy. *European Business Journal, 13,* 37-44. http://dx.doi.org/10.2139/ssrn.319642

Coen, D., & Katsaitis, A. (2013). Chameleon pluralism in the EU. *Journal of European Public Policy, 20,* 1104-1119. http://dx.doi.org/10.1080/13501763.2013.781785

Costello, R., & Thomson, R. (2013). The distribution of power among EU institutions. *Journal of European Public Policy, 20,* 1025-1039. http://dx.doi.org/10.1080/13501763.2013.795393

Crombez, C. (1997). The Co-Decision Procedure in the European Union. *Legislative Studies Quarterly, 22,* 97-119. http://dx.doi.org/10.2307/440293

Delreux, T., & Happaerts, S. (2016). *Environmental Policy and Politics in the European Union.* Basingstoke: Palgrave.

Diez, T. (2005). Constructing the Self and Changing Others. Millenium: *Journal of International Studies, 33,* 613-636. http://dx.doi.org/10.1177/03058298050330031701

Dür, A., Bernhagen, & Marshall, D. (2015). Interest Group Success in the European Union. *Comparative Political Studies, 48,* 951-983. http://dx.doi.org/10.1177/0010414014565890

EEB. (2013a). *The EEB's response to the European Commission's Consultation on a 2030 framework for climate and energy policies.* Retrieved July 15, 2016, from http://www.eeb.org/EEB/?LinkServID= D49AB73B-5056-B741-DB84A81BCCF2201E&showMeta=0

EEB. (2013b). *Annual Report 2013.* Retrieved July 15, 2016, from http://www.eeb.org/EEB/?LinkServID=5CB28CB5-5056-B741-DBF934A519D00261

Egeberg, M. (2006). Executive politics as usual. *Journal of European Public Policy, 13,* 1-15. http://dx.doi.org/10.1080/13501760500380593

Eising, R. (2001). Interessenvermittlung in der Europäischen Union. In Reutter, W., & P. Rütters (eds.). *Verbände und Verbandssysteme in Westeuropa.* Opladen: Leske + Budrich.

Eising, R. (2007). The access of business interests to European Union institutions. *Journal of European Public Policy, 14,* 384-403. http://dx.doi.org/10.1080/13501760701243772

ERT. (2013). *Energy Impacts on Industrial Competitiveness.* Retrieved July 2, 2016, from www.ert.eu/sites/ert/files/generated/files/document/energy_impacts_on_industrial_competitiveness.pdf

EUROFER. (2013). *Green Paper on a 2030 framework for climate and energy policies.* Retrieved July 4, 2016, from http://ec.europa.eu/energy/en/consultations/consultation-climate-and-energy-policies-until-2030

European Parliament. (2014). *European Parliament resolution of 5 February 2014 on a 2030 framework for climate and energy policies (2013/2135(INI)).* Retrieved July 7, 2016, from http://www.europarl. europa.eu/sides/getDoc.do?pubRef=-//EP//NONSGML+TA+P7-TA-2014-0094+0+DOC+PDF+V0//EN

Evans, S. (2014). *Who wants what from the EU 2030 climate framework.* Carbon Brief 17.10.2014. Retrieved July 4, 2016, from https://www.carbonbrief.org/analysis-who-wants-what-from-the-eu-2030-climate-framework

Fagan-Watson, B., Elliott, B., & Watson, T. (2015). *Lobbying by Trade Associations on EU Climate Policy.* London: Policy Studies Institute. Retrieved July 2, 2016, from http://miljo-utveckling.se/files/2015/04/PSI-Report_Lobbying-by-Trade-Associations-on-EU-Climate-Polic y.pdf

Falkner, R. (2008). *Business Power and Conflict in International Environmental Politics.* Basingstoke: Palgrave.

Fischer, S. (2014). *The EU's New Energy and Climate Policy Framework for 2030.* SWP Comments 55. Retrieved July 1, 2016, from www.swp-berlin.org/fileadmin/contents/products/comments/2014C55_fis.pdf

Fischer, S., & Geden, O. (2015). *EU climate policy.* Energypost 21.05.2015. Retrieved July 7, 2016, from http://www.energypost.eu/eu-climate-policy-time-come-earth/

FoEE. (2013). *Submission on the European Commission's Green Paper on a 2030 framework for climate and energy policies.* Retrieved July 15, 2016, from crowdsourcing.simpolproject.eu/static/staticdata/gpc/consultations/friends_of_the_ earth_ europe.pdf

Frank, A. (1978). *Dependent Accumulation and Underdevelopment.* London: Macmillan.

Fuchs, D. (2007). *Business Power in Global Governance.* Boulder: Lynne Rienner.

Fuchs, D. (2013). Theorizing the Power of Global Companies. In Mikler, J. (Ed.), *Handbook of Global Companies.* Hoboken: Wiley-Blackwell.

Fuchs, D., & Graf, A. (2015). Interessenvertretung in der globalisierten Welt. In Zimmer, A., & L. Speth (Eds.), *Lobbywork..* Heidelberg: Springer.

Fuchs, D., Gumbert, T., & Schlipphak, B. (Forthcoming). Euroscepticism and Big Business. In Startin, N., & S. Usherwood (Eds.), *The Routledge Handbook of Euroscepticism.* London: Routledge.

Fuels Europe. (2013). *EUROPIA Response tot he public Consultation on a 2030 framework for climate and energy policies.* Retrieved July 22, 2016, from http://ec.europa.eu/energy/en/consultations/consultation-climate- and-energy-policies-until-2030

Geiger, S. (2005). *Europäische Governance.* Marburg: Tectum.

Gill, S. (1995). Globalisation, Market Civilisation, and Disciplinary Neoliberalism. *Millenium, 24,* 399-423. http://dx.doi.org/10.1177/03058298950240030801.

Graf, A., & Fuchs, D. (2014). Macht – ihr diskursives Regierungspotenzial. In Wullweber, J., A. Graf & M. Behrens (Eds.), *Theorien der Internationalen Politischen Ökonomie.* Wiesbaden: VS.

Grande, E. (2001). *Institutions and Interests.* Working Paper 1/2001, TU München, Lehrstuhl für Politische Wissenschaft.

Grande, E. (2003). How the Architecture of the EU Political System Influences Business Associations. In

Greenwood, J. (Ed.), *The Challenge of Change in EU Business Associations*. Houndmills: Palgrave.

Grant, W. (2013). Business. In Jordan, A., & C. Adelle (Eds.), *Environmental Policy in the EU*. London: Routledge.

Grant, W., Matthews, D., & Newell, P. (2000). *The effectiveness of European Union environmental policy*. Basingstoke: Palgrave.

Greenwood, J. (1998). Regulating Lobbying in the European Union. *Parliamentary Affairs, 51*, 587-599. http://dx.doi.org/10.1093/pa/51.4.587

Greenwood, J. (2011). *Interest Representation in the European Union*. Basingstoke: Palgrave.

Hall, M. (2012, October 12). Oettinger tells Volkswagen he relaxed new CO2 targets. *EurActiv*.

Holzscheiter, A. (2005). Discourse as Capability. *Millenium, 33*, 723-746. http://dx.doi.org/10.1177/03058298050330030301

Hüller, T. (2010). Playground or Democratisation? *Swiss Political Science Review, 16*, 77–107. http://dx.doi.org/10.1002/j.1662-6370.2010.tb00153.x

IOGP. (2013). OGP Response to the Green Paper "A 2030 framework for climate and energy policies". Retrieved July 15, 2016, from http://www.gasnaturally.eu/uploads/Modules/Publications/ogp_response_to _2030_green_paper_-_june_2013.pdf

Karns, M., & Mingst, K. (2004). *International organizations*. Boulder: Rienner.

Kassim, H., Connolly, S., Dehousse, R., Rozenberg, O., & Bendjaballah, S. (2016). Managing the House. *Journal of European Public Policy*. http://dx.doi.org/10.1080/13501763.2016.1154590

Knill, C., & Liefferink, D. (2007). *Environmental Politics in the European Union*. Manchester: Manchester University Press.

Knodt, M., Quittkat, C., & Greenwood, J. (Eds.). (2012). *Functional and Territorial Interest Representation in the EU*. London: Routledge.

Kohler-Koch, B. (2000). Unternehmensverbände im Spannungsfeld von Europäisierung und Globalisierung. In Bührer, W., & E. Grande (Eds.), *Unternehmerverbände und Staat in Deutschland*. Baden-Baden: Nomos.

König, T., Lindberg, B., Lechner, S., & Pohlmeier, W. (2007). Bicameral Conflict Resolution in the European Union. *British Journal of Political Science, 37*, 281-312. http://dx.doi.org/10.1017/s0007123407000142

Krämer, L. (2013). The European Court of Justice. In Jordan, A., & C. Adelle (Eds.), *Environmental Policy in the EU*. London: Routledge.

Levy, D., & Newell, P. (Eds.). (2005). *The Business of Environmental Governance*. Cambridge: MIT.

Manners, I. (2002). Normative Power Europe. *Journal of Common Market Studies, 40*, 235-58. http://dx.doi.org/10.1111/1468-5965.00353

Marazzi, C. (1995). Money in the World Crisis. In Bonefeld, W., & J. Holloway (Eds.), *Global Capital, National State and the Politics of Money*. New York: St. Martin's.

Mazey, S., & Richardson, J. (1997). Policy Framing. *West European Politics, 20*, 111-133. http://dx.doi.org/10.1080/01402389708425207

Neslen, A. (2012, July 19). EU energy chief warms to offshore oil and shale gas. *EurActiv*.

Neslen, A. (2016, April 20). EU dropped climate policies after BP threat of oil industry 'exodus'. *The Guardian*.

Newell, & Levy, D. (2006). The Political Economy of the Firm in Global Environmental Governance. In May, C. (Ed.), *Global Corporate Power*. Boulder: Rienner.

Nollert, M. (1997). Verbändelobbying in der Europäischen Union. In Alemann, U. von, & B. Weßels (Eds.), *Verbände in vergleichender Perspektive*. Berlin: sigma.

Nollert, M. (2016). High-level Lobbying und Agenda Setting. In Wendt, B., M. Klöckner, S. Pommrenke, & M. Walter (Eds.), *Wie Eliten Macht organisieren*. Hamburg: VSA.

Oberthür, S., & Dupont, C. (2011). The Council, the European Council and international climate policy. In Wurzel, R., & J. Connelly (Eds.), *The European Union as a Leader in International Climate Change Politics*. London: Routledge.

Partzsch, L., & Fuchs, D. (2012). Philanthropy: Power with in International Relations. *Journal of Political Power*,

5, 359-376. http://dx.doi.org/10.1080/2158379x.2012.735114

Phillips, L. (2010, July 4). Battling the 'Multilateral Zombie'. EU Observer. Retrieved July 4, 2010, from https://euobserver.com/environment/29354

Rasmussen, M. (2012). Is the European Parliament still a policy champion for environmental interests? *Interest Groups and Advocacy, 1*, 239-259. http://dx.doi.org/10.1057/iga.2012.12

Ronit, K., & Schneider, V. (Eds.). (2000). *Private Organizations in Global Politics*. London: Routledge.

Sanchez Salgado, R. (2014). Rebalancing EU Interest Representation? *Journal of Common Market Studies, 52*, 337-353. http://dx.doi.org/10.1111/jcms.12092

Schön-Quinlivan, E. (2013). The European Commission. In Jordan, A., & C. Adelle (Eds.), *Environmental Policy in the EU*. London: Routledge.

Selin, H., & Van Deveer, S. (2015). *European Union and Environmental Governance*. London: Routledge.

Shepsle, K. (1979). Institutional Arrangements and Equilibrium in Multidimensional Voting Models. *American Journal of Political Science, 23*, 27-59. http://dx.doi.org/10.2307/2110770

Skovgaard, J. (2014). The limits of entrapment. *Journal of Common Market Studies, 51*, 1141-57. http://dx.doi.org/10.1111/jcms.12069

Smith, A. (2014). How the European Commission's Policies Are Made. *Journal of European Integration, 36*, 55-72. http://dx.doi.org/10.1080/07036337.2013.809344

Smith, M. (2008). All Access Points are Not Created Equal. *The British Journal of Politics & International Relations, 10*, 64-83. http://dx.doi.org/10.1111/j.1467-856x.2007.00318.x

Steinebach, Y., & Knill, C. (2016). Still an entrepreneur? *Journal of European Public Policy.* http://dx.doi.org/10.1080/13501763.2016.1149207

Strange, S. (1998). *Mad Money*. Ann Arbor: University of Michigan Press.

Teffer. (2014, October 16). Leaked papers show EU disagreement on climate goals. *EU Observer*. Retrieved July 10, 2016, from https://euobserver.com/news/126112

Thomson, R. (2015). The distribution of power among the institutions. In Richardson, J., & S. Mazey (Eds.), *European Union*. London: Routledge.

Thomson, R., & Hosli, M. (2006). Who has Power in the EU? *Journal of Common Market Studies, 44*, 391–417. http://dx.doi.org/10.1111/j.1468-5965.2006.00628.x.

Tsebelis, G., & Garrett, G. (1996). Agenda setting power, power indices, and decision making in the European Union. *International Review of Law and Economics, 16*, 345-361. http://dx.doi.org/10.1016/0144-8188(96)00021-x

Unbehaun, S. (2016). *A Façade of Unity. Transatlantic Policy Symposium 2016.* Retrieved July 22, 2016, from http://tapsgeorgetown.com/wp-content/uploads/2016/01/Sarah-Unbehaun-EU-climate-and-energy-divisions.pdf

Van Renssen, S. (2014, March 5). *EU deeply divided over 2030 climate and energy policy.* Energypost. Retrieved July 4, 206, from http://www.energypost.eu/eu-divided-2030-climate-energy-policy/

Waele, H. de. (2010). The Role of the European Court of Justice in the Integration Process. *Hanse Law Review, 6*, 3-26.

Walter, A. (2001). NGOs, Business, and International Investment. *Global Governance, 7*, 51-73.

Wasserfallen, F. (2010). The Judiciary as Legislator? *Journal of European Public Policy, 17*, 1128-1146. http://dx.doi.org/10.1080/13501763.2010.513559.

Wessels, W., & Höing, O. (2013). The European Commission's position in the post-Lisbon institutional balance. In Monar, J., & M. Chang (Eds.), *The European Commission in the post-Lisbon era of crises*. Frankfurt: Peter Lang.

WWF. (2013). *Response to the European Commission public consultation on a 2030 climate and energy package.* Retrieved June 10, 2016, from Awsa.panda.org/.../clean_wwf_response_to_2030_green_paper_on_climate_and_ energy.pptx

Ydersbond, I. (2016). Where is power really situated in the EU? FNI Report 3/2016.

Notes

Note 1. Retrieved July 07, 2016, from http://ec.europa.eu/clima/policies/strategies/2020/index_en.htm

Note 2. Retrieved July 07, 2016, from http://ec.europa.eu/clima/policies/strategies/2030/index_en.htm. The energy efficiency target is not binding on either the European or the national level, and the renewables target only on the European level, but not on the national level.

Note 3. These actors also include the European Court of Justice as well as a range of other institutions, of course, which are not central to our case but have received substantial attention in the literature in general (e.g. Wasserfallen, 2008; De Waele, 2010; Krämer, 2013).

Note 4. Burns underlines that the augmented practice of informal meetings potentially excludes smaller groups such as the Greens in the EP or marginalized interests such as environmental NGOs: "[...], if committee meetings and plenaries are increasingly used to rubber stamp agreements negotiated by a few actors behind closed doors, the scope to have one's views considered during the legislative process is much lower" (Burns, 2013: 143). This, in fact, has great ramifications for the often mentioned democratic deficit in the EU, considering that the EP should also represent the views of small groups.

Note 5. Furthermore, the principle of collegiality implies that "each decision made has to be endorsed and defended publicly by all Commissioners" (Barnes, 2011: 98).

Note 6. Steinebach and Knill (2016) analyze the EU's policy outputs from 1980 to 2014 in the area of clean air and water protection and find that the EU "has entered a period of almost complete regulatory inactivity after the year 2010" (11).

Note 7. Levy and Newell (2005) also include the "revolving door", i.e. (often high ranking) politicians and buraucrats moving to the private sector and vice versa, in their concept of instrumental power.

Note 8. Scholars speak of "single issue maximizing" as a strategy in this context, i.e. the creation of temporary, issue based coalitions accompanied by professional campagins (Kohler-Koch, 2000).

Note 9. Considering different policy domains, different types of legitimacy (input/output) and different types of information (technical/political), Coen and Katsaitis (2013) conclude that "policy domains that place higher utility on output legitimacy demand more technical expertise and as a result show a stronger presence of business interests and associations" (1117).

Note 10. This, in turn, means that special interests have continued access to and get involved in the policy process much earlier than traditionally thought and with much more room for shaping the pursuing policy process (Grande 2001).

Note 11. The Green 10 consist of i.a. the World Wide Fund for Nature European Policy Office (WWF EPO), European Environment Bureau (EEB), Friends of the Earth Europe (FoEE), Greenpeace European Unit, Climate Action Network Europe (CAN-E) and Birdlife International. It is important to mention that all members of the Green 10, except for Greenpeace, receive financial support from the European Commission. For a further understanding of the effects of the Commission's funding on the EU system of interest representation see Sanchez-Salgado (2014).

Note 12. Structural power can also result in rule-setting power, in so far as an actor's resources and/or competencies may put it in the position to create collectively binding rules for others. This latter form of structural power has come to be associated with business actors in the context of private governance initiatives, which the EU also has encouraged and relied on in a range of policy fields. Given that the focus of the present analysis is on climate targets set by the EU institutions, business' rule-setting power will not be further discussed here.

Note 13. The WWF wanted a reduction of GHG emissions of at least 55 %, a renewables target of 45 % and an energy efficiency target of 40 % (WWF, 2013). EEB demanded a 60%-45%-50%-trias (EEB, 2013a), FoEE a 80%-marked increase-50%-trias (FoEE, 2013).

Note 14. CoE is the overarching coalition for the groups working for greater energy efficiency in Europe (Ydersbond, 2016: 61).

Note 15. EREC is the umbrella organization for the European renewables industry.

Note 16. In the Barroso-Kommission II (2009-2014) there was a separate Environment Commissioner, Energy Commissioner, Climate Action Commissioner and Maritime Affairs and Fisheries Commissioner.

Note 17. This is particularly noteworthy, as DG ENV and DG CLIMA are considered relatively small to begin with. The staff numbers of DG ENV and DG CLIMA relative to the output in this domain and in comparison to other DGs are limited and scholars view their administrative capacity, therefore, as restricted (Delreux and Happaerts, 2016: 63). Whereas DG ENV employs 500 officials and DG CLIMA 154 officials, DGs like Enterprise and Industry employ 900 officials and DG Agriculture and Rural Development 1000 officials (ibid.). Delreux and Happaerts argue that human and financial resources equate to institutional power in an institution like the Commission and that DG ENV and DG CLIMA therefore have an inferior status (Schön-Quinlivan, 2013: 104).

Note 18. The interviewee linked these developments also to a subsequent deterioration in the relationship between the Commission and the Environment Committee in the European Parliament resulting from the fact that the Juncker Commission only started talking about sustainable development after pressure from the Environment Committee.

4

Study on the Impact of Sand-Clay Bond in Geo-grid and Geo-Textile on Bearing Capacity

Mohammadehsan Zarringol[1] & Mohammadreza Zarringol[1]

[1] Geotechnical Engineering Department, University of Guilan, Rasht, Iran

Correspondence: Mohammadehsan Zarringol, Geotechnical Engineering Department, University of Guilan, Rasht, Iran. E mail: Ehsan.zaringol@gmail.com.

Abstract

This paper aims to determine the impact of sand-clay bond in geo-grid and geo-textile on bearing capacity. In doing so, we examined clay-geo-synthetics, sand-geo-synthetics and clay-sand-geo-synthetics samples using direct shear tests. The friction between clay and reinforcement was provided by encapsulated-sand system.

This method is used to transfer the tensile force mobilized in geo-synthetics from sand to clay and improve the strength parameters of clay. This study indicated that the provision of a thin layer of sand at both sides of the reinforcement significantly improved the shear strength of clay soil.

Bond coefficient computations indicated that the shear strength of clay-geo-synthetics samples was higher than non-reinforced clay. The increased strength was due to the impact of open meshes of geo-synthetics which provided some degree of resistance bearing. To determine the share of resistance bearing provided by geo-synthetic transverse members in the entire direct shear strength, we conducted a series of tests on geo-synthetics-reinforced samples with and without transverse members. The resistance bearing provided by geo-synthetic transverse members was almost 10% of total shear strength. The results indicated that encapsulated geo-grid and geo-textile sand system increased the bearing capacity of clay, with geo-grid being more efficient than geo-textile.

Keywords: geo-synthetics, tensile force, reinforced soil, transverse members, bond coefficient

1. Introduction

Soft clay soils have low bearing capacity and tend to settle under the influence external loads. In order to prevent excessive settlement, they must be improved before construction process. This paper aims to determine the impact of geo-synthetic coarse-grained reinforcement on bearing capacity of soft clay.

The advantages of grain materials include good drainage, high friction resistance, and good stability against humidity and time [Elias, V., Barry, P. E., & Christopher, R. 1997]. Reinforced-soil structures are often constructed by coarse-grained materials because fine-grained soils have poor drainage and friction resistance and tend to change under the influence of humidity and time [Lin, H., & Atluri, S. N. 2001]. Normally, membranous and confinement impacts are considered in soil reinforcement. In membranous impact, foundation and soil move downward and reinforcement layers (geo-grid and geo-textile) are gradually tensioned. To bear the load, the deformed reinforcement exerts a force upward. The reinforcement under tension needs a specific amount of settlement to mobilize membranous effect. Furthermore, the reinforcement must have sufficient length and stiffness. The upward force is developed at the conjunction of reinforcement and failure surface [Safari, E., Ghazizade, M. J., Abduli, M. A., & Gatmiri, B. 2014; Bhatia, S. K., Khachan, M. M., Stallings, A. M., & Smith, J. L. 2014].

Patra et al. conducted a series of studies on geo-grid reinforced sand with eccentric loading and reported that the increased layers of geo-grid resulted in the increased bearing capacity [Patra, C. R., Das, B. M., Bhoi, M., & Shin, E. C. 2006].

Alawaji conducted a study on settlement and bearing capacity of geo-grid reinforced sand located on weak soil. He reinforced the sand using a geo-grid layer and found that the optimal depth of geo-grid placement for reaching the maximum bearing capacity was 10% of foundation diameter. This result was in line with the studies of other researchers.

Alawaji conducted a series of studies to determine the impact of geo-grid diameter change on load-settlement behavior of samples. He reported that the increased diameter of geo-grid resulted in a significant reduction of settlement in equal capacities [Alawaji, H. A. 2001].

Radhey et al. examined the foundations placed on geo-grid reinforced soil and found that the first layer of reinforcement must be close to foundation with an optimal depth range of 0.5B-2B. They also reported that the length of reinforcement must be between 2B and 8B [harma, R., Chen, Q., Abu-Farsakh, M., & Yoon, S. 2009].

Nazir and Sawwaf conducted a study on the behavior of rectangular foundation located on geo-grid reinforced sand. They reported that the increased soil density resulted in the increased impact of geo-grid layers on bearing capacity [El Sawwaf, M., & Nazir, A. K. 2010].

Basudhar et al. investigated the behavior of circular foundation located on geo-textile reinforced sand. They found that the increased number of geo-textile layers resulted in the increased bearing capacity. They also reported that the increased density of sand led to failure in higher settlements [Basudhar, P. K., Saha, S., & Deb, K. 2007].

This paper aims to determine the impact of sand-clay bond in geo-grid and geo-textile on bearing capacity. In doing so, we conducted a comprehensive study on different parameters of sand and clay in various geo-synthetic layers. This is the first paper to make a comprehensive study on this subject.

2. Research Method

2.1 Shear Strength

The shear strength between reinforcement and soil is provided by 1) the shear strength in the contact area between reinforcement and soil and 2) the shear strength between soils in open meshes of the geo-synthetics.

Equation 1 represents total direct shear strength (F_t) [Sreekantiah, H. R., & Unnikrishnan, N. 1992]:

$$F_t = \sigma_n A[\alpha_{ds} tan(\delta + (1 - \alpha_{ds}) tan\phi) \tag{1}$$

ϕ: Internal friction angle of soil in direct shear

δ: Apparent friction angle of shear surface of reinforcement and soil

α_{ds}: The ratio of shear surface of reinforcement to total shear surface

σ_n: Vertical stress in shear surface and A: total shear surface

Many researchers have recognized bond coefficient as an important parameter in reinforced-soil design [Liu, C. N., Ho, Y. H., & Huang, J. W. 2009; Wang, Z., & Richwien, W. 2002; GHIASIAN, H., & Jahannia, M. 2004]. They have defined bond coefficient as the ratio of soil-soil contact area strength to soil-reinforcement contact area strength. In this definition, soil-soil contact surface strength is the direct shear strength of the soil, provided that shear surface is soil-reinforcement shear surface [Das, B. M., Cook, E. E., Shin, E. C., Yen, S. C., & Puri, V. K. 1993].

If the bond coefficient is bigger than unit number, a powerful bond exists between soil and geo-synthetics and soil-reinforcement contact area strength is higher than soil-soil contact area strength. If it is smaller than 0.5, there is a weak bond between soil and geo-synthetics [Jewell, R. A., & Wroth, C. P. 1987].

2.2 Bond Coefficient

So far, many researchers have studied the friction behavior of soil-geo-synthetics contact area, with an emphasis on the parameters of humidity percentage, soil type and density, geometry, stiffness of geo-synthetics, and vertical stress [Sridharan, A., Murthy, B. S., Bindumadhava, & Revanasiddappa, K. 1991]. They have compared the friction angle mobilized in soil-geo-synthetics contact area with internal friction angle of soil and defined factor C_i (effective bond coefficient) in the form of equation 2:

$$C_i = \frac{C_a + \sigma_n tan\delta_a}{C + \sigma_n tan\phi} \tag{2}$$

C_a: Bond between soil and geo-synthetics

δ_a: Apparent friction angle of contact area

C: Soil bond

ϕ: Internal friction angle of soil

σ_n: Vertical stress

If the tested soil is sand, bond coefficient is simplified as follows:

$$C_i = \frac{\tan\delta_a}{\tan\phi} \qquad\qquad (3)$$

Bond coefficient of the reinforcement depends on geo-synthetic surface, bearing capacity of transverse members, bearing capacity of surrounding soil, type of soil, and the length of buried sample. In soil-reinforcement direct shear bond mechanism, the shear strength of contact area is a combination of soil-soil direct shear strength and soil-reinforcement direct shear strength. If bond coefficient is smaller than 0.5, a weak bond exists between soil and geo-synthetics. If it is bigger than 1, a strong bond exists between soil and geo-synthetics.

2.3 Materials

We carried out the tests using kaolinite clay and silica sand. The sand was a mixture of three types of uniform sand with grain sizes of 0.7-1.2, 1-2 and 2-4 mm. Figure 1 illustrates the information of clay-sand mixture based on ASTM standards. Based on unified soil classification system (USCS), clay is classified in CL group (clay with low plastic property) and sand is classified in SW group (finely grained sand). The tests were carried out using a large direct shear device with some modifications for connecting the geo-synthetics to lower shear box by clamp (according to D5321 ASTM) [Unnikrishnan, N., Rajagopal, K., & Krishnaswamy, N. R. 2002]. Figure 2 illustrates the compaction curves.

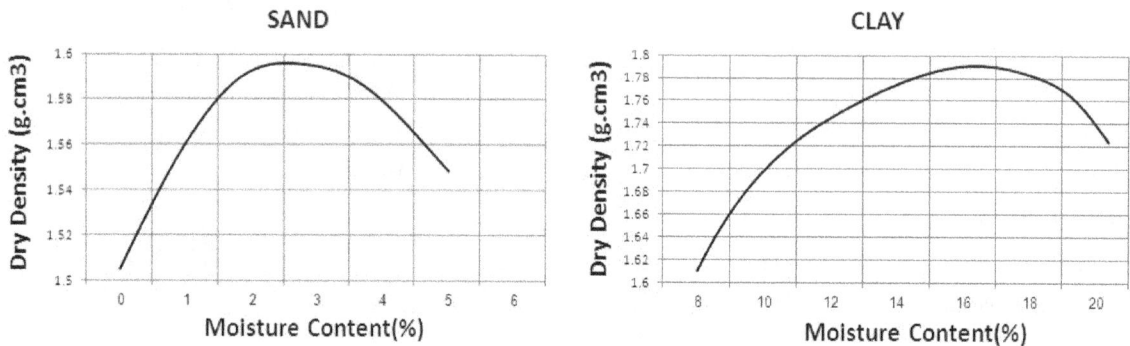

Figure 1. Clay-sand compaction curve

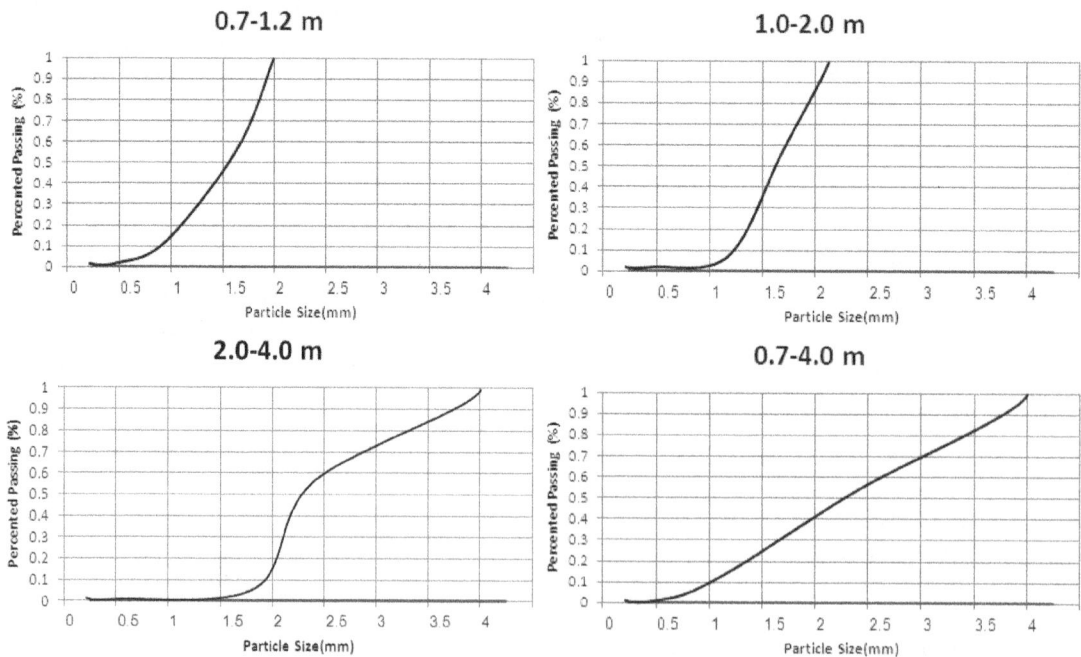

Figure 2. Sand gradation curve

2.4 Direct Shear Test

We performed soil compaction test inside the shear box using a hammer designed for this purpose. The test was carried out with and without transverse members. The samples were prepared as follows:

We completely mixed the clay and sand in the needed amount with optimal moisture content. Then we divided the mixture into two equal parts to fill the lower and upper parts of the shear box. The first part was poured in the lower shear box until reaching the maximum density obtained from proctor standard density [Mohiuddin, A. 2003; Biswas, A., Ansari, M. A., Dash, S. K., & Krishna, A. M. 2015; Bhat, S., & Thomas, J. 2015]. The number of required blows for reaching the intended density was determined by trial and error method and preliminary tests. The thickness of clay in lower box was 10 cm minus half of the sand layer for covering geo-synthetics at the middle of shear box. The needed sand was divided into two parts based on the required thickness. One part was poured into the lower shear box and the other part was poured into the upper shear box. The clamp was connected to the lower part of the shear box and was fixed by screws. After placing the geo-synthetics at the shared boundary of the box, we poured the remaining sand over the geo-synthetics at the upper box. The upper shear box was filled and compressed with the clay just like the lower shear box. After placing the loading plate on the sample, we let the force be equally distributed for 20 minutes. The test was started with the shear force rate of 1mm/min.

2.5 Test Method

We carried out direct shear tests to simulate the bond mechanism between soil and geo-synthetic reinforcement in contact area. In order to match other researches, we selected vertical stresses of 25, 50 and 75 kilopascal and shear force rate of 1mm/min [Zhang, C., Jiang, G., Liu, X., & Buzzi, O. 2016].

All tests were performed by strain control method according to ASTM D5321. We measured the shear force by load cell (LVDT$_s$) and measured vertical and horizontal displacements by a transducer connected to computer. Next, we depicted the curves based on shear stress and displacement and determined shear strength parameters. After completing each test, the shear box soil was sampled to control moisture content and dry weight of the samples.

2.6 Layering Method

For this purpose, we used three types of geo-grid and two types of geo-textile with the dimensions of 300×300 mm. The geo-grids and geo-textiles were placed at the middle of sand layer and at the shared boundary between two layers of soft clay and sand (Table 1). To prepare the clay, we increased its moisture content to 26% by adding water. The uniaxial resistance of clay in moisture content of 26% was 0.21 kg/cm^2, which is in the range of soft clay. The thickness of clay in test box was 15 cm, which was compressed by hand and compactor in three 5 cm layers after being fully mixed. The sand was placed on the clay in one 5 cm layer and was compressed by compactor and plastic hammer to reach the density of 55%. The geo-grid and geo-textile layers were used in three modes: 1) between two layers of sand and clay, 2) at the middle of sand layer, and 3) at both places [Demir, A., Yildiz, A., Laman, M., & Ornek, M. 2014; Wang, Z., Jacobs, F., & Ziegler, M. 2014].

After preparing the sample in test box, we placed the box in the device and put the loading plate exactly at the middle of sand surface. The force speed was set on 1.27 mm/min and the amount of force and settlement was measured by two gauges. The final settlement value was 25 mm and 50 readings were recorded. As the basic test, we tested a clay layer with the thickness of 20 cm. Next, we performed a few other tests. In one test, we placed a non-reinforced sand layer with the thickness of 5 cm on a clay bed. In another test, we used a reinforced sand layer with three types of geo-grid and two types of geo-textile.

Table 1. The used symbols

CLAY						
Flow limit%	Plasticity limit%	Plasticity index %	Optimal moisture%	Maximum dry density (g/cm3)	Bond (kPa)	Internal friction angle°
26.5	4.5	22	17	1.78	11.7	22.5

SNAD				
Curvature coefficient (Cc)	Optimal moisture%	Maximum dry density (g/cm3)	Bond (kPa)	Internal friction angle°
1.69	4	1.6	12.1	33.7

GeoGrid								
	Weight (kg/m2)	Thickness (mm)	Mesh shape	Color	Tensile strength (KN/m)	Material	Mesh size (mm)	Product name
GEO_GRID_1	0.73	2.2	Square	Black	8.2	HDPE	8×8	CE121
GEO_GRID_2	0.7	2.2	Square	Black	7.6	HDPE	10×10	CE161
GEO_GRID_3	0.42	2.4	Square	Orange	5.5	HDPE	15×15	SQ15L

Geotextile				
	Permeability (cm/sec)	Tensile strength (KN/m)	Thickness (mm)	Weight (gr/cm3)
GEO_TEXTILE_1	0.23	18	2.6	300
GEO_TEXTILE_2	0.28	35	4.5	600

Location of geo-synthetic layers	
SAND_CENTER	Middle of sand layer
SAND_CLAY_MID	Between two layers of sand and clay
SAND_CLAY_CENTER_MID	At the middle and between two layers of sand and clay

3. Results

3.1 Clay

Figure 3 illustrates the shear stress-displacement curves obtained from direct shear tests for non-reinforced and geo-synthetic reinforced clay. As you can see, non-reinforced and reinforced clays have shown stiffening behaviors. Shear surface was formed more quickly due to deformation of compressed soil. Moreover, reinforced clay curves indicate that shear failure has been formed more quickly in the reinforced samples. This behavior denotes that large-scale direct shear test is much more efficient for geo-technical structures of the reinforced soils [Ze, L., & Yong, L. 2014; Infante, D. U., Martinez, G. A., Arrúa, P., & Eberhardt, M. 2016; Liu, C. N., Yang, K. H., & Nguyen, M. D. 2014]. The curves indicate that shear strength in geo-synthetic-soil contact area is higher than in soil-soil contact area. This is probably due to the impact of geo-synthetic transverse elements which have provided some degree of resistance bearing along the shear surface.

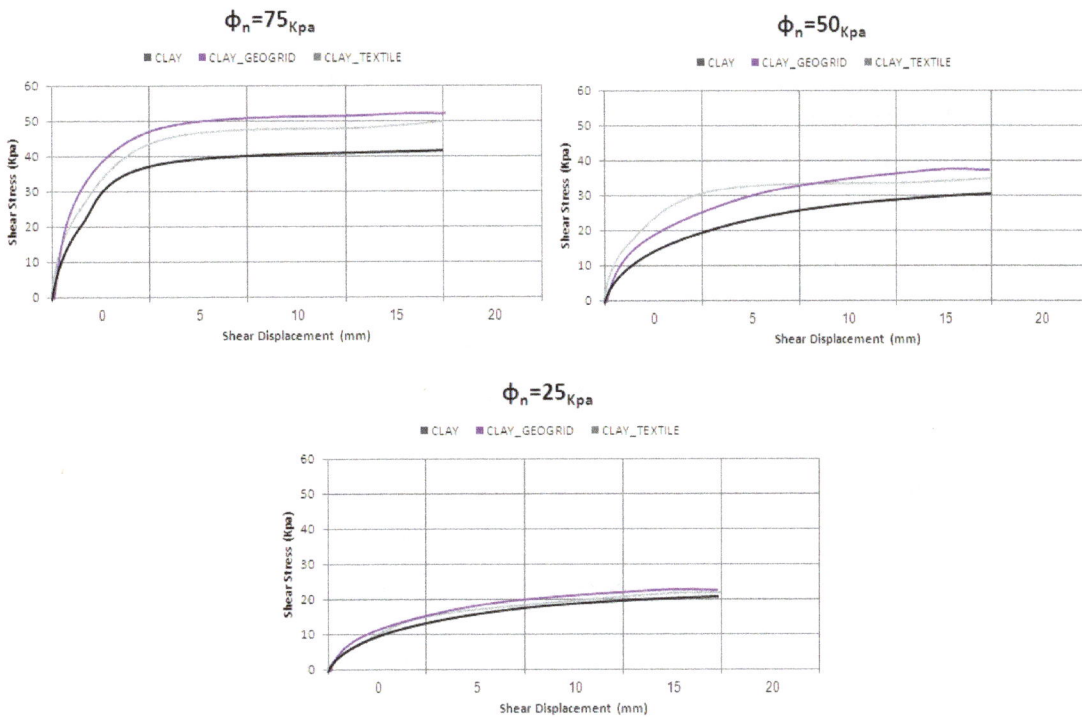

Figure 3. Shear stress curves – shear displacement of clay samples

3.2 Sand

Figure 4 illustrates the results of direct shear tests for non-reinforced and geo-synthetic-reinforced sand. As you can see, geo-synthetic reinforcement has increased the shear strength of the sand. The shear strength of both reinforced and non-reinforced sands has significantly and linearly increased after slight increase of displacement (1 mm). After reaching the maximum shear strength with a displacement of about 3 mm, shear strength has declined. In high horizontal displacements (about 4 mm), shear strength has become stable. Both reinforced and non-reinforced sands showed similar behaviors, with slight difference in the amount of shear strength. As you can see in the figure, the tensile strength of geo-synthetics has not been fully mobilized.

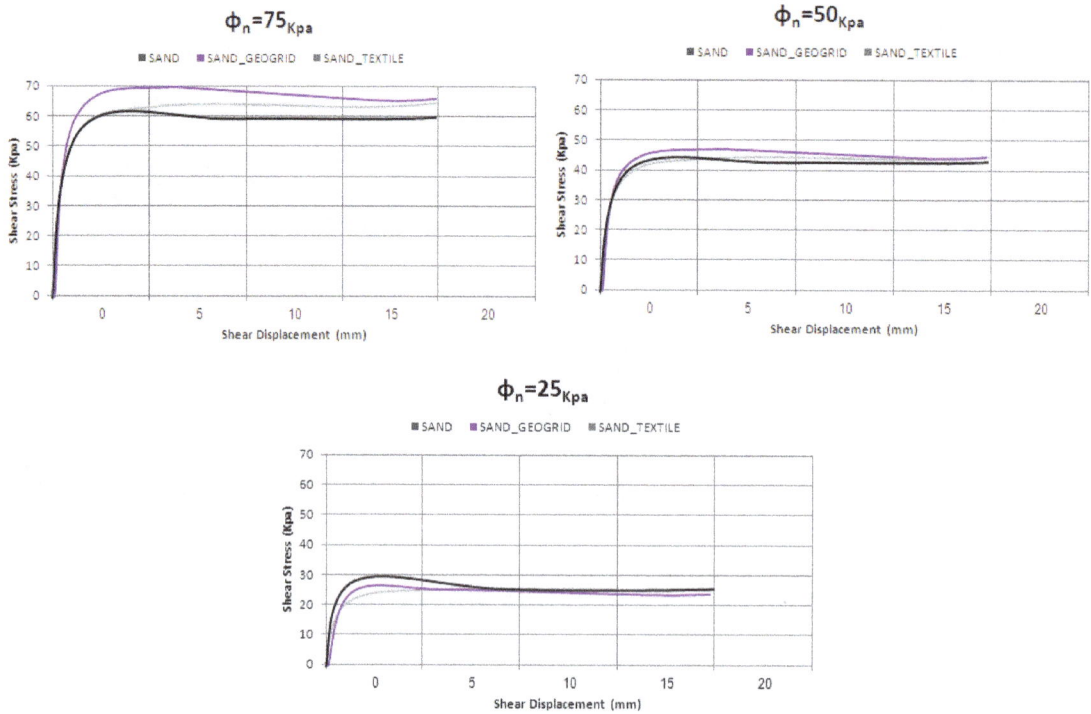

Figure 4. Shear stress-displacement curves of sand samples

3.3 Rupture Envelope

Table 2 contains the rupture envelope for non-reinforced and geo-synthetic-reinforced clay. As you can see, geo-synthetic reinforcement has increased the bond and slightly reduced the friction angle in contact area.

Table 2 contains the rupture envelope for non-reinforced and geo-synthetic-reinforced sand. As you can see, geo-synthetic reinforcement has significantly increased the apparent friction angle, but apparent bond has not changed. The increased apparent friction angle of the reinforced sand may be explained by the locking of sand particles in open meshes of geo-synthetics.

Table 2 represents the rupture envelopes of geo-synthetics-reinforced samples with and without transverse members. As you can see, geo-synthetic transverse members have increased the apparent bond, but have not significantly affected the apparent friction angle.

Table 2. Rupture envelopes

							Shear Strengh (Kpa)			
normal stress (Kpa)	SAND	SAND_GEOGRID	SAND_TEXTILE	CLAY	CLAY_GEOGRID	CLAY_TEXTILE	CLAY_SAND_GEOGRID with transverse members	CLAY_SAND_GEOGRID without transverse members	CLAY_SAND_TEXTILE with transverse members	CLAY_SAND_TEXTILE without transverse members
25	28	29	28	18	20	19	25	22	24	22
50	42	44	43	30	34	39	49	40	47	39

3.4 Transverse Members

To determine the impact of geo-synthetic transverse members in the entire direct shear strength of contact area, we carried out a series of tests on geo-synthetics-reinforced clay samples with and without transverse members. You can see the results in figure 5. Figure 6 illustrates the results for contact area. The results indicated that geo-synthetic transverse members increased the resistance bearing by about 10%.

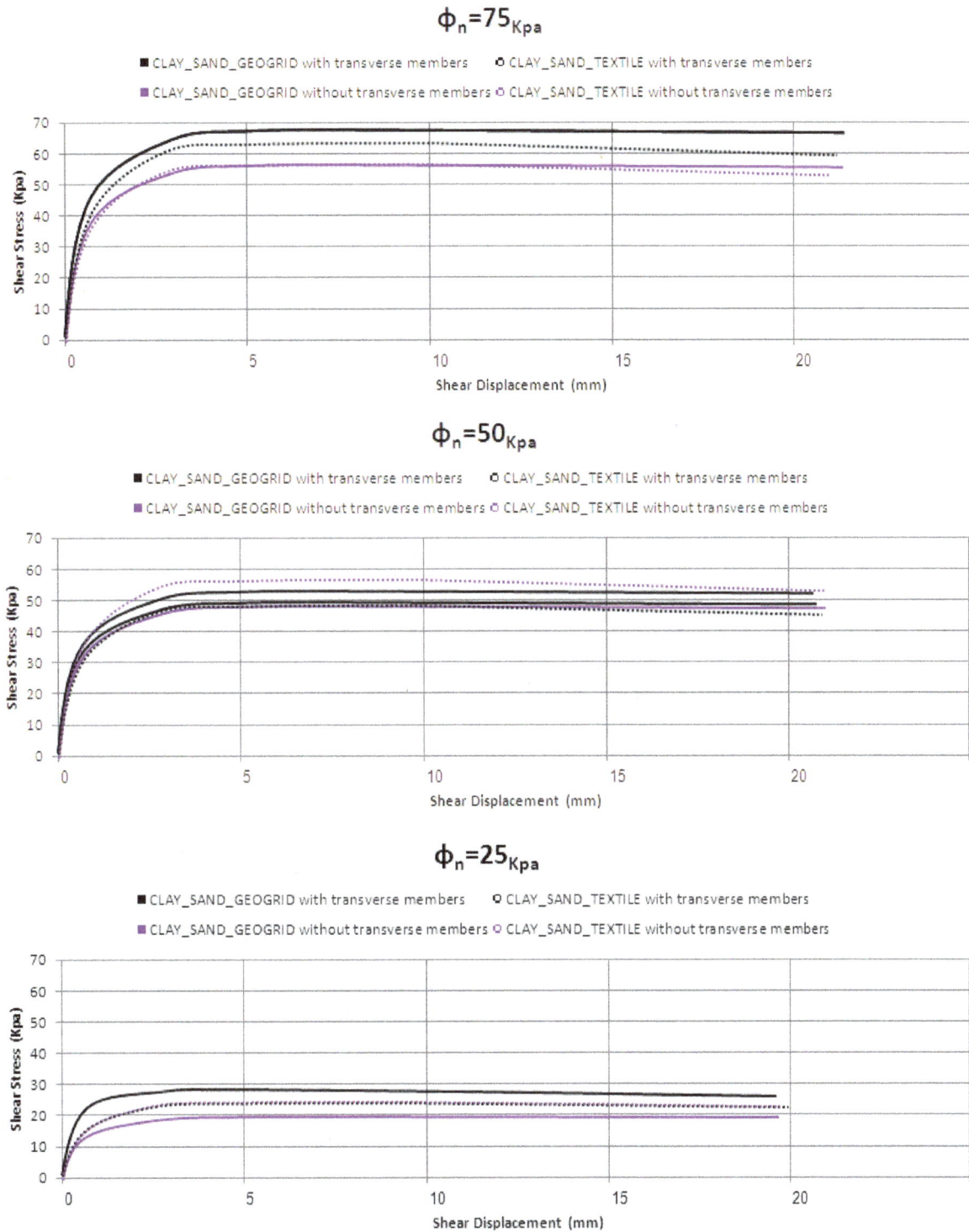

Figure 5. Shear stress-displacement curves for clay-sand-geo-synthetics samples without and without transverse members

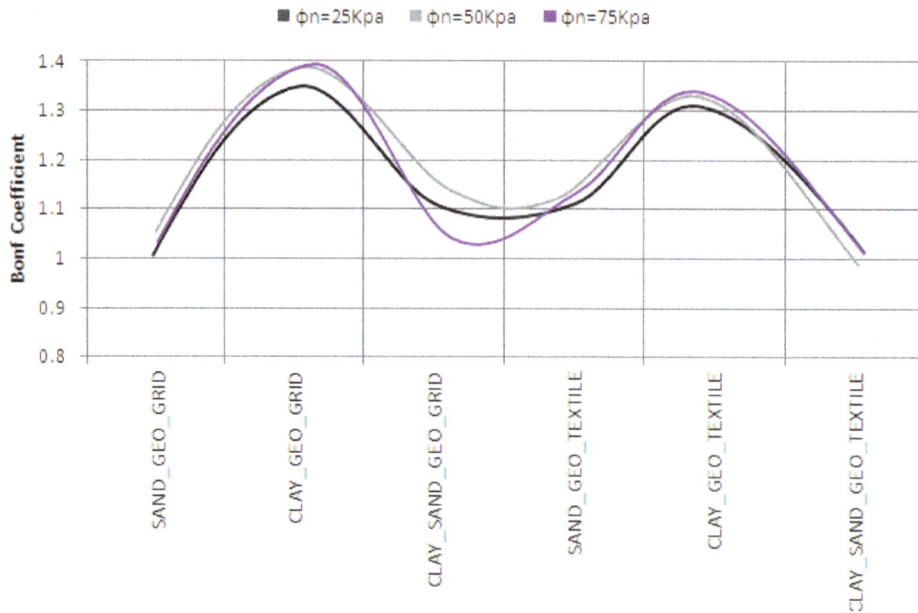

Figure 6. Bond coefficient changes of geo-synthetics compared to vertical stress

3.5 Layering

Figure 7 illustrates pressure-settlement diagram for clay samples with geo-synthetics-reinforced sand layer. The diagrams indicate that the sand layer with one geo-grid or geo-textile layer increases the bearing capacity and reduces the settlement. The use of two geo-grid or geo-textile layers at below and middle of sand layer increases the bearing capacity more than when only one geo-grid or geo-textile layer is used. It should be noted that the increased bearing capacity is more visible in high settlements, because geo-grid and geo-textile strengths are fully mobilized and the settlement is prevented [Keskin, M. S. 2015; Moreira, A., Vieira, C. S., das Neves, L., & Lopes, M. L. 2016; Aldeeky, H., Al Hattamleh, O., & Alfoul, B. A. 2016].

When load is applied, the circular plate and the reinforced soil move downward and the reinforcement layers (geo-grid and geo-textile) located in the sand are gradually tensioned. Consequently, the reinforcements are deformed due to their stiffness and exert a force upward to bear the load. Therefore, a specific amount of settlement is needed to mobilize the membranous impact in the reinforcement under tension [Shirlal, K. G., & Mallidi, R. R. 2015].

The comparison of pressure-settlement diagrams for different geo-grids indicated that the better coarsely-grained particles are locked in geo-grid meshes, the less settlement will occur. Since the coarse-grained particles are sand, the mesh size of 8×8 mm would be more efficient than the mesh size of 15×15 mm. It should also be noted that the tensile strength of GEO_GRID_1 is higher than GEO_GRID_3. Therefore, geo-grid is more efficient than geo-textile in the improvement of bearing capacity. Where the reinforcement layers are located at the middle of the sand, the type of geo-grid or geo-textile does not affect the bearing capacity. The best mode is the use of two reinforcement layers at the middle and below the sand layer. The worst mode is the use of one reinforcement layer below the sand layer.

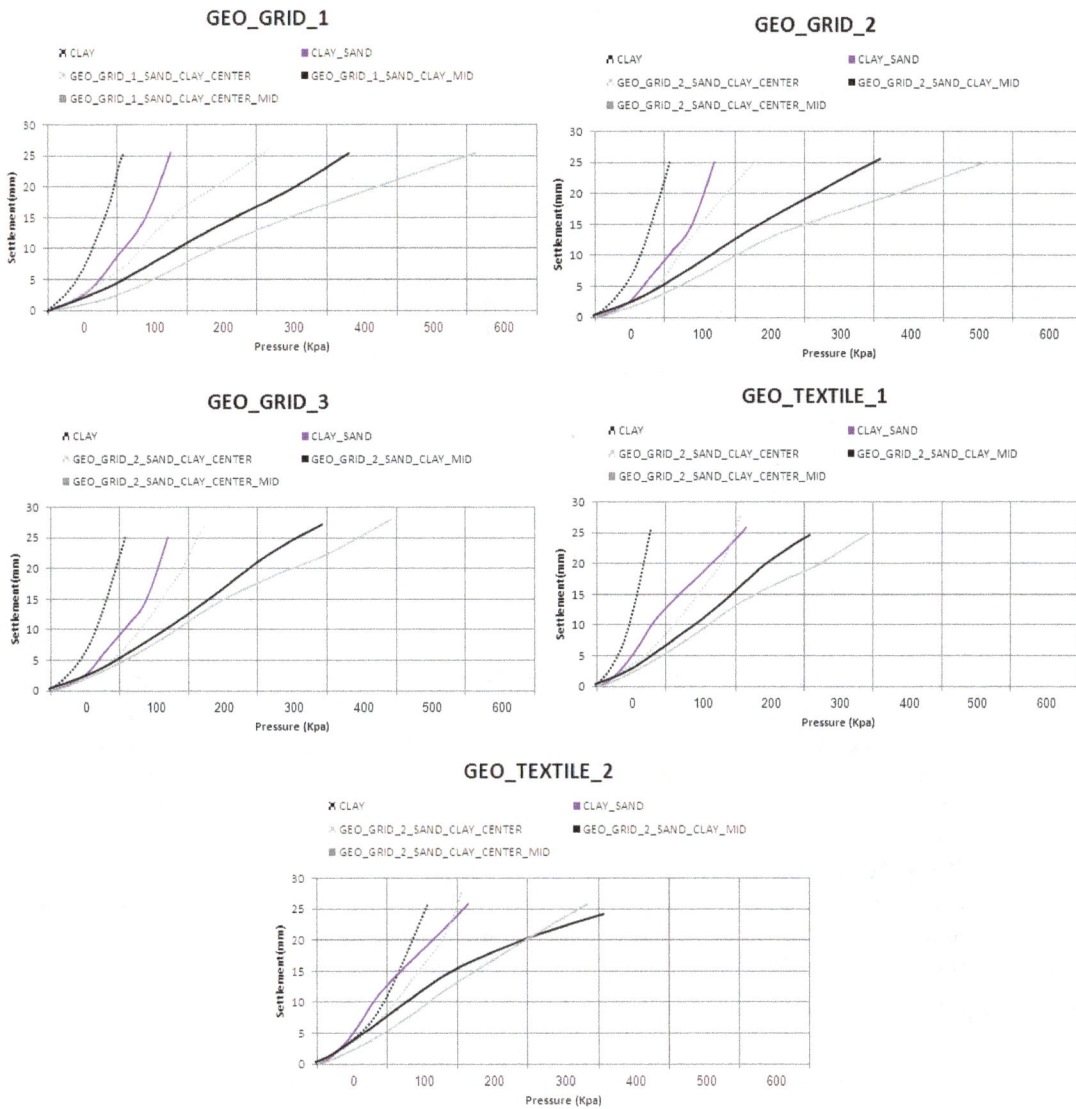

Figure 7. Pressure-settlement diagram for the reinforced samples

Figure 8 illustrates the pressure changes in settlement percentages of 5 and 50 for all modes (the ratio of settlement to loading plate diameter). In low settlement percentages, sand layer increases the bearing capacity by about two times. In high settlement percentages, since the clay bed is gradually more tensioned, the bearing capacity is declined by 1.6 times in settlement percentage of 50% compared to where no sand layer exists. In other words, the clay bed bears more pressure in higher settlement percentages. As you can see, geo-textile negatively affects the bearing capacity in low settlement percentages, because geo-textile is tensioned more slowly in low settlement percentages and increases the settlement due to its thickness. According to Figure 8, when geo-textile is located below the sand layer (B), the bearing capacity is negatively affected until the settlements of 20-25 mm.

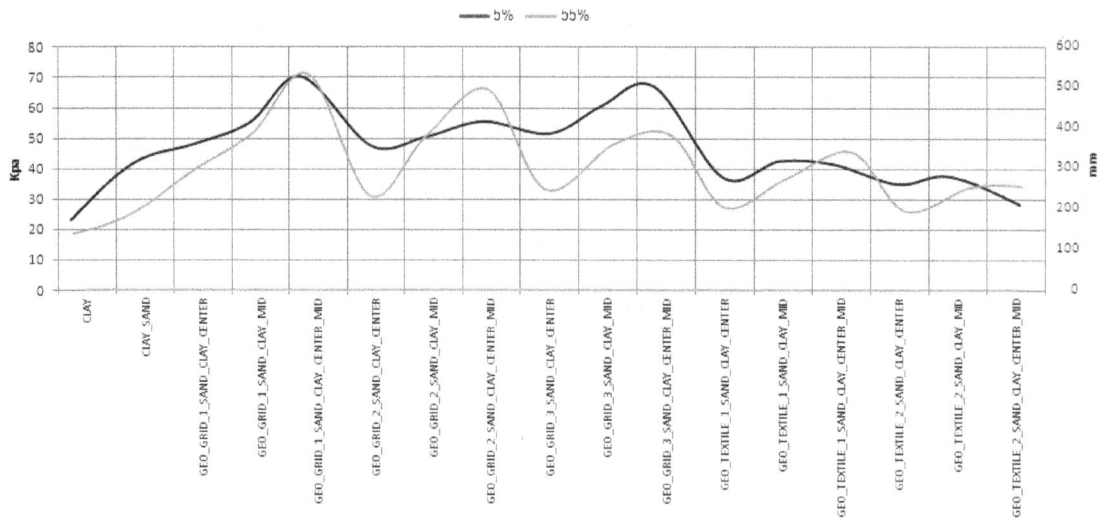

Figure 8. Comparison of the pressures (S/D=5% and S/D=55%)

4. Conclusion

We examined the bond between clay-geo-synthetics, sand-geo-synthetics and clay-sand-geo-synthetics using large-scale direct shear tests. In order to determine the impact of geo-synthetic transverse members on total strength of contact area, we repeated the tests without the transverse members. Studies on geo-synthetic surface friction indicated that the friction of contact area between soil and geo-synthetic materials may be smaller than internal friction angle of the same soil. This is likely to develop a weak surface along the contact area between geo-synthetic and soil. Figure 6 illustrates the changes of bond coefficient. As you can see, the shear strength of total contact area of the reinforced soil is higher than that of non-reinforced soil. The comparison of curves indicates that bond coefficient of sandwich method is higher than reinforced clay and reinforced sand. The reinforced sand shows the increased vertical stress, the reinforced clay shows the reduced vertical stress, and clay-geo-synthetic-sand sample shows an ascending trend [Deb, K., & Konai, S. 2014; Yang, K. H., Yalew, W. M., & Nguyen, M. D. 2015].

We reached the following conclusions based on test results:

Geo-synthetic reinforcement of clay increases the shear strength, which is mainly explained by the impact of open meshes of geo-synthetics. The reinforcement significantly improved the bond, but its impact on apparent friction angle was not significant. Geo-synthetic reinforcement of sand increased the shear strength of contact area, with a significant impact on internal friction angle. The provision of a thin layer of sand at both sides of the reinforcement significantly improved the apparent friction angle of the clay soil. The geo-synthetic transverse members developed the resistance bearing and constituted about 10% of total shear strength.

Test results indicated that shear surface of the reinforced soil is formed more quickly and its total strength remains unchanged after displacements of 2-4 mm. The bond coefficient was 1.10 for clay, 1.04 for sand and 1.11 for clay-sand-geo-synthetics sample. The increased coefficient denotes the significant impact of burying the geo-synthetics in thin layers of sand as clay reinforcement.

The placement of a sand layer on soft clay bed increases the bearing capacity and reduces the settlement. In low settlements, the sand layer bears almost half of the pressure. As the load and settlement increase, soft clay bed bears more load than sand layer does. The reinforcement of sand layer by geo-grid and geo-textile increases the bearing capacity. The use of two reinforcement layers at the middle and below the sand layer is more efficient than the use of one reinforcement layer.

As regards the bearing capacity, the placement of reinforcement at the middle of the sand layer is more efficient than when it is placed below the sand layer. Also, geo-grid is more efficient than geo-textile in the improvement of bearing capacity, because the soil below the circular plate moves downward in high settlement percentages and the reinforcement layers (geo-grid and geo-textile) are gradually tensioned. Consequently, geo-grid and geo-textile strength is fully mobilized and prevents the settlement.

In low settlement percentages, geo-textile negatively affects the bearing capacity because geo-textile is tensioned more slowly than geo-grid and increases the settlement due to its thickness. In all settlement percentages, the highest bearing capacity belongs to the sample in which two GEO_GRID_1 layers have been used at the middle and below the sand layer.

References

Alawaji, H. A. (2001). Settlement and bearing capacity of geogrid-reinforced sand over collapsible soil. *Geotextiles and Geomembranes, 19*(2), 75-88. https://doi.org/10.1016/S0266-1144(01)00002-4

Aldeeky, H., Al Hattamleh, O., & Alfoul, B. A. (2016). Effect of Sand Placement Method on the Interface Friction of Sand and Geotextile. *International Journal of Civil Engineering, 14*(2), 133-138. https://doi.org/10.1007/s40999-016-0019-0

Basudhar, P. K., Saha, S., & Deb, K. (2007). Circular footings resting on geotextile-reinforced sand bed. *Geotextiles and Geomembranes, 25*(6), 377-384. https://doi.org/10.1016/j.geotexmem.2006.09.003

Bhat, S., & Thomas, J. (2015). Use of Polymer Geogrid Composite to support rail track over weak saturated clay subgrade–a case study.

Bhatia, S. K., Khachan, M. M., Stallings, A. M., & Smith, J. L. (2014). Alternatives for the Detection of Residual Polyacrylamide in Geotextile Tube Dewatering—Streaming Current Detection and China Clay Settling Rate Methods. *Geotechnical Testing Journal, 37*(4). https://doi.org/10.1520/GTJ20130162

Biswas, A., Ansari, M. A., Dash, S. K., & Krishna, A. M. (2015). Behavior of geogrid reinforced foundation systems supported on clay subgrades of different strengths. *International Journal of Geosynthetics and Ground Engineering, 1*(3), 1-10. https://doi.org/10.1007/s40891-015-0023-5

Chong, S. Y., Kassim, K. A., Chiet, K. T. P., & Tan, C. S. (2015). Static Response on Lime Column and Geotextile Encapsulated Lime Column (GELC) Stabilised Marine Clay Under Vertical Load. *Jurnal Teknologi, 77*(11).

Das, B. M., Cook, E. E., Shin, E. C., Yen, S. C., & Puri, V. K. (1993). Bearing capacity of strip foundation on geogrid-reinforced clay.

Deb, K., & Konai, S. (2014). Bearing capacity of geotextile-reinforced sand with varying fine fraction. *Geomechanics and Engineering, 6*(1), 33-45. https://doi.org/10.12989/gae.2014.6.1.033

Demir, A., Yildiz, A., Laman, M., & Ornek, M. (2014). Experimental and numerical analyses of circular footing on geogrid-reinforced granular fill underlain by soft clay. *Acta Geotechnica, 9*(4), 711-723. https://doi.org/10.1007/s11440-013-0207-x

El Sawwaf, M., & Nazir, A. K. (2010). Behavior of repeatedly loaded rectangular footings resting on reinforced sand. *Alexandria Engineering Journal, 49*(4), 349-356. https://doi.org/10.1016/j.aej.2010.07.002

Elias, V., Barry, P. E., & Christopher, R. (1997). Mechanically Stabilized Earth Walls and Reinforced Soil Slopes Design and Construction Guidelines: FHWA Demonstration Project 82, Reinforced Soil Structures WSEW and RSS. US Department of Transportation, Federal Highway Administration.

Ghiasian, H., & Jahannia, M. (2004). Influence of encapsulated geogrid-sand system on bearing capacity and settlement characteristics of reinforced clay.

Infante, D. U., Martinez, G. A., Arrúa, P., & Eberhardt, M. (2016). Behavior of Geogrid Reinforced Sand Under Vertical Load. *International Journal, 10*(21), 1862-1868. https://doi.org/10.21660/2016.21.5168

Jewell, R. A., & Wroth, C. P. (1987). Direct shear tests on reinforced sand. *Geotechnique, 37*(1), 53-68. https://doi.org/10.1680/geot.1987.37.1.53

Keskin, M. S. (2015). Model studies of uplift capacity behavior of square plate anchors in geogrid-reinforced sand. *Geomechanics and Engineering, 8*(4), 595-613. https://doi.org/10.12989/gae.2015.8.4.595

Lin, H., & Atluri, S. N. (2001). The meshless local Petrov-Galerkin (MLPG) method for solving incompressible Navier-Stokes equations. *CMES- Computer Modeling in Engineering and Sciences, 2*(2), 117-142.

Liu, C. N., Ho, Y. H., & Huang, J. W. (2009). Large scale direct shear tests of soil/PET-yarn geogrid interfaces. *Geotextiles and Geomembranes, 27*(1), 19-30. https://doi.org/10.1016/j.geotexmem.2008.03.002

Liu, C. N., Yang, K. H., & Nguyen, M. D. (2014). Behavior of geogrid–reinforced sand and effect of reinforcement anchorage in large-scale plane strain compression. *Geotextiles and Geomembranes, 42*(5), 479-493. https://doi.org/10.1016/j.geotexmem.2014.07.007

Mohiuddin, A. (2003). Analysis of Laboratory and Field Pull-Out Tests of Geosynhthetics in Clayey Soils

(Doctoral dissertation, Faculty of the Louisiana State University and Agricultural and Mechanical College in partial fulfillment of the requirements for the degree of Master of Science in Civil Engineering in The Department of Civil and Environmental Engineering by Ather Mohiuddin BE, Osmania University).

Moreira, A., Vieira, C. S., das Neves, L., & Lopes, M. L. (2016). Assessment of friction properties at geotextile encapsulated-sand systems' interfaces used for coastal protection. *Geotextiles and Geomembranes, 44*(3), 278-286. https://doi.org/10.1016/j.geotexmem.2015.12.002

Patra, C. R., Das, B. M., Bhoi, M., & Shin, E. C. (2006). Eccentrically loaded strip foundation on geogrid-reinforced sand. *Geotextiles and Geomembranes, 24*(4), 254-259. https://doi.org/10.1016/j.geotexmem.2005.12.001

Safari, E., Ghazizade, M. J., Abduli, M. A., & Gatmiri, B. (2014). Variation of crack intensity factor in three compacted clay liners exposed to annual cycle of atmospheric conditions with and without geotextile cover. *Waste management, 34*(8), 1408-1415. https://doi.org/10.1016/j.wasman.2014.03.029

Sharma, R., Chen, Q., Abu-Farsakh, M., & Yoon, S. (2009). Analytical modeling of geogrid reinforced soil foundation. *Geotextiles and Geomembranes, 27*(1), 63-72. https://doi.org/10.1016/j.geotexmem.2008.07.002

Shirlal, K. G., & Mallidi, R. R. (2015). Physical Model Studies on Stability of Geotextile sand Containers. *Procedia Engineering, 116*, 567-574. https://doi.org/10.1016/j.proeng.2015.08.327

Sreekantiah, H. R., & Unnikrishnan, N. (1992). Behaviour of geotextile under pullout. In Proc. of the Indian Geotechnical Conference, Calcutta (pp. 215-228).

Sridharan, A., Murthy, B. S., Bindumadhava, & Revanasiddappa, K. (1991). Technique for using fine-grained soil in reinforced earth. *Journal of geotechnical engineering, 117*(8), 1174-1190. https://doi.org/10.1061/(ASCE)0733-9410(1991)117:8(1174)

Unnikrishnan, N., Rajagopal, K., & Krishnaswamy, N. R. (2002). Behaviour of reinforced clay under monotonic and cyclic loading. *Geotextiles and Geomembranes, 20*(2), 117-133. https://doi.org/10.1016/S0266-1144(02)00003-1

Wang, Z., & Richwien, W. (2002). A study of soil-reinforcement interface friction. *Journal of Geotechnical and Geoenvironmental Engineering, 128*(1), 92-94. https://doi.org/10.1061/(ASCE)1090-0241(2002)128:1(92)

Wang, Z., Jacobs, F., & Ziegler, M. (2014). Visualization of load transfer behaviour between geogrid and sand using PFC 2D. *Geotextiles and Geomembranes, 42*(2), 83-90. https://doi.org/10.1016/j.geotexmem.2014.01.001

Yang, K. H., Yalew, W. M., & Nguyen, M. D. (2015). Behavior of Geotextile-Reinforced Clay with a Coarse Material Sandwich Technique under Unconsolidated-Undrained Triaxial Compression. *International Journal of Geomechanics, 16*(3), 04015083. https://doi.org/10.1061/(ASCE)GM.1943-5622.0000611

Ze, L., & Yong, L. (2014). Test Study on the Interface Friction Characteristics of Coal Gangue-Geogrid-Sand Layered System. *Industrial Construction, 4*, 22.

Zhang, C., Jiang, G., Liu, X., & Buzzi, O. (2016). Arching in geogrid-reinforced pile-supported embankments over silty clay of medium compressibility: Field data and analytical solution. *Computers and Geotechnics, 77*, 11-25. https://doi.org/10.1016/j.compgeo.2016.03.007

5

Land Use Dynamics and Wetland Management in Bamenda: Urban Development Policy Implications

Balgah Sounders Nguh[1] & Jude Ndzifon Kimengsi[2]

[1] University of Buea, Buea, Cameroon

[2] Catholic University of Cameroon (CATUC), Bamenda, Cameroon

Correspondence: Jude Ndzifon Kimengsi, Catholic University of Cameroon (CATUC), P.O. Box 782 Bamenda, Cameroon. E-mail: ukjubypro2@yahoo.com

Abstract

Wetland ecosystems in the world have been affected by changing land uses brought about by rapid urbanization. The thrust of this study therefore is to examine the trend of land use dynamics and their implications on wetland management. Using land use maps for two periods – 1984 and 2014, aided by the administration of 75 semi-structured questionnaires, we exploited the rate of change of land uses and their effects on wetland management as well as the urban development policy implications for Bamenda. A positive relationship (0.5) was observed for land use change and wetland degradation. Furthermore, the results from land use analysis showed that between 1984 and 2014, significant changes were observed for residential land use which increased in surface area from 42% as of 1984 to 53% in 2014. In addition, agricultural land use increased from 11% to 34%. Conversely, the surface area covered by wetlands reduced from 27% in 1984 to 6% in 2014. The conclusion drawn is that in the face of further wetland degradation, the current trend of land use dynamics can be checked by the application of zoning laws to control the changes witnessed in the land uses (residential and agricultural land uses). In addition, the Bamenda City Council should promote public awareness through sensitization on wetland resources and should actively encourage the participation of the public, local government authorities and institutions in sustainably managing wetlands.

Keywords: land use, dynamics, wetland management, implications, Bamenda

1. Introduction

Very few environments exist on earth today which has not witnessed significant alterations or transformations by humanity for one reason or the other (Balgah, 2007). Land use, the way human employ the land and its resources (Balgah, 2007) continue to witness significant transformations. This occurs especially within urban centres and introduces a challenge to reconcile the often-competing demands of land to accommodate urban functions and environmental protection (UN-Habitat, 2009).

Urbanisation and land use changes in developing countries presents formidable challenges. Of particular concern are the risks of immediate and surrounding environment, its effects on natural resources, health conditions, social cohesion and on individual rights. Each year, cities attract new migrants who, together with the increasing native population, expand the number of squatter settlement and shanty towns, exaggerating the problem of urban congestion and sprawl and hampering local authorities' attempts to improve on basic infrastructures and deliver essential services (Cohen, 2006).

Wetland surface areas are estimated at 12.8 million km^2 with a global annual economic value worth US$ 70 billion (WWF 2004). This value is declining with the ever increasing human pressure on the world's wetlands and justifies the fact that since the 1900s, more than 50% of world wetlands have been lost to other uses like agriculture and/or infrastructural development. It is important to mention that one of the key drivers of wetland degradation is urbanisation which is characterised by infrastructural development.

Wetlands are among the most valuable and productive ecosystems on earth (Castaineda and Herro, 2008) which are affected by land use dynamics leading to their degradation (Tiner et al., 2002). Public usage of wetlands is the root cause of wetland loss. Negative views towards wetlands potentially results from misunderstanding of the value and services that they provide for the society and inadequate public policy (Xie et al., 2010). This has led to their conversion to intensive agricultural, industrial and residential lands (Grillas *et al.*, 2004). It has been reported that a large percentage of wetlands have been lost in the last century apparently due to drainage and land clearance as a consequence of land use change - agriculture, urban and industrial development activities (Williams *et al* 2009).

Cameroon, like many other countries in the Inter-tropical zone, is home to a number of significant wetlands such as the Waza logone floodplain, the Limbe and Wouri estuaries, Bakassi and Rio del Rey Creeks, the Ndop plain, Bamendjim dam, Mape dam, Menchum river basin, Mboh and Santchou Floodplains and crater lakes like the lake Oku, Awing, Wum and Barombi; and in the south, they exist around the forested swamps. Bamenda is home to major wetlands such as in Ngomgham, Mulang and Menda-Nkwen. As the town continues to witness rapid multiplication of land uses due to her primacy status, wetland encroachment and degradation has been aggravated. Emerging as a city in the colonial days of the British, French and Germans from around the 19th Century, Bamenda, due to land use dynamics, has transcended from being a traditional monoculture village to becoming a complex heterogeneous city offering many services to its inhabitants as well as to its hinterlands (Nyambod, 2010). The multiplication of urban functions occurs at the expense of wetland conservation – it precipitates the colonisation of wetlands by agricultural, residential and commercial land uses, among others.

2. Problem Statement

Wetlands occupy a central position as far as the earth's natural resource base is concern – they offer numerous ecosystem services as spelt out by the 1971 Ramsar Convention (Ramsar Information Sheet 2009-2012). Land use change, a result of urban development, affects the management of wetlands. The causal mechanisms associated with land-use change remain relatively poorly understood, in part because of the complexity of urban systems. Consequently, urban planners and policy makers are often faced with the difficult task of making land-use decisions without sufficient analyses or vision (Sun *et al.* 2009).

This is the case with the town of Bamenda which is witnessing rapid urbanisation characterised by the multiplication of her major land uses – agricultural, settlement and administrative land uses. Bamenda is home to major wetlands such as in Ngomgham, Mulang and Menda-Nkwen. As the town continues to witness rapid multiplication of land uses due to her primacy status, wetland encroachment and degradation has been aggravated. In other words, this rapid pace of urbanisation in Bamenda ultimately affects its wetland ecosystem as urban development is encroaching onto the wetlands to secure space for multiple urban functions. Anthropogenic activities such as settlement, conversion of wetlands into farm land, waste dump sites reduces the ability of wetlands to build resistance and resilience leading to eventual collapse.

Previous studies on land use dynamics have focused on its connection with population growth and their effect on the environment (Kimengsi, 2011; Lambi & Balgah, 2010; Balgah *et al.*, 2008; Balgah, 2007; Balgah 2005), and land use conflicts (Kimengsi, 2009; Kimengsi, 2008). In addition, the hydro-geomorphological implications of urbanization have also been researched upon including the causes and effects of land use changes (Kometa, and Ndi, 2012). However, their implications for wetland management have received little attention. This is particularly necessary at a time when the Bamenda City Council has embarked on moves towards conserving, restoring and revitalising wetland environments in Bamenda against the backdrop of increasing land use dynamics precipitated by human activities – agriculture, settlement, commerce and waste disposal. The purpose of this study is to examine the trend of land use dynamics exploring two periods – 1984 and 2014, and their implications for wetland management. The study equally seeks to assess the rate of change of land uses and their effects on wetland management on the one hand, and the policy implications for urban development on the other hand.

3. Literature Review

Land use changes involve the transformation of diverse land use activities due to population increase and economic development constitute one of the main stressors of wetland ecosystems (Zorrilla-Miras *et al.*, 2014; Tijani *et al.*, 2011; Ripken, 2009). Added to these forces is the political interference which manifest in cases where governments contradict environmental regulations to encroach and or/watch without interference such encroachments onto wetlands. These eventually lead to the degradation and collapse of wetland ecosystems as its resilience is decreased (Ajibola, *et al.*, 2010). Land use dynamics introduces direct and indirect impacts on wetland quality. Direct impact occurs when a wetland is degraded, filled, drained or otherwise altered by activities occurring within the wetland boundary. Examples of direct impact include drainage of wetlands for agricultural use by constructing drainage ditches or installing underground drainage tiles and filling wetlands to provide useable land on which to build. Indirect impacts are caused by increase storm water and pollution generated by land development within the wetland contributing drainage area (Tiffany *et al.*, 2006).

In this regard, the wise use of *Wetlands becomes imperative. The* Millennium Ecosystem Assessments (MEA's) (2005) explains that "wise use of wetlands refers to the maintenance of their ecological character, achieved through the implementation of ecosystem approaches, within the context of sustainable development. Managing wetlands properly requires adequate control of land use activities. Considerations for managing wetlands should include strategies to reduce anthropogenic stresses and those to increase resistance and resilience to climate change. As a first step towards protection, there is a need to make an inventory and classification of wetlands in terms of their physical settings (Ndenecho and Fonteh, 2012). Wetland management requires intense monitoring and increased interaction and co-operation among various agencies such as state departments concerned with the environment, soil, agriculture, forestry, urban planning and development, natural resource managers, public interest groups, citizens and research institutions. Management strategies should involve protection of wetlands by regulating inputs, using water quality standards (WQS) promulgated for wetlands and such surface waters to promote their normal functioning. Equally, monitoring restoration endeavours should include both structural and functional attributes. Monitoring of attributes at the population, community, ecosystem and landscape level is appropriate in this regard and community training (Ramachandra, 2001). In view of these, the institutional support frameworks and regulations, municipal zoning, community ownership and a change in management techniques in the face of extensive pressure from land use multiplications are imminent (Ripkens, 2009; Kometa, 2009)

4. The Study Area

Bamenda is located between latitude 5°56' and 5°58' north of the Equator and longitude 10.09° and 10.11° east of the Greenwich Meridian (Bamenda City Council, 2014). It is the capital of the North West Region with headquarters in Mezam Division. It covers a surface area of about 37,560 km² (Bamenda City Council, 2014). Bamenda is bounded to the west and southwest by Momo Division and Bali sub-division respectively. To the north, it is flanked by Bafut sub division, to the north east by Tubah sub division and to the south by Santa sub division. Within Bamenda, the study sites chosen include Mulang, Ngomgham, Mbelem, Menda and Ntenesoh (Figure 1).

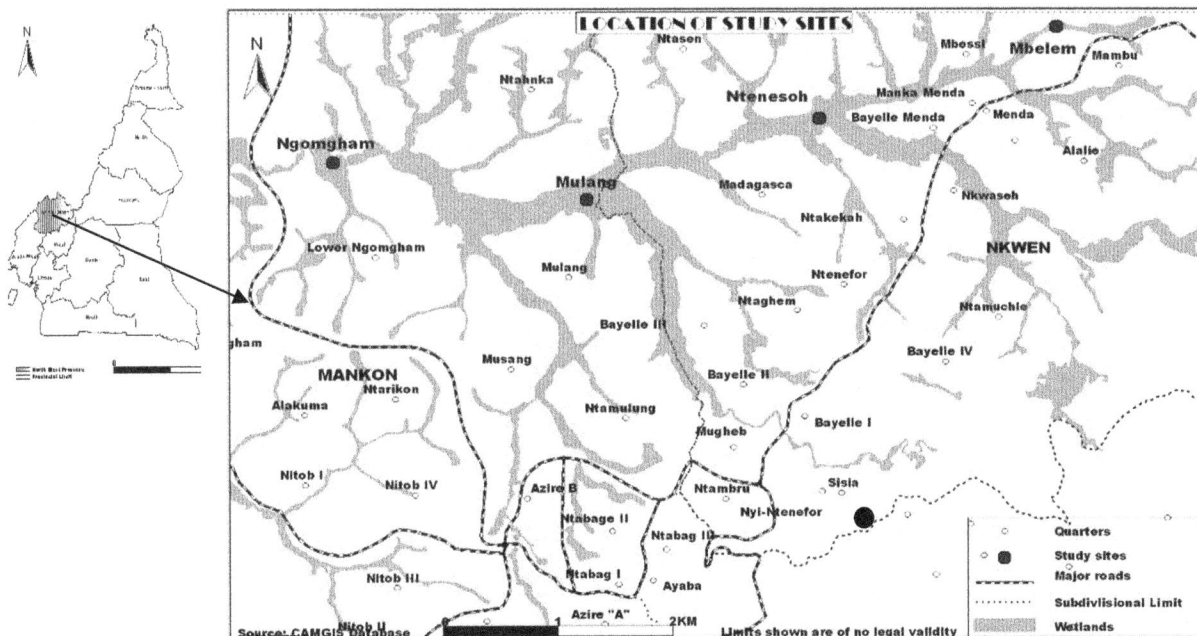

Figure 1. Location of the study sites

Sources: Adapted from CAMGIS, (2013)

5. Materials and Methods

Five targeted sites were identified for the study, they include, Mulang, Ngomgham, Mbelem, Menda and Ntenesoh.

Table 1. Description of the study sites

Zone	Site	Site Description
Mulang	A	This is the most extensive wetland area in Bamenda. This area is experiencing gradual human encroachment. It is covered by savannah vegetation and raffia palms.
Ntenesoh	B	This area is characterised by intensive wetland agriculture. Observable vegetation species here is the raffia palm. This area is also highly settled.
Ngomgham	C	This area is also highly settled and there is equally the practice of urban agriculture.
Mbelem	D	This area is highly encroached mainly by filling of wetlands with soil for housing construction.
Menda	E	This area is characterized by the practice of urban agriculture.

Source: Field work, 2015

These sites were chosen because they were judged to have extensive wetlands whose management is affected by changing land uses. The target population of these areas is estimated at about 2000 inhabitants. From this target population, a 5% sample was drawn involving 100 inhabitants. There was bias in the distribution of the questionnaires as Mulang and Ngomgham which were characterised by intense land use activities on wetlands received 30 questionnaires each while Menda, Ntenesoh and Mbelem received 15, 15 and 10 questionnaires respectively. Random sampling was employed in the distribution of the questionnaires. The targeted respondents consisted of land use actors (agriculturalists, real estate operators, council officers and government representatives). In addition, interviews with municipal and planning authorities, farmers as well as quarter heads were conducted. Data on population evolution for Bamenda between 1976 and 2014 were obtained from the Bamenda City Council including estimates from the quarter heads of the five targeted neighbourhoods.

To determine changes in land uses and wetland degradation, Remotely sensed data (optical multispectral images) was used in the classification of the study area is from Landsat (MSS, TM and ETM+) sensors (Table 2). The data for multispectral images ranged from visible (VIS; 0.45 – 0.9 μm), near-infrared (NIR; 0.76 – 0.98 μm), mid-infrared/short wave infrared (MIR/SWIR; 1.55 – 2.35 μm) to thermal infrared (TIR; 10.4 – 12.5). These images were obtained from the Global Land Cover Facility (GLCF, 2005). These images were processed using Geographic Information System (GIS) and Remote Sensing (RS) softwares.

Table 2. Satellite data acquired, uses and their applications

Sensor	Year	Date of each scene	Operational	Applications
Landsat MSS	1975	1975-06-06	1972 – 1984	Supervised classification to discriminate between settlement and other features.
Landsat TM	1986	1986-01-10	1972–present	Supervised classification to discriminate between settlement and other features.
Landsat ETM+	2000 2002	2000-06-01 2002-01-30	1999–present	Supervised classification to discriminate between settlement and other features

In analysing the optical multispectral images presented in Table 2, different stages were involved based on the feature of interest. These images were acquired already geo-referenced in the WGS84 ellipsoid in UTM32N with the exception of Landsat ETM+ 2002. This image was geo-referenced in ENVI using the Landsat ETM+ 2000 image as reference following the option of image-to-image registration since they came from the same sensor. After geo-referencing, since these images came in separate bands for a single year, these bands were stacked together to give them the same spatial sizes that facilitated false colour composite combinations and other analysis. Two types of false colour composites were made using a combination of bands (321 and 742) to be able to clearly discriminate between settlement and the other observed features on these images. For the Landsat TM and ETM+ spectral bands, bands 1-3 provide increased penetration of water bodies and correspond to the reflectance of green vegetation (Biradar et al., 2003). Band 3 however exhibit more contrast than bands 1 and 2 because the effect of the atmosphere is reduced here (Biradar et al., 2003). Band 4 is also responsive to the amount of vegetation biomass present in a scene and useful for identification of vegetation types, emphasizing soil-crop and land-water contrasts. Band 5 on its part helps to reduce the effect of thin clouds and smoke while band 7 is particularly effective in identifying zones of hydrothermal alteration in rocks because of its increased wavelength (2.08-2.35 μm; Biradar et al., 2003).

Band 321 was used to better observe road infrastructure and other less detailed features. Band 742 of Landsat was used for vegetation analysis. Further image analyses carried out using ENVI included change detection, supervised and unsupervised classification. Results from the supervised classification were the ones represented in the study to show land use dynamics over the years. After classification, these images were saved in geotiff format and imported to ArcGIS where the maps were produced in the final form presented in this thesis. A limiting factor common in all the images used in the classification process is the presence of cloud cover. In cases where this cloud covered settlement, these areas were also classified as settlement. It is impossible to remove these clouds from the images, thus making classification in the tropics very difficult. A general limitation to Landsat images is the fact that they are acquired at intervals of 16 days at best. This time lag between image acquisition and the availability to users, which normally takes from 1-14 days and the high cost incurred in case of on-request data over specific target areas, prevent true real-time monitoring of settlement changes. The land use and land cover maps for 1984 and 2014 were developed by CAMGIS. Also supported by the data from the questionnaires, change maps were used to evaluate the percentage change of land uses between 1984 and 2014 (Appendix I). They were further compared to show the degree of changing land uses and consequent wetland encroachment. A total of 75 semi-structured questionnaires were administered to obtain respondents opinion on the driving forces of land use change and their effects on wetlands. The data was analysed using the Stata 11.1 Statistical Package in which a correlation was established between land use change and wetland degradation in Bamenda. The correlation analysis was done in which the variables of land use change were correlated with those of wetland degradation at 1% and 5% levels of significance (Appendix II).

6. Results and Discussion

Land use dynamics in Bamenda

The land use situation of the Bamenda municipality in 1984 (Figure 2) showed that residential (42%) and forest (19%) land uses were the most dominant (Figure 3a).

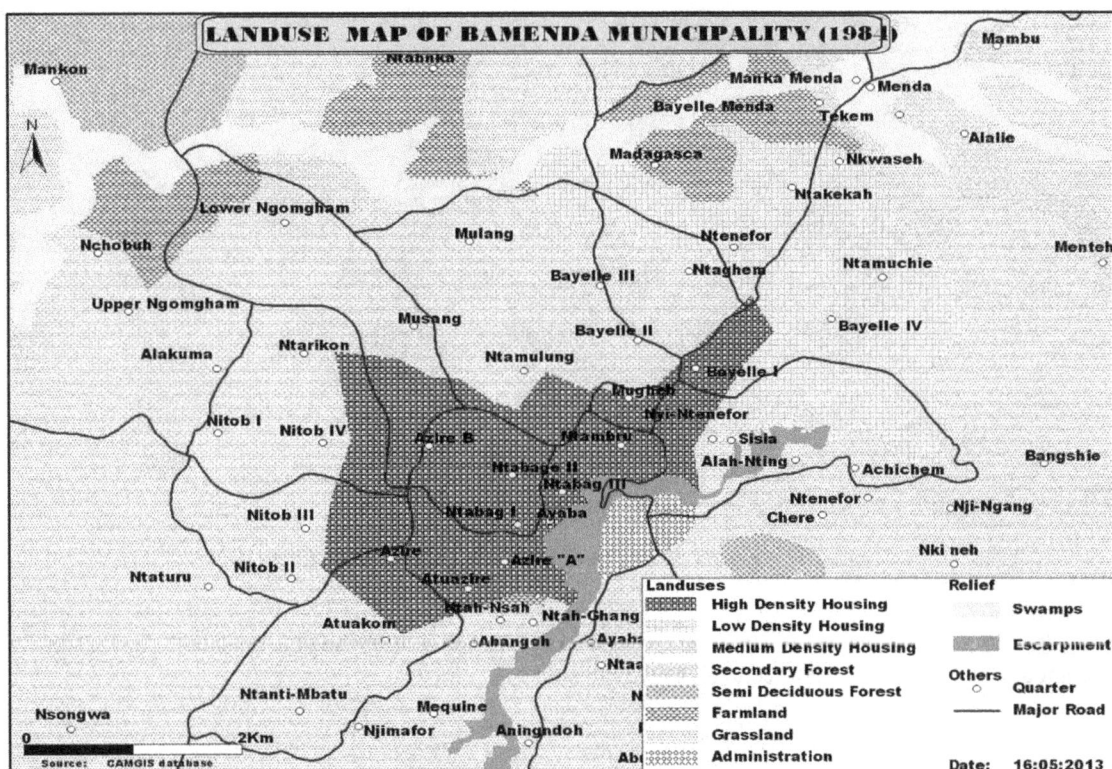

Figure 2. land use of the Bamenda municipality in 1984

During this period, wetlands occupied 27% of the total land use surface area in Bamenda. As a regional headquarter, Bamenda was exposed to significant land use transformation which was brought about by the influx of migrants to work there.

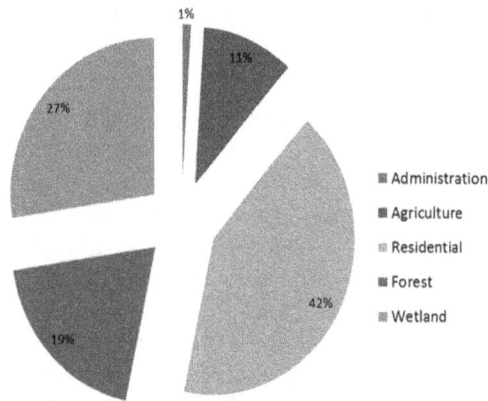

Figure 3a. Proportion of land uses in 1984   Figure 3b. Proportion of land uses in 2014

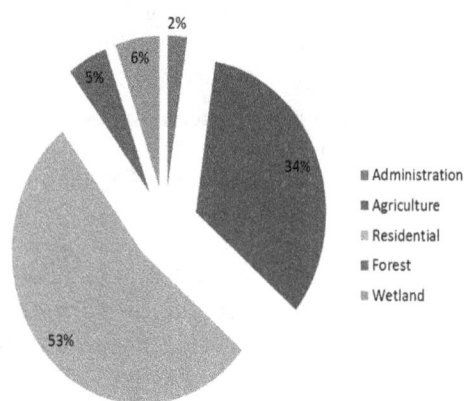

By 2014, significant changes were observed for the major land uses – residential land use moved from 42% and now occupies 53% of the total land use area (Fig 3b). This is followed by agricultural land use which occupies 34%. Conversely the surface area covered by wetlands reduced from 27% in 1984 to 6% in 2014. It is evident that residential and agricultural land uses have encroached and converted these wetlands (Figure 4 & Appendix I). Such changes signal the colonisation of wetland environments.

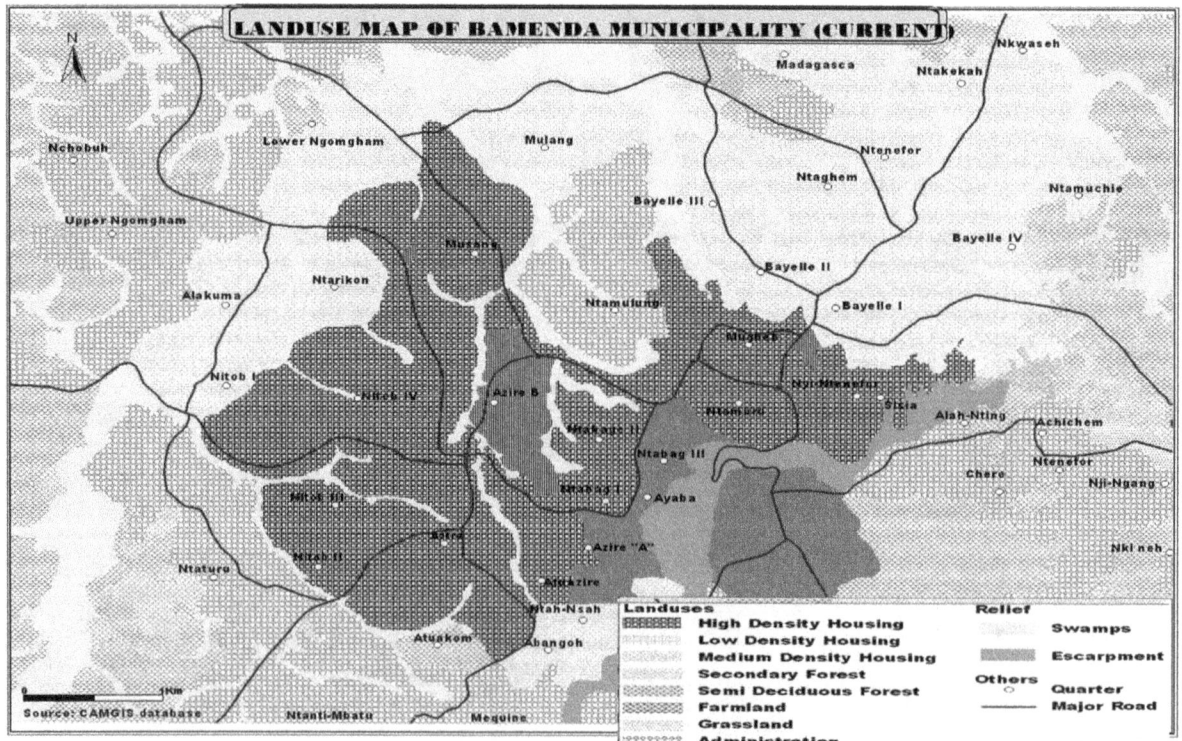

Figure 4. Land use situation of Bamenda municipality (2014)

The key driving force of land use dynamics and wetland degradation in Bamenda is population growth (Figure 5). Such demographic growth has resulted to a change in land uses.

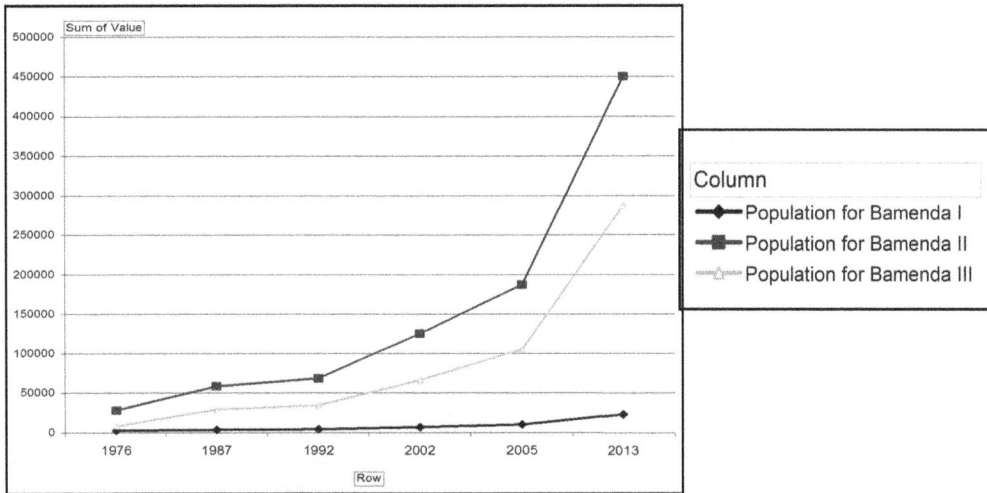

Figure 5. Population growth in Bamenda (Bamenda City Council, 2014)

Population increase has been accompanied by an increase in land speculation resulting to an increase in land value around the city. This has pushed many people to acquire cheap and marginal lands - wetlands. The quest for land from the wetlands is as a result of the necessity to build more houses to accommodate the fast increasing population. In the process of construction, the wetland ecosystem is destroyed through the cutting of trees and the killing of the habitats of some wetlands. Thereafter, the area is drained and backfilled with ground for buildings to be erected. Still in the same way with the rapid growing population these wetlands are over exploited for farmlands and other beneficial activities. Deforestation and wetland exposure (26.3%), increase in construction of houses (25%), drainage and backfilling (17.5%), over exploitation of wetlands (16.3%) and dumping of household waste (15%) were the key forces operating through land use change to cause wetland degradation as revealed from the questionnaires (Table 3).

Table 3. Activities that degrade the wetland ecosystem

Locality	Causes of wetland degradation					
	Increase in housing construction	Dumping of household waste	Over exploitation of wetlands	Drainage and filling	Deforestation and exposure of wetlands	**Total**
Ngomgham	3.8%	6.3%	1.3%	2.5%	3.8%	17.5%
Mulang	6.3%	3.8%	3.8%	5.0%	6.3%	25.0%
Mbelem	8.8%	5.0%	8.8%	5.0%	10.0%	37.5%
Ntenesoh	6.3%		2.5%	5.0%	6.3%	20.0%
Total	25.0%	15.0%	16.3%	17.5%	26.3%	100.0%

Source: Field work, 2015

Mbelem is the most degraded with a degradation rate of 37.5%, closely followed by Mulang with a degradation rate of 25.0%, Ntenesoh with a rate of 20.0% and finally Ngomgham with a degradation rate of 17.5%. Using the Stata 11.1 Statistical Package, a correlation was established between land use change and wetland degradation in Bamenda. The correlation analysis was done in which land use change (mainly agricultural and residential) was correlated with those of wetland degradation (reduction in size of wetlands, waste disposal and drainage and filling of wetlands) at 1% and 5% levels of significance. The correlation results showed a positive relationship between land use dynamics and wetland exploitation and degradation (Appendix II).

Previous studies on land use dynamics and wetland degradation observed agricultural and settlement expansion as the key factors influencing wetland conversion and subsequent degradation (Zorrilla-Miras *et al.*, 2014; Dahl & Johnson 1991; Dahl, 1993). Liu *et al* (2004) observed a similar situation for the Small Sanjiang Plain (SSP) which was formerly the largest wetland complex in China, located in the Northeastern part of Heilongjiang Province, China. Between 1950 and 2000, significant changes were observed for this wetland environment which was largely attributed to human activities. This led to a rapid decline in waterfowl and plant species with the loss and fragmentation of natural wetlands and wetland ecosystem degradation; greater variation in wetland water levels as the result of land-use changes over the years; and a decrease in floodplain area that caused increased flooding peak flows and runoff. In addition, Nyamasyol and Kihima (2014) observed that in Kenya, land use changes over the past three decades characterized by a noticeable increase in the size of farmland, settlement, and other lands and a decline in forestland, grassland, wetland, and woodland have affected the Kimana Wetland ecosystems. Murungweni (2013) using GIS equally observed a decline in the quality of urban wetlands in the Monavale Wetland in Harare.

7. Conclusion and Recommendations

Wetlands in Bamenda provide varied functions and services to the inhabitants ranging from flood control, provision of wetland products, wetland agriculture, habitat for wetland species, natural filter and reserve lands for vegetation. Due to rapid urbanisation these wetlands have been affected through reclamation, deforestation and exposure of wetlands, use of wetlands as dump sites and over exploitation of wetlands.

Between 1984 and 2014, significant changes were observed on land uses in Bamenda wherein residential land use increased from 42% in 1984 to 53%, and from 11% to 34% for agricultural land use. The surface area covered by wetlands reduced from 27% in 1984 to 6% in 2014. Population increase, coupled with an increase in land speculation result to an increase in land value around the city. This has pushed many people to acquire cheap and marginal lands - wetlands.

Based on the above observations, the following recommendations are suggested:

The Bamenda City Council should map out risk zones as well as structure the urban space into various functions with wetlands serving as protected area and green spaces. Penalties should be levied on people who go against this zonation by encroaching into wetlands. Efforts by the city council to discourage encroachment needs to be complemented by other stakeholders involved in environmental management. In addition, there is a need to integrate wetland management in the urban planning strategy of Bamenda. Urban planning in Bamenda should consider wetland environments as natural conservation sites, landscape planning and water resource management.

Given the current trend of land use change and its effects on wetlands, urban development policy should consider the proper zonation to ensure that the multiplication of land uses is contained. The current trend of land use dynamics can be checked by the application of zoning laws which control the changes witnessed in the main land uses (residential and agricultural land uses).

The wetland environments should be carved out and effectively declared as non-encroachment areas. The ecosystem services of these wetlands should be promoted. In addition, the Bamenda City Council should promote public awareness and understanding of wetland resources and actively encourage participation of the public, local government authorities and institutions in sustainable wetland management. This can be achieved by disseminating awareness on the importance of wetlands through leaflets, posters, radio, television and other media as well as periodically monitoring public responds or view on the need to conserve wetlands.

The city council should attempt to value the wetlands by indicating the cost of wetland degradation and the benefits of wetland conservation.

The city council should train extension staff of relevant ministries at District level to equip them with knowledge and skills to facilitate their supervisory role. They should also establish a mechanism and develop capacity for carrying out Environment Impact Assessment on proposed wetland development projects.

Since wetlands are a multi-sectoral resource, there is need to create and establish an appropriate institutional arrangement for their management. Although there are sectoral laws that refer to some aspects of wetlands such as water, or land or prevention of pollution, there is no comprehensive law for management of wetlands as an ecological entity. This can be realised by enacting a national law for regulating the management of wetland resources and encourage local authorities to make bye-laws for the proper management of wetlands.

The Bamenda City Council should establish fully "Protected Wetland Areas" of important biological diversity. No modification, drainage or other impacts will be entertained for the so-protected wetlands. The restoration program should involve all aspects of the ecosystems, including habitat restoration, elimination of undesirable species, and restoration of native species from the ecosystem perspective with holistic approach designed at watershed level.

Acknowledgements

The authors acknowledge the assistance of Bridget Ngeminiy Fru and Kongbime Elvis Shiangong for their support during the data collection process.

References

Ajibola, M. O., Adewale, B. A., & Isasan, K. C. (2012). Effects of Urbanisation on Lagos Wetlands. *International Journal of Business and Social Science, 3*(17).

Balgah, S. N. (2005). Landuse and Land Cover Dynamics in Buea and Tiko Sub-Divisions, Cameroon. Unpublished PhD Thesis, University of Buea, Department of Geography.

Balgah, S. N. (2007). Population Growth and land use dynamics in Buea Urban Area, Cameroon. *Loyola Journal of Social Sciences, XXI*(1), 103-115.

Balgah, S. N., Ndjib, G., & Ngwa, S. N. (2008). Monitoring land use and land cover dynamics in Buea and Tiko sub divisions in Cameroon: A Natural Resource Management, Published in Annals of the Faculty of Arts, Letters and Social Sciences, University of Yaounde 1, Vol. 1, No 8, Nouvelle Serie 2008.

Biradar, C. M., Thenkabail, P. S., Gangodagamage, C., & Islam, A. (2003). Landsat Enhanced Thematic Mapper (ETM+) Mosaic. Nominal 2000's Mosaic for the Limpopo River Basin. International Water Management Institute (IWMI), Sunil Mawatha, Sri Lanka. Retrieved from http://ww.iwmi.org

Castañeda, C., & Herrero, J. (2008). Assessing the degradation of saline wetlands in an arid agricultural region in Spain. *Catena, 72*, 205-213. http://dx.doi.org/10.1016/j.catena.2007.05.007

Cohen, B. (2006): Urbanization in Developing Countries: Current Trends, Future Projections, and Key Challenges for Sustainability. *Technology in Society, 28*, 63–80. http://dx.doi.org/10.1016/j.techsoc.2005.10.005

Dahl, T. E. (1993). Monitoring wetland changes - the U.S. Wetlands Status and Trends Study. In M. Moser, R. C. Prentice, & J. Van Vessem (Eds.), *Waterfowl and Wetland Conservation in the 1990's: A Global Perspective* (pp. 170-174). The International Waterfowl and Wetlands Research Bureau. IWRB Special Publication No. 26.

Dahl, T. E., & Johnson, C. E. (1991). Status and trends of wetlands in the contenninous United States, mid-1970's to mid-1980's. U.S. Department of the Interior, Fish and Wildlife Service, Washington, D.C. 28 pp.

Fombe, L. F., & Balgah, S. N. (2007). *The Urbanisation Process in Cameroon: Process, Pattern and Implications*. Nova Science Publishers, Inc. New York.

Grillas, P., Gauthier, P., Yavercovski, N., & Perennou, C. (2004). Mediterranean Temporary Pools: (1) Issues Relating to Conservation, Functioning and Management. Tour du Valat, Arles.

Kimengsi, J. N. (2008). The Contribution of Pamol Plantation and Its Associated Environmental Impacts to the Development of Ekondo-Titi Sub-Division, South West Province of Cameroon. Unpublished Master of Science Thesis, FSMS, Department of Geography, University of Buea.

Kimengsi, J. N. (2009). Pamol Industrial Growth and Land Use Conflicts in Ekondo-Titi Sub-Division, South West Region of Cameroon. Proceedings of the Second Post Graduate Seminar on Conflict Prevention Management and Resolution, Faculty of Social & Management Sciences, University of Buea, 28th Jan, 2009.

Kimengsi, J. N. (2011). Population Growth, Land Use Change and Forest Degradation in Ndian Division of Cameroon: Drivers and Policy Options. 17th International Interdisciplinary Conference on the Environment, 29th June to 3rd July, 2011, Kona, Hawaii. Retrieved from http://ieaonline.org/wp-content/uploads/2013/06/IEA-Program-And-Abstracts-2011-062411.pdf

Kometa, S. (2009). Wetland exploitation and its environmental impact: the case of the Bamenda-Bafoussam axis of the Western Highlands of Cameroon. Proceedings of the postgraduate Seminar on the conflict Prevention, Management and Resolution. University of Buea. Pg 25-26.

Kometa, S. S., & Ndi, R. A. (2012). The Hydro-Geomorphological Implications of Urbanization in Bamenda Cameroon. *Journal of Sustainable Development, 5*(6). http://dx.doi.org/10.5539/jsd.v5n6p64

Lambi, C. M. (2001). Changing Landuse in the Batie Hills of the West Province of Cameroon. *Journal of Applied Social Sciences (JASS), 2*(1/2).

Lambi, C. M., & Balgah, S. N. (2010). Reflections on the Changing Landuse Patterns in Some Grassland Rural Environments of the North West Region of Cameroon. *African Journal of Social Sciences (AJOSS), 1*(1).

Lambi, C. M., & Takang, R. T. (2010). Land Use Dynamics on the Eastern Slopes of Mount Cameroon. *African Journal of Social Sciences (AJOSS), 1*(3), 20-35.

Liu, H., Zhang, S., Li, Z., Lu, X., & Yang, Q. (2004). Impacts on wetlands of large-scale land-use changes by agricultural development: the Small Sanjiang Plain, China. *Ambio, 33*(6), 306-10. Retrieved from http://www.ncbi.nlm.nih.gov/pubmed/15387064#

Millennium Ecosystem Assessment. (2005). *Ecosystems and Human Well-being.* Retrieved from http://www.unep.org/maweb/documents/document.354.aspx.pdf

Murungweni, F. M. (2013). Effect of Land Use Change on Quality of Urban Wetlands: A Case of Monavale Wetland in Harare. *Geoinfor Geostat: An Overview S1.*

Ndenecho, E. N., & Fonteh, M. L. (2012). Freshwater and Coastal Resource Management in Cameroon: Building Resistance and Resilience to Climate Change, Bamenda, Angwecams printers.

Nyamasyo, K. S., & Kihima, B. O. (2014). Changing Land Use Patterns and Their Impacts on Wild Ungulates in Kimana Wetland Ecosystem, Kenya. *International Journal of Biodiversity, 10.* http://dx.doi.org/10.1155/2014/486727

Nyambod, E. M. (2010). Environmental Consequences of Rapid Urbanisation: Bamenda City, Cameroon. *Journal of Environmental Protection, 1,* 15-23. Retrieved from http://www.SciRP.org/journal/jep

Ramachandra, T. V. (2001). *Restoration and Management strategies of Wetlands in Developing Countries.* Electronic Green Journal, UELA Library UC Los Angeles.

Ripkens, C. (2009). Resilience and Vulnerability of wetlands, Germany, Msc thesis.

Sun, Z., Deal, B., & Pallathucheril, V. G. (2009). The Land-use Evolution and Impact Assessment Model: A Comprehensive Urban Planning Support System. URISA Journal, 21(1). Retrieved April 20, 2012, from http://downloads2.esri.com/campus/uploads/library /pdfs/119199.pdf

Tiffany, W., Tomliason, J., Schueler, T., Cappiela, K., Kitchell, A., & Hirschman, D. (2006). *Direct and Indirect Impacts of Urbanization on Wetlands Quality.*

Tijani, M. N., Olaleye, A. O., & Olubanjo, O. O. (2012). *Impact of Urbanization on Wetland Degradation. A Case Study of Eleyele Wetland.* Ibadan, South West Nigeria.

Tiner, R. W., Bergquist, H. C., DeAlessio, G. P., & Starr, M. J. (2002). Geographically Isolated Wetlands: a Preliminary Assessment of their Characteristics and Status in Selected Areas of the United States. U.S. Department of the Interior, Fish and Wildlife Service, Northeast Region, Hadley, MA.

UN-Habitat. (2009). *Harmonious Cities: State of the World's Cities2008/2009.* London. Retrieved September 30, 2009, from www.clc.org.sg/pdf/un-habitat20Report%20Overview.pdf

Williams, B., Walsh, C., & Boyle, I. (2009). *The Development of the Functional Urban Region of Dublin: Implications for Regional Development Markets and Planning.* Retrieved from http://www.uep.ie/pdfs/fur_markets_WilliamsWalshBoyle.pdf

WWF. (2004). The Economic Values of the World's Wetlands. Prepared with support from the Swiss Agency for the Environment, Forests and Landscape (SAEFL) Gland/Amsterdam, January 2004. Retrieved from http://www.unwater.org/downloads/wetlandsbrochurefinal.pdf

Xie, Z., Xu, X., & Yan, L. (2010). Analyzing qualitative and quantitative changes in coastal wetland associated to the effects of natural and anthropogenic factors in a part of Tianjin, China. *Estuarine Coastal and Shelf Science, 86,* 379-386. http://dx.doi.org/10.1016/j.ecss.2009.03.040

Zorrilla-Miras, P., Palomo, I., Gómez-Baggethun, E., Martín-López, B., Lomas, P. L., & Montes, C. (2014, February). Effects of land-use change on wetland ecosystem services: A case study in the Doñana marshes (SW Spain). *Landscape and Urban Planning, 122,* 160-174. http://dx.doi.org/10.1016/j.landurbplan.2013.09.013

Appendix I. Land use and surface area for Bamenda (1984 & 2014)

Domain	1984 Surface area (M^2)	Surface area (M^2) 2014	% change
Administration	511270.7	1822113	256.39
Farmland (Agriculture)	5566952	25988842	366.84
Residential (High density)	591832	6420252	90.78
Residential (Low density)	6150645	8054526	30.95
Residential (Medium density)	15158850	26793169	67.75
Forest	10135329	3827929	-164.77
Wetland	14350072	4275093	-235.67

Source: CAMGIS 2014

Appendix II. Correlation results on land use change and wetland degradation in Bamenda

	Dependent Variables		
	Wetland reduction	Waste disposal	Drainage & filling
Independent Variables	Coefficient	Coefficient	Coefficient
Agriculture	0.6*	0.4*	0.5**
Residential	-0.66**	0.3*	0.6**
Observations			

* significant at 1%; ** significant at 5%

Source: Analysis of questionnaire using Stata 11.1 Package

6

Profit Efficiency of Rice Farmers in Cambodia
The Differences between Organic and Conventional Farming

Rada Khoy[1], Teruaki Nanseki[2] & Yosuke Chomei[2]

[1] Graduate School of Bioresource and Bioenvironmental Sciences, Kyushu University, Fukuoka, Japan

[2] Faculty of Agriculture, Kyushu University, Fukuoka, Japan

Correspondence: Teruaki Nanseki, Faculty of Agriculture, Kyushu University, Hakozaki 6-10-1, Higashiku, Fukuoka 812-8581, Japan. E-mail: nanseki@agr.kyushu-u.ac.jp

Abstract

This article highlights some important issues regarding the relative profit efficiency of organic and conventional farming in selected study areas of Cambodia, by estimating pool and separate profit frontiers of the two groups and accounting for the self-selection problem. We identify the relationship between the efficiency score from each frontier with farmers' characteristics. The results indicate that farmers cannot manage their rice farming effectively in larger fields and fail to optimize their labor input and costs owing to limited skills and knowledge in rice production. Organic fertilizers can help to increase farmers' rice income, while chemical fertilizers are less effective in doing so. Interestingly, being an organic farmer had no effect on farmers' income elasticity when we conducted pool frontier estimation. However, these results were rejected by an LR test that was favorable to the estimation of a separate frontier, which suggested a better efficiency score if farmers adopted organic farming. We found some significant factors influencing the efficiency score, including *education, own-tractor,* and *credit use* (negative correlation) and *selling, other farming,* and *number of poultry* (positive correlation). *Off farm* was negatively correlated with the efficiency score in organic farming, but positively correlated in matched conventional.

Keywords: profit efficiency, organic rice, conventional rice, stochastic production frontier, propensity score matching, Cambodia

1. Introduction

To help mitigate environmental problems, sustainable farming systems have existed for over half a century. Organic farming is regarded as one of the most environmentally friendly practices that can solve some environmental deterioration issues, leading many countries to adopt this farming practice. However, it is still questionable whether organic farming can be adopted on a global level, or can help to increase farmers' income. These questions remain the main concerns in the production of organic products in the developing world.

Able to produce organic rice and skeptical about the excessive use of farm chemicals, Cambodian farmers have adopted organic rice practices since 2003 (Cambodian Organic Agriculture Association [COrAA], 2011). During the first few years, Cambodian rice farmers produced organic rice with surprising success, and many organic rice cooperatives became established throughout the main rice production areas in Cambodia. However, not surprisingly, after the support from NGOs was terminated, organic rice farming diminished in scale, and many organic rice farmers reverted to conventional farming, although some studies, for example: Taing (2008), and Sa (2011), documented that organic farming could increase farmers' rice yield and profit.

Many studies about organic practice, for example: Cary and Wilkinson (1997), Musshoff and Hirschauer (2008), Sheeder and Lynne (2009) and Ponti, Rijk, and Ittersum (2012), have acknowledged that financial concerns are the main motivating factors behind the increased adoption of organic farming. Generally, organic products often obtain price premiums (Nieberg & Offermann, 2003). However, as argued by Imbens and Wooldridge (2009), the better performance of technology adopters might result from differences in their characteristics, rather than being adopters or non-adopters, implying that a selection bias exists among farmers. This could affect farmers' adoption decision and, hence, performance. To solve the selection bias problems of organic rice farming adoption in Cambodia, Khoy, Nanseki, and Chomei (2015, 2016) employed two approaches, propensity score matching and endogenous switching regression, to evaluate the impact of the adoption. Their studies suggested that Cambodian rice farmers could benefit from adopting organic rice farming in terms of rice yield and profit.

Even organic rice farming has been introduced for years; information regarding production practices is very limited. In particular, none of the studies focus on the respective efficiency of organic and conventional rice farming. Some studies, Taing (2008) and Sa (2011), have tried to directly compare the yield and profit differences between organic and conventional rice farmers, but didn't account for selection bias. Khoy et al. (2015, 2016) accounted for selection bias in their studies by applying propensity score matching and endogenous switching regression. However, their work did not describe the profit efficiency of organic and conventional farmers. Thath (2014) studied the cost efficiency of Cambodian rice farmers by comparing different rice production zones, but this study did not explicitly analyze organic rice farming. Self-selection remains the main issue although some articles have documented that the organic movement is the potential practice of profit gains compared to conventional rice production, especially for smaller farms. Some farmers could obtain higher profit when adopting organic rice farming, while many would not receive this benefit in terms of their conditional issues. Furthermore, many farmers are reluctant to begin this new practice because they believe their present farming has suited them, and they have become accustomed to it. As Khoy et al. (2016) demonstrated, Cambodian farmers adopted organic rice farming based on their comparative advantage, suggesting farmers who possessed relative advantage with organic farming adopted the new practice, and those who were suited to conventional stayed with the old practice. Hence, the detail about the relative profit efficiency between organic and conventional farmers needs to be examined.

This study aims to assess the profit efficiency of organic and conventional farmers and its' determinants by accounting for selection bias. The article highlights two important aspects of the profit efficiency of organic and conventional rice farming in Cambodia. First, pooled and separate profit frontiers between organic and conventional farmers that account for the self-selection problem were estimated. Second, the relationship between the efficiency score from each profit frontier and farmers' characteristics was identified.

2. Method

2.1 Study Site and Data Collection

This study was conducted in two provinces, Takeo and Kampot province. We purposely selected three targeted districts from two provinces, because there are organic rice cooperatives located in those districts, they border one other, and they possess similar social demographic and agro-ecosystems. The districts are Srer Cheng Organic Agriculture Development Cooperative, situated in Chum Kiri district, Kampot Province; Chhuk Organic Agriculture Development Cooperative, in Chhuk district, Kampot province; and Trapaing Sronger Agriculture Development Cooperative in Tram Kak District, Takeo Province. Random organic and conventional farmers were selected from each cooperative and district. Data was collected by face-to-face interviews for the 2013 wet season rice production. A total of 221 respondents were interviewed, of which 84 organic respondents were selected from organic cooperatives and 137 were randomly selected from conventional farmers in the same study areas. Among all respondents, 36, 21, and 27 organic farmers, and 64, 49, and 24 conventional farmers, were selected from Chum Kiri, Chhuk, and Tram Kak districts, respectively.

2.2 Analytical Framework

This article discusses some issues arising from a comparison of profit efficiency between organic and conventional rice farmers in Cambodia, by estimating both pool and separate profit frontiers using stochastic production frontiers, controlling for farmers' selection bias. We employed propensity score matching to control for farmers' observable characteristics. We then ran a regression of the efficiency score generated from each frontier with farmers' characteristics.

In microeconomic theory, the production or profit frontier explains the maximum output resulting from a set of production inputs and technology. While some inputs are decided by farmers, some are exogenously generated by fixed technology provided to farmers. This would add some constraints and/or advantages to the production performance of farmers (Mayen, Balagtas, & Alexander, 2010). In our study, organic farmers in particular adopted a set of technologies that would affect their performance. Thus, the production inputs of organic and conventional farmers may be different. To account for technology differences between organic and conventional farmers, Mayen et al. (2010) included a treatment variable (organic or conventional) to estimate the production frontier and discussed whether the correlation coefficient of the treatment variable had a positive or negative effect on farmers. We believe that estimating the production frontier separately would result in a better conclusion, as organic and conventional might be two completely different groups in terms of the allocation of production inputs. This simply means that the production inputs for the production frontiers of organic farmers may differ from those of conventional farmers (Bravo-Ureta, Greene, & Solís, 2012). We conduct a LR test proposed by Greene (2007), to confirm our assumption of a technology difference.

In addition to the technology difference issue, farmers themselves decided whether to adopt this technique or not, which resulted in selection bias among farmers. To accurately evaluate the impact of organic farming adoption and its correlated factors on profit efficiency levels, we applied a multi-step framework to account for potential selection problems in the estimation of the Stochastic Production Frontier (SPF) model. Monteiro (2010) demonstrated that in order to obtain an accurate estimation of adoption impact, we have to set a control group that has characteristics that are as similar as possible to those in the treated group. Propensity Score Matching (PSM) has become a common approach that can balance the observed characteristics of the control group to resemble those in the treated group. In other words, this approach can generate the counterfactual situation and mitigate potential selection bias associated with observable characteristics (Rosenbaum and Rubin 1983). PSM is used in some recent studies such as Bravo-Ureta, Almeida, Solís, and Inestroza (2011), Bravo-Ureta et al. (2012), Cerdán-Infantes, Maffioli, and Ubfal (2008) and Mayen et al. (2010) to access the impact of technology adoption.

For this paper, we estimated the pool and separate profit frontier by employing a SPF approach in unmatched and matched samples generated by PSM, to correct for biases from observed characteristics. We then measured and compared efficiency scores from each frontier between organic and conventional farmers, before and after matching. We firstly estimated the profit frontier of the pool unmatched sample by including an adopter variable (organic or conventional farmer) as an input variable, and we also estimated the separate profit frontiers of organic and conventional farmers. In the second step, the pool and separate profit frontiers were re-estimated by using a matched sample produced by PSM. After assessing the different profit efficiencies, we identified the relationship between the efficiency score of each frontier and farmers' social economic characteristics, by employing OLS regression.

2.3 Empirical Models

2.3.1 Stochastic Production Frontier (SPF)

The SPF framework was used to estimate profit frontiers and the profit efficiency score. This approach can deal with the stochastic nature of agricultural processes. The SPF model is written as:

$$ln y_i = \beta x_i + v_i - u_i \tag{1}$$

where y_i denotes the output (we used profit as the output), x_i (in logarithm) is a vector of the production inputs (described in table 1), β is a vector of parameters to be estimated, v_i is a two-sided stochastic term that accounts for statistical noise, and u_i is a non-negative stochastic term representing inefficiency.

We measured the efficiency score suggested by Battese and Coelli (1988). Because the output is in natural logarithmic form, the efficiency score is specified as:

$$TE_i = y_i/exp^{(\beta x_i + v_i)} = exp^{(\beta x_i + v_i - u_i)}/exp^{(\beta x_i + v_i)} = exp^{(-u_i)} \tag{2}$$

2.3.2 Propensity Score Matching (PSM)

PSM is a two-step procedure (Becker & Ichino, 2002). Firstly, farmers' propensity scores were determined by estimating the probability model in probit or logit, specified as:

$$Y(1;0) = \beta_0 + \beta_1 X_1 + \beta_2 X_2 + \cdots \beta_n X_n \tag{3}$$

where Y is a binary dependent variable (1=Organic farmer; 0=Conventional farmer), β is the regression coefficient to be estimated, and X is an independent variable to be explained (described in table 1). The propensity score of each farmer is then estimated based on the following equation:

$$P_{score} = 1/[1 + e^{-(\beta_0 + \beta_1 X_1 + \beta_2 X_2 + \cdots \beta_n X_n)}] \tag{4}$$

Secondly, each farmer in the organic group is matched up to a conventional farmer with similar propensity score values, by using some comparison techniques, in order to estimate the average treatment effect (ATE). Here, we adopted single nearest neighbor matching (NNM) to measure average treatment effect on treated (ATT).

2.3.3 OLS Regression

We employed OLS regression to identify the relationship between farmers' efficiency score and their characteristics, before and after the matching procedure. The OLS model is written as:

$$y_i = \beta_0 + \beta_i x_i + \varepsilon_i \tag{5}$$

where y is the dependent variable (efficiency score), x is the independent variable to be explained (described in table 1), β is the regression coefficient to be estimated, and ε is an error term.

2.4 Description of Data Variables

Table 1 describes all the variables used in each model. It shows the variable name, definition, and unit of each variable. In SPF, *rice income* regarded as profit was used as the output variable. We have used *production land, labor input, organic fertilizer, chemical fertilizer, other cost,* and a dummy *adopter* variable as production inputs in pool frontier analysis. We excluded the *adopter* variable in the separate frontier estimation. After estimating SPF, an efficiency score was estimated and used as a dependent variable in the OLS regression. We included independent variables such as *age, gender,* and *education* for farmers' characteristics; *farming labor, rice plots, rice field, selling, other farming, number of cows, number of poultry,* and *membership* for farm characteristics; and *off farm, own-tractor,* and *credit-use* for economic characteristics.

Table 1. Definitions of variables to be used in each approach

Variable	Definition	Unit
	Stochastic production frontier model	
Rice income[a]	Total rice income per hectare (excluding family labor cost)	$/ha
Production land	Organic rice production land (for conventional: rice field size produced Phka Rumduol Rice variety)	Ha
Labor input	Total labor employed in rice production per hectare	Man-day
Organic fertilizer	Total organic fertilizer applied in rice production per hectare	Kg
Che. fertilizer	Total chemical fertilizer applied in rice production per hectare	Kg
Other cost	Total production cost excluded labor and fertilizer cost	US$
Adopter	= 1 if farmer produces organic rice	Dummy
	OLS regression	
TE score	TE score estimated from SPF	0-1
Age	Age of household head	Years
Gender	= 1 if household head is male	Dummy
Education	Years of schooling of household head	Year
Farming labor	Number of family labors available for rice farming	Person
Rice plots	Numbers of rice plots farmers owned	Number
Rice field	Total rice field size farmers owned	Ha
Selling	= 1 if farmers sell their rice	Dummy
Other farming	= 1 if farmers have other farm activities besides rice	Dummy
No. of cows	Numbers of cows they owned	Number
No. of poultry	Numbers of poultry they raised	Number
Membership	= 1 if farmers belong to any agricultural related group	Dummy
Off farm	= 1 if farmers have off-farm job	Dummy
Own-tractor	= 1 if farmers have two-wheel tractor	Dummy
Credit-use	= 1 if farmers loan credit	Dummy
	Propensity score matching	
Adopter	= 1 if farmer produces organic rice	Dummy
Age	Age of household head	Years
Gender	= 1 if household head is male	Dummy
Education	Years of schooling of household head	Year
Farming labor	Number of family labors available for rice farming	Person
House size	The square meter of house farmers owned	M^2
Rice plots	Numbers of rice plots farmers owned	Number
Rice field	Total rice field size farmers owned	Ha

Other farming	= 1 if farmers have other farm activities besides rice	Dummy
No. of cows	Numbers of cows they owned	Number
No. of poultry	Numbers of poultry they raised	Number
Off farm	= 1 if farmers have off-farm job	Dummy
Own-tractor	= 1 if farmers have two-wheel tractor	Dummy
Credit-use	= 1 if farmers loan credit	Dummy

Note. a: Rice income = (Yield * Price) – (Fixed cost + Variable cost); Family labor cost is not included in variable cost; It was regarded as profit.

We have specified some variables to be included in PSM for balancing the characteristics between organic and conventional farmers. The balanced variables are *age, gender, education, farming labor, rice plots, rice field, selling, other farming, number of cows, number of poultry, off farm, own-tractor,* and *credit-use.* We believe that these variables potentially influence farmers' propensity to adopt organic rice farming.

3. Results and Discussions

3.1 Descriptive Results Before and After Matching

Table 2 presents the descriptive statistics and statistical significance tests of two farmer groups, before and after matching.

Table 2. Descriptive results and statistical significant test between organic and conventional farmers

	Unmatched				Matched			
	Pool	Organic	Con.		Pool	Organic	Con.	
Variable	M (221)	M (84)	M (137)	Diff.	M (121)	M (84)	M (37)	Diff.
Age	46.15	47.35	45.42	1.92	46.69	47.35	45.19	2.16
Gender	0.90	0.94	0.88	0.06	0.94	0.94	0.95	-0.01
Education	5.90	7.11	5.17	1.94***	7.00	7.11	6.76	0.35
Farming labor	2.79	2.85	2.76	0.09	2.83	2.85	2.81	0.03
House size	38.21	39.35	37.51	1.84	39.13	39.35	38.63	0.73
Rice plots	2.57	2.82	2.42	0.41***	2.73	2.82	2.51	0.31
Rice field	1.02	1.17	0.94	0.23***	1.12	1.17	1.01	0.15
Selling	0.80	0.96	0.69	0.27***	0.93	0.96	0.86	0.10**
Other farming	0.29	0.44	0.19	0.25***	0.42	0.44	0.38	0.06
No. of cows	2.60	3.12	2.28	0.83***	3.02	3.12	2.81	0.31
No. of poultry	81.41	121.74	56.68	65.06	133.98	121.74	161.76	-40.02
Membership	0.52	0.98	0.25	0.73***	0.74	0.98	0.19	0.79***
Off farm	0.21	0.26	0.18	0.08	0.26	0.26	0.27	-0.01
Own-tractor	0.19	0.25	0.15	0.10*	0.21	0.25	0.14	0.11
Credit use	0.24	0.19	0.26	-0.07	0.17	0.19	0.11	0.08
Production land	0.59	0.47	0.66	-0.19***	0.52	0.47	0.63	-0.16**
Labor input	254.50	282.70	237.21	45.49**	274.55	282.70	256.03	26.67
Org. fertilizer	1586.87	2115.29	1262.88	852.42***	2016.54	2115.29	1792.34	322.95
Che. fertilizer	83.92	0.00	135.38	-135.38	34.12	0.00	111.59	-111.59
Other cost	172.78	168.65	175.31	-6.66	162.63	168.65	148.96	19.69
Yield	2.86	3.32	2.58	0.75***	3.12	3.32	2.68	0.65***
Rice income	603.13	973.77	375.88	597.90***	827.43	973.77	495.17	478.60***

Note. *, **, *** significant at 10%, 5%, and 1% respectively; Con. is conventional; Diff. is difference; M is mean; Values in parenthesis represent the numbers of sample in each group

Before matching, many variables were statistically different between organic and conventional farmers. It indicates that the education of organic farmers was 1.94 years higher than that of conventional farmers. Organic farmers also owned statistically more numbers of rice plots, and possessed larger rice field size vis-à-vis conventional farmers. We found that 96 percent of organic farmers had sold their products, which was 27 percent higher than conventional farmers. Organic farmers also had a higher percentage of other farming activity, and raised more cows. In addition, almost all organic farmers (98 percent) belonged to some agricultural groups (*membership*) compared to only 25 percent of conventional farmers. Organic farmers also had a bigger proportion of owning tractor versus conventional. Based on unmatched results, our testing implies that organic farmers possess better characteristics vis-à-vis conventional farmers. They have higher education that could aid the adoption of new technology because they can access much information through various sources. Organic farmers possess larger farms, greater farming skills, and more machinery, favorable conditions for them to adopt organic farming.

Production inputs and outputs of organic and conventional farmers are also presented in table 2. The results from the unmatched sample show that organic farmers allocate statistically fewer hectares of their land (production land) to organic farming compared to conventional farmers for the phka rumduol rice variety. The results suggest that organic farming is more labor intensive, organic farmers employing 45.49 man-day/ha of labor input, which is significantly higher than conventional farming, because organic farmers need to employ more labor to meet organic farming requirements. Not surprisingly, organic farmers applied more organic fertilizer to their farm and obtained a significantly higher yield and income compared to conventional farmers. Nevertheless, as argued earlier, this improved performance in rice farming may be due to the better characteristics of organic farmers rather than being conducting organic farming per se. Hence, we used PSM to control for characteristic differences so as to obtain unbiased results.

After we conducted the matching approach, the difference between the two groups was minimized. For all the variables included in PSM, only the variable selling was still significant, while the other variables showed no significant difference. For variables excluded in PSM, membership, production land, yield, and rice income still showed significant differences. Surprisingly, there is no significance difference for application of organic fertilizer between the two groups, suggesting matched conventional farmers have knowledge about the advantages of organic fertilizer. The reduction in significant difference between the two groups could be because PSM minimizes the heterogeneity and the matched sample became more homogeneous in term of observed variables used in the analysis. As shown in figure 1, compared to all conventional farmers, the propensity score of matched conventional farmers is similar to that of organic farmers. This suggests that our proposed matching technique was fairly successful.

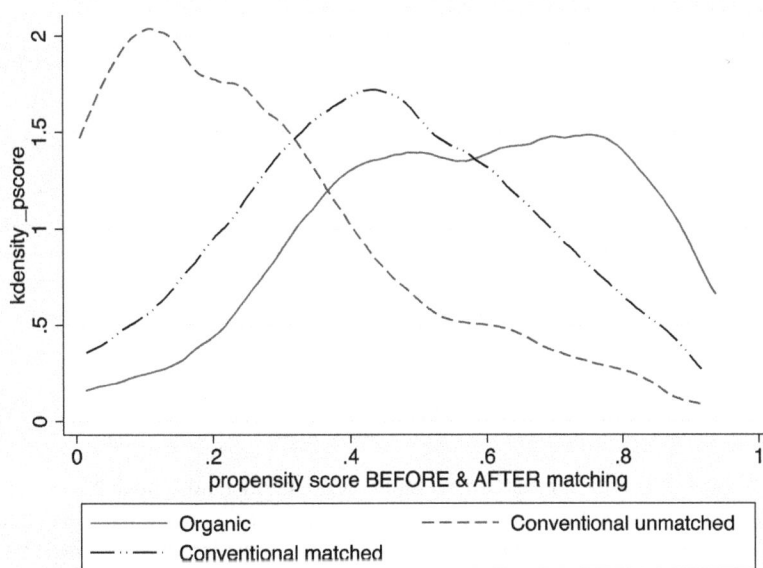

Figure 1. Kernel density of propensity score matching

3.2 Stochastic Production Frontier Results

Table 3 gives stochastic production frontier results of unmatched and matched sample. It also shows the pool and separate frontier results. For the unmatched pool sample, we found that production land, labor input, and other costs were negative and statistically significant with output, while organic fertilizer was positive and significant. A negative relationship between production land and output indicates that producing rice in a larger field size does not increase farmers' profit elasticity, as Cambodian farmers are not able to manage the large field effectively due to limited production techniques and skills. The result was consistent with Islam, Sipilainen, and Sumelius (2011) with respect to the profit efficiency of rice farmers in Bangladesh, but it was inconsistent with Kiatpathomchai (2008) who assessed the economic efficiency of rice production in Thailand, and Thath (2014) who focused on the cost efficiency of rice farming in Cambodia. Labor input and other costs were negative and significant suggesting that sample farmers failed to manage input effectively. The optimal use of labor input and cost are necessary to increase farmers' efficiency. Aung (2011) suggests a different result for rice farmers in Myanmar in the case of labor input. The results show that applying organic fertilizer helps to increase profit elasticity, as it helps to increase the yield and minimize the external input cost. Conversely, applying chemical fertilizer would decrease farmers' rice income even there was no significant correlation. This result was consistent with that of Aung (2011), but contrasted with that of Costantin, Martin, and Rivera (2009) in the case of the Brazilian grain crops. Asian Development Bank (2014) pointed out that applying both organic and inorganic fertilizer could increase rice farmers' production and value of production. As expected, compared to a conventional farmer, adopting organic practices (an adopter) could increase profit elasticity, the variable adopter having a positive and statistically significant correlation coefficient. However, this result might be due to farmers' particular characteristics rather than being organic or conventional. We will discuss this matter in more depth below in the matched results.

Table 3. Pool and separate stochastic production frontier analysis of unmatched and matched sample

Variable	Unmatched pool sample		Organic		Conventional		Matched pool sample		Matched conventional	
Log (Rice Income)	Coef.	Std. Err.	Coef.	Std. Err.	Coef.	Std. Err.	Coef.	Std. Err.	Coef.	Std. Err.
Production land	-0.269***	0.063	-0.279***	0.098	-0.258***	0.086	-0.347***	0.077	-0.315**	0.149
Labor input	-0.368***	0.089	-0.314**	0.136	-0.449***	0.119	-0.475***	0.108	-0.711***	0.174
Organic fertilizer	0.029*	0.016	0.123**	0.060	0.025	0.018	0.099***	0.036	0.076	0.050
Chemical fertilizer	-0.009	0.045			-0.022	0.061	-0.220***	0.076	-0.174	0.162
Other cost	-0.157***	0.041	-0.076	0.051	-0.217***	0.060	-0.126***	0.050	-0.356**	0.140
Adopter	0.270***	0.086					-0.175	0.144		
Constant	3.895***	0.250	3.527***	0.329	4.246***	0.347	4.228***	0.292	5.164***	0.851
Variance of u	0.141***	0.019	0.092***	0.026	0.154***	0.027	0.078*	0.040	0.007	0.593
Variance of v	0.133***	0.012	0.104***	0.017	0.154***	0.017	0.133***	0.021	0.168***	0.032
Lambda	1.061***	0.028	0.885***	0.039	0.997***	0.038	0.583***	0.059	0.043	0.619
	N = 221		N = 84		N = 137		N = 121		N = 37	
	Chi^2 = 190.440***		Chi^2 = 21.170***		Chi^2 = 28.840***		Chi^2 = 137.730***		Chi^2 = 52.680***	
	Log like. = 56.859		Log like. = 48.299		Log like. = 18.532		Log like. = 55.298		Log like. = 13.383	

Note. *, **, *** significant at 10%, 5%, and 1% respectively; Coef. is coefficient; Std. Err. is standard error

In the separate frontier result for the unmatched sample, production land, and labor input were negative and significantly associated with rice income for both organic and conventional groups. As stated earlier in the pool analysis, this suggests that both groups of farmers cannot manage their larger sized rice farms properly and fail to allocate labor to their farms efficiently. Our results show that organic fertilizer is positively and significantly correlated with output suggesting organic substance could increase income elasticity for organic farmers. For conventional farmers, results also suggest a positive correlation between organic fertilizer and rice income and a negative relationship between chemical fertilizer and rice income, but in both cases, the correlation is not statistically significant. Hence, organic fertilizer can increase rice income for both organic and conventional farmers, while chemical fertilizer lowered conventional farmers' rice income. Other costs were negatively related with output for both groups, but it was only statistically significant for conventional farmers.

Following the matching process, we also estimated SPF in pool and separate frontiers. In pool estimation, there were surprising changes of statistical significance. Production land, labor input, and other costs showed no difference to the unmatched pool estimation in terms of sign and statistical significance of its correlation coefficient. However, organic fertilizer, and chemical fertilizer became highly significant with the same sign. This strongly suggests that organic fertilizer can help to increase farmers' rice income, while chemical fertilizer lowers their income. Interestingly, the variable adopter was negative and has no statistical significance, which suggests that organic farming had no effect on farmers' rice income. This result is consistent with (Mayen et al., 2010) who conducted tests on dairy farms in the United States. Therefore, the greater efficiency of organic farmers was most likely due to their characteristics, rather than due to conducting organic or conventional farming, when estimated in the pool frontier. These results were rejected by a LR test suggested by Greene (2007), which produced better results for the separate frontier estimation for organic and conventional farmers. We estimated a LR test based on the following equation:

$$LR = 2*[\ln L_P - (\ln L_O + \ln L_C)] \tag{6}$$

where $\ln L_P$, $\ln L_O$, and $\ln L_C$ denote the log-likelihood values obtained from the pool frontier, organic frontier, and conventional frontier, respectively, in both unmatched and matched sample. The estimated LR tests were -19.944 in the unmatched sample and -12.768 in the matched sample. This rejected the null hypothesis for the equality of the pool and separate frontier model, with 0.01 significance in both cases (unmatched and matched sample). This confirms that the production inputs included in the estimation varied across the two groups of farmers, and the negative sign of the LR test offered the indicator favorable to the separate frontier estimation. Hence, we can infer that using pool estimation for the profit frontier of organic and conventional farmers leads to overstatement of the efficiency of conventional farmers. By allowing the variable adopter in pool estimation before and after matching, we cannot accurately confirm the efficiency of both organic and conventional farmers. We will compare and discuss the farmers' level of efficiency in the next section. For the result in the matched conventional frontier, we find similar results to those in the unmatched conventional, in terms of both sign and significance of correlation coefficient.

3.3 Efficiency Score of Farmers

We present and compare the average efficiency scores of organic and conventional farmers in table 4. In pool estimation, farmers had an average efficiency score of 0.877 for the unmatched sample, and this increased to 0.928 after we matched the sample and estimated the production frontier. The increase in average score after matching suggests that the extent of poor farmers' characteristics had been reduced. The matched sample was also more homogenous. There was significant difference between organic and conventional farmers for the unmatched sample, but no significant difference in the matched sample, and matched PSM (ATT) resulted from single nearest neighbor matching. Again, this suggests that organic farming had no effect on farmers' efficiency when we estimated it in a pooled frontier.

Table 4. The estimation and comparison of efficiency score between organic and conventional farmers

Variable	Pool estimation				Separate estimation			
	All	Organic	Conventional	Diff.	All	Organic	Conventional	Diff.
Unmatched	0.877	0.891	0.869	0.022*	0.832	0.915	0.780	0.135***
Matched	0.928	0.929	0.925	0.005	0.878	0.915	0.792	0.123***
Matched PSM[a]	ATT	0.891	0.894	-0.003	ATT	0.915	0.799	0.116***

Note. *, **, *** significant at 10%, 5%, and 1% respectively; Diff. is difference; a: we estimate average treatment effect on treated (ATT) by employing single nearest neighbor matching

However, when we estimated the profit frontier differently, results indicate that, on average, farmers had a 0.832 efficiency score in the unmatched sample, lower than those in the pool estimation. In the matched sample, the average score was 0.878, which is also lower than the pool estimation. The lower average efficiency score in the separate estimation is due to the decrease in efficiency score for conventional farmers. In Contrast to pool frontier estimation, there is a highly significant difference between organic and conventional farmers in the unmatched sample, matched sample, and matched PSM. The highly significant difference even in the matched sample and matched PSM suggest that organic farming would help to increase farmers' profit efficiency when we estimate in a separate frontier. When two groups were estimating in the pool frontier, the efficiency score of conventional farmers increased, leading to no significant difference between the two groups. This was because organic farmers allocated higher production inputs to their smaller production land, while conventional farmers allocated lower inputs to their larger field. On the other hand, estimating the separate frontier allowed us to calculate efficiency scores for conventional farmers accurately, because the score of all conventional farmers was estimated for their most efficient farm. Eventually, after conducting a LR test, we could obtain better results in the separate frontier.

3.4 Determinants of Profit Efficiency Score

In this section, we will explain how farmers' characteristic affects their efficiency score. The relationship between efficiency score and farmers' characteristics was shown in table 5. All the results of efficiency score estimated from the unmatched pool sample, unmatched separate frontier (organic and unmatched conventional), matched pool sample, and matched conventional, were used as the dependent variable regressed with some independent variables.

Table 5. Relationships between efficiency score and farmers' characteristics by OLS regression estimation

Variable	Unmatched pool sample		Organic		Conventional		Matched pool sample		Matched conventional	
	Coef.	Std. Err.	Coef.	Std. Err.	Coef.	Std. Err.	Coef.	Std. Err.	Coef.	Std. Err.
Age	0.001	0.001	0.000	0.001	0.000	0.001	0.000	0.000	-0.001	0.003
Gender	0.024	0.020	0.010	0.028	0.006	0.036	0.004	0.015	-0.152	0.108
Education	-0.004**	0.002	-0.001	0.002	-0.009**	0.004	-0.002	0.001	-0.016**	0.007
Farming labor	0.004	0.006	0.005	0.006	0.017	0.014	0.003	0.003	0.020	0.029
Rice plots	0.009	0.007	-0.007	0.007	0.016	0.015	-0.004	0.004	0.003	0.031
Rice field	0.003	0.013	0.011	0.014	0.002	0.027	0.009	0.007	0.013	0.043
Selling	0.060***	0.016	0.056	0.038	0.095***	0.029	0.026*	0.014	0.028	0.064
Other farming	0.026*	0.014	0.019	0.016	0.037	0.031	0.018**	0.008	-0.006	0.050
No. of cows	0.000	0.005	-0.002	0.005	-0.011	0.011	0.000	0.003	-0.017	0.019
No. of poultry	0.000	0.000	0.000	0.000	0.000**	0.000	0.000	0.000	0.000**	0.000
Membership	0.010	0.013	0.017	0.045	0.024	0.034	0.007	0.008	0.074	0.069
Off farm	-0.022	0.014	-0.030*	0.016	0.012	0.030	-0.009	0.008	0.173***	0.052
Own-tractor	0.000	0.016	-0.035**	0.016	0.032	0.038	-0.016*	0.009	0.037	0.067
Credit use	-0.026*	0.015	-0.032*	0.018	-0.034	0.034	-0.011	0.010	0.037	0.087
Constant	0.766***	0.036	0.842***	0.061	0.697***	0.069	0.890***	0.024	0.987***	0.163

Note. *, **, *** significant at 10%, 5%, and 1% respectively; Coef. is coefficient; Std. Err. is standard error

Our results showed that education was negatively correlated with efficiency score in the unmatched pool sample, with the conventional and matched conventional indicating that farmers with higher educational levels obtain a lower efficiency score. Generally, more educated farmers often had other jobs, in addition to their farming activity, that could result in a lower efficiency score. This was consistent with Thath (2014), but inconsistent with Aung (2011).

The category of selling was positively associated with the efficiency score in the unmatched pool sample, organic unmatched conventional, and matched pool sample. This suggests business oriented farmers may increase efficiency due to their motivation in gaining profit from rice production. Other farming was positively associated with the efficiency score in the unmatched and matched pool sample, indicating that farmers with other farming activity may be more accessible to organic resources, and have higher skill and knowledge levels in farming activities. Number of poultry was positively correlated with efficiency score in organic, conventional and matched conventional. This was because farmers who raised more poultry may have an additional organic resource to their farm, and had greater knowledge of farming.

Off farm is negatively correlated with efficiency score in organic, but positively associated with efficiency score in matched conventional. Organic farmers with off farm jobs may focus more on their off farm job rather than rice farming, which would result in poor management in farming practice. However, matched conventional was found to be more efficient when they have an off farm job. This suggests that, in the matched conventional sample, those farmers with an off farm job were able to manage their business activities more effectively. Own-tractor is negatively associated with efficiency score in the organic and matched pool sample. With a tractor, farmers may increase their production cost if it helped to increase productivity and intensity of adoption of organic farming. Credit use is negatively associated with efficiency score in the unmatched pool sample and the organic sample. It is often associated with poorer farmers who have easy access to credit funding, and as a result, would get lower performance in farming.

4. Conclusions and Implication

This study contributes some important findings regarding the relative profit efficiency of organic and conventional farming, by highlighting pool and separate profit frontiers between organic and conventional farmers, accounting for the self-selection problem, and identifying the relationship between efficiency score from each profit frontier and farmers' characteristics.

Our results show that organic farmers possess better characteristics versus conventional farmers. This necessitated we control for those differences to access an accurate estimate for the profit efficiency of both groups. After we conducted the matching approach, the difference between the two groups was minimized, indicating that our proposed matching technique was fairly successful.

The tests of the stochastic production frontier indicate that farmers cannot manage their rice farming effectively if they produce in a larger field owing to their limited skills and knowledge in rice production. In addition, with higher labor input and other input costs, farmers have lower income elasticity, suggesting farmers have little knowledge in the optimization of their farm inputs. Organic fertilizer helps to increase farmers' rice income for both groups, while chemical fertilizer was found to be less effective in doing so. Furthermore, being an organic farmer would result in higher rice income in the unmatched pool sample, but it had no effect after matching. However, these results were rejected by a LR test that was favorable to the separate frontier estimation.

In comparing efficiency scores of pool estimation, average efficiency score had increased from 0.877 for the unmatched sample to 0.928 for the matched sample. There was significant difference between organic and conventional farmers for the unmatched sample, but no significant difference in the matched sample and matched PSM (ATT), suggesting that organic farming had no effect on farmers' efficiency. In contrast, the organic group had a higher efficiency score compared to those in conventional, for both unmatched sample, matched sample and matched PSM, suggesting that organic farming helps to increase farmers' profit efficiency when we estimate the profit frontier separately. We believe that estimating a separate frontier allowed us to calculate the efficiency score for conventional farmers accurately, as the production practice was different across the two groups, this being confirmed by the LR test.

We found some factors significantly influence farmers' efficiency score. Education, own-tractor, and credit use are negatively correlated with efficiency score. While selling, other farming, and number of poultry are positively correlated with efficiency score. Off farm is negatively correlated with efficiency score in organic, but positively correlated with efficiency score in matched conventional.

From this study we would suggest all relevant organizations should introduce an effective technique that would help farmers to manage their rice farming on a larger scale, and allocate their labor input and cost more efficiently, by encouraging farmers to further apply organic fertilizer, raise more livestock, and engage with other cropping systems. Farmers should be supported to commercialize themselves to get benefit from rice production, by encouraging them to grow either market demand variety or organic rice, together with mixed farming systems, which is more sustainable to increase their profit efficiency.

References

Asian Development Bank. (2014). *Improving rice production and commercialization in Cambodia: findings from a farm investment climate assessment.* Philippines: Asian Development Bank.

Aung, N. M. (2011). Agricultural efficiency of rice farmers in Myanmar: a case study in selected areas. *IDE Discussion Paper No. 306,* Institute of Developing Economies.

Battese, G. E., & Coelli, T. J. (1988). Prediction of firm-level technical efficiencies with a generalized frontier production function and panel data. *Journal of Econometrics, 38,* 387–399. https://doi.org/10.1016/0304-4076(88)90053-X

Becker, S. O., & Ichino, A. (2002). Estimation of average treatment effects based on propensity scores. *The Stata Journal, 2*(4), 358-377.

Bravo-Ureta, B. E., Almeida, A., Solís, D., & Inestroza, A. (2011). The economic impact of MARENA's investments on sustainable agricultural systems in Honduras. *Journal Agricultural Economics, 62,* 429–448. https://doi.org/10.1111/j.1477-9552.2010.00277.x

Bravo-Ureta, B. E., Greene, W., & Solís, D. (2012). Technical efficiency analysis correcting for biases from observed and unobserved variables: an application to a natural resource management project. *Empirical Economics, 43,* 55–72. https://doi.org/10.1007/s00181-011-0491-y

Cambodian Organic Agriculture Association. (2011). *Organic agriculture and food processing in Cambodia: status and potentials.* Phnom Penh: Cambodian Organic Agriculture Association. Retrieved from http://www.coraa.org/page.php?id=8

Cary, J., & Wilkinson, R. (1997). Perceived profitability and farmers' conservation behavior. *Journal of Agricultural Economics, 48*(1), 13–21. https://doi.org/10.1111/j.1477-9552.1997.tb01127.x

Cerdán-Infantes, P., Maffioli, A., & Ubfal, D. (2008). *The impact of agricultural extension services: the case of grape production in Argentina.* Office of Evaluation and Oversight, Inter-American Development Bank.

Costantin, P. D., Martin, D. L., & Rivera, E. B. B. R. (2009). *Cobb-douglas, translog stochastic production function and data envelopment analysis in total factor productivity in Brazilian agribusiness.* Simpoi-Anais.

Greene, W. (2007). *Econometric analysis* (6th ed.). Prentice Hall, New Jersey.

Imbens, G. W., & Wooldridge, J. M. (2009). Recent developments in the econometrics of program evaluation. *Journal of Economic Literature, 47,* 5–86. https://doi.org/10.1257/jel.47.1.5

Islam, Z. K. M., Sipilainen, T., & Sumelius, J. (2011). Access to microfinance: does it matter for profit efficiency among small scale rice farmers in Bangladesh? *Middle-East Journal of Scientific Research, 9*(3), 311-323.

Khoy, R., Nanseki, T., & Chomei, Y. (2015). Impacts of organic rice farming on production performance in Cambodia: an application of propensity score matching. *Japanese Journal of Farm Management, 53*(2), 85-90.

Khoy, R., Nanseki, T., & Chomei, Y. (2016). Assessment of the premium on rice yield and rice income from adoption of organic rice farming for Cambodian farmers: an application of endogenous switching regression. *Journal of Agricultural Economics and Development, 5*(2), 33-44.

Kiatpathomchai, S. (2008). *Assessing economic and environmental efficiency of rice production systems in southern Thailand: an application of data envelopment analysis* (Doctoral dissertation, Giessen, Germany). Retrieved from http://geb.uni-giessen.de/geb/volltexte/2008/6373/

Mayen, C., Balagtas, J., & Alexander, C. (2010). Technology adoption and technical efficiency: organic and conventional dairy farms in the United States. *American Journal of Agricultural Economics, 92,* 181–195. https://doi.org/10.1093/ajae/aap018

Monteiro, N. (2010). Using propensity matching estimators to evaluate the impact of privatization on wages. *Applied Economics, 42,* 1293–1313. https://doi.org/10.1080/00036840701721281

Musshoff, O., & Hirschauer, N. (2008). Adoption of organic farming in Germany and Austria: An integrative dynamic investment perspective. *Agricultural Economics, 39,* 135–145. https://doi.org/10.1111/j.1574-0862.2008.00321.x

Nieberg, H., & Offermann, F. (2003). The profitability of organic farming in Europe. In: *Organization for Economic Co-operation and Development (OECD). Organic Agriculture: Sustainability, Markets and Policies.* Wallingford: CABI Publishing. p. 141–152.

Ponti, D. T., Rijk, B., & Ittersum, M. K. (2012). The crop yield gap between organic and conventional agriculture. *Agricultural Systems, 108,* 1-9. https://doi.org/10.1016/j.agsy.2011.12.004

Rosenbaum, P., & Rubin, D. (1983). The central role of the propensity score in observational studies for causal effects. *Biometrika, 70,* 41–55. https://doi.org/10.1093/biomet/70.1.41

Sa, K. (2011). Organic rice farming systems in Cambodia: socio-economic impact of smallholder systems in Takeo Province. *International Journal of Environmental and Rural Development, 2*(1), 115-119.

Sheeder, R., & Lynne, G. (2009). Empathy conditioned conservation: "walking in the shoes of others" as a conservation farmer. *Agricultural and Applied Economics Association's Annual Meeting,* Milwaukee, WI, July 26–28.

Taing, K. (2008). *Economic analysis of organic-culture rice in rural household economy: case studies in Tram Kork and Chumkiri District* (Unpublished master's thesis). Royal University of Agriculture, Phnom Penh, Cambodia.

Thath, R. (2014). Factors affecting cost efficiency of Cambodian rice farming households. *Forum of International Studies, 4*(2), 18-38.

7

Key Factors for Sustainable Industrial Cities

Ingy M. El Barmelgy[1] & Motaz S. Aly[2]

[1] Architecture Department, Faculty of Engineering, Cairo University, Egypt

[2] General Organizations for Physical Planning, Ministry of Housing, Cairo, Egypt

Correspondence: Ingy M. El Barmelgy, 12 hanora tower, El barmelgy street of Al bahr El Azam street, Giza, Egypt. E-mail: i_barmelgy@yahoo.com

Abstract

The industry is one of the main pillars of a strong economic city. Unfortunately, third world countries' industrial cities face environmental threats to the point that sustainable environments are considered a luxury (Pugh, C, 2013). According to a report issued by the Ministry of Agriculture and Land Reclamation in 2015 Egypt lost approximately 8618 acres of the finest farmland in the Delta and the Nile Valley as a result of urban sprawl on farmland to take advantage of employment opportunities and services in cities (Ministry of Agriculture land protection and land reclamation, 2015). The paper attempts to monitor different cases in the Egyptian context, trying to conclude the effective factors for their environmental and urban form as a result of industrial use. The Aim is to conclude key factors for sustainable industrial cities. The paper's results are based on a designed questionnaire that is analysed using SPSS. The questionnaire is completed with the help of experts and executives in order to specify the main factors in the sustainable urban form regarding industrial cities. It is followed by cluster analysis to determine the positive or negative effects of each element in relation to the rest of the elements concluding the most effective factor affecting the environment in every group as a tool to help the urban planning decision makers (environmental - urban - economic and social).

Keywords: environment, industrial cities, pollution, sustainable form, sustainability, urban planning

1. Introduction

This paper attempts to solve the dilemma between sustainable forms and industrial bases through specifying the main factors clearly as a tool for urban planners. There are several factors affecting the urban environment, especially within the industrial cities where the industrial use causes a lot of environmental threats. These threats include dumping industrial sewage into waterways, air pollution with toxic gases and hazardous waste for some industries, and negative impacts on surrounding residential areas and the urban form generally (Tan Yigitcanlar& M. Kamruzaman, 2015).

The main objective of this paper is to determine the key factors affecting the urban environment. For practice, to determine these factors, a theoretical approach is adopted to form a clear review of the factors through studying Industrial Development Authority standards, Millennium Development Goals of the United Nations (United Nations [UN], 2013), Studies and indicator theory. This is followed by an analysis of the EEAA standards and international cities such as the Joshua city in China (Institute of design and urban planning (Tongji), 2014), Skikda city in Algeria (Institute for Urban and management techniques, 2011), the city of Glasgow in the UK (Glasgow city plan & vision, 2014), and the policies of some countries such as France, England and China (World Bank, 2014).

The finalized factors are then revised and matched with the analysis of six Egyptian cities; Kafr El-Dawar (Ministry of Housing (MOH, 2013) - Kafr El Zayat (MOH, 2014) - Hawamdeya (MOH, 2014) - Mahalla Al Kubra (MOH, 2013) - Talkha (MOH, 2012) - Deshna (MOH, 2015).

After reviewing and studying different sources, the paper will identify these factors with respect to developing countries. To arrange these factors, a designed questionnaire for experts and executives is adopted, afterwards, the most and least influential factor in the environment as a result of the presence of the industry is identified by using the statistical analysis program (SPSS) as well as the most influential factor in each group (environmental, urban, economic and social).

2. Overview

All throughout Egypt, the cultural heritage factor is an important factor. Damietta has been famous for the wood industries (furniture - Arabesque - fishing boats) since the Ottoman conquest of Egypt, while Assiut is famous for spinning wool, carpets and Kilims since ancient times. This cultural heritage acts as a great comparative advantage in the authenticity either in small industries with similar or more sophisticated to urban communities with greater history and their neighbouring communities. This must be taken into account when proposing a small rural industry, as it should have historical roots in the city to keep pace with the current market needs (Allam, A, Khalid, 1995).

2.1 Overview of Egypt's Situation as Developing Country

Egyptian cities are suffering from the random spread of residential areas where these areas are held in the absence of both urban planning and the protection of agricultural land. Such areas may exist as part of the city or on the edges of the agricultural land; they lack proper planning and suffer from facilities and services deficiency. Mostly, they suffer from environmental pollution in addition to the low level of social, economic and environmental life (General Organization for Physical Planning (GOPP), 2014).

One of the main causes of the spread of these areas is the presence of attractive industries for employment, leading to endemic workers on the outskirts of the city.

This problem is common in cities that have an agricultural boundary and include concentrated labour-intensive industries in them such as the textile industry, which exists in cities such as: Kafr El Dawar, Kafr El Zayat, Hawamdeya, and Mahalla al-Kubra. These cities rely on a large number of employees where most of them are from the city and the neighbouring villages. It is some of those employees that decided to move next to their work, creating these expanding residential areas (Aly, M Sayed, 2014).

Also, this problem appears in coastal cities such as Port Said and Suez, where the industries exist and as a result of increased population growth rates, lack of an appropriate residential balance exists due to unplanned growth (Papa, Rocco, 2016).

The presence of industrial use in these cities has a significant impact on their population densities compared to those of the rest of the cities as illustrated (tab.1) (Source author using data from, Egypt Census Report (ECR) and Egyptian Urban Observatory (EUO), 2016). Shaded cities are the cities that have based industries where increased population density of 100 people / acre and up to 160 people / acre like Girga City as the cities that do not have a based industrial of the average population density of 70 people / acre.

Table 1. Illustrate high densities in industrial cities compared to other cities

GOVERNMENT	CITY	Density Per/acre
GHARBYA	Kafr El-Dawar	120
GHARBYA	Kafr El-Zayat	72
GIZA	Hawamdia	113
GHARBYA	Mahalla al-Kubra	90
PORT SAID	Port Said	21
SUEZ	Suez	20
DAQAHLYA	Talkha	119
QENA	Deshna	75
SOHAG	Girga	160
MINYA	Abo Qorqas	30
DAMYAT	Faraskor	112
SHARQYA	Belbis	92
GIZA	Badrashen	35
BEHERA	Edko	88
MINYA	Maghagha	143

MONOFYA	Berket Al-Sabeaa	69
DAQAHLYA	Bani Ebeid	82
DAQAHLYA	Nabaruh	99
SHARQYA	Faqus	71
SHARQYA	Kafr Saqr	111
QALYOBIA	Tokh	124
QALYOBIA	Kafr Shokr	87
QALYOBIA	Qaha	155
KAFR AL SHEKH	Desok	136
GIZA	El- Ayyat	89

2.2 Factors Affecting the Physical Growth of the Egyptian Cities

Urban extensions of cities are a real dilemma for land management as many factors affect the urban extension; these factors change depending natural factors, social factors, political, administrative and economic, etc.

Yet, the main factor which encourages the extension of the Egyptian cities is the economic factor. This is illustrated (tab. 2) (Source: author using data from, ECR, 2016 and EUO, 2015). The shaded rows show cities with an industrial base. The urban cluster of cities with an industrial base increased compared to those without an industrial base and this is a result of increased demand for residential and service land in these cities as a result of labour mobility for housing. For example, Kafr Al-Dawar city has an area of 2190 acres, Hawamdeya 1223.64 acres, and Mahalla Al-kobra 1291 acres (Source author using data from (GOPP, 2015), while the rest of the cities in the same category have an average area of about 1,000 acres. This confirms the effect of the presence of industries on the urban cluster of cities (ECR, 2016).

Table 2. Shows the population and areas in some industrial cities in Egypt

CITY	Area (Acre)	Inhabitants 2016
Kafr El-Dawar	2190.7	262,748
Kafr El-Zayat	1193	85,313
Hawamdeya	1223.64	137,722
Mahalla al-Kobra	1291	115,777
Port Said	32000	672,072
Suez	31621	629,616
Talkha	798	94,991
Deshna	877	65,459
Girga	706	112,920
Abo Qorqas	580	17,354
Faraskor	350	39,257
Belbis	1820	169,036
Badrasheen	2254	78,679
Edko	1366.7	119,952
Maghaghah	530	75,700
EL-Hamam	2582	72,393
Berket El- Sabaa	581	40,089
Belqas	1401	115,117

Dekernes	1214	81,570
Baniebed	441	36,065
Nabaroo	463	45,652
Faqos	1310	93,357
Kafr Saqr	334	37,109
Tokh	406	50,351
Kafr Shokr	326.9	28,300
Qaha	209.5	32,381
Desok	971	132,272
EL-Ayyat	467.22	41,528

It has an impact on attracting construction in the direction of economic activities due to the availability of work and the low prices of residential units.

This advantage for the working group seeking residence is considered as a component of attracting extensions in the desired directions by establishing factories, economic projects and other activities near the areas in order to drive them away from the random extensions of cities. When industries appear in one of the cities, margins grow along with the residential suburbs industry associated with it, even if it was outside the maximum chamber taking over agricultural lands (H, Al Omar et al., 2002).

The ratio of the industrial area to the residential cluster in the cities which have an industrial base is too high in cities like Kafr El-Dawar, Talkha or Deshna (Fig.1) (author). It's more than the residential area, which makes us wonder if it's a residential city with industries in its margins or an industrial city with labour camps. Generally, the industrial ratio is too high in all cities which give an indicator about the impact of the industry on the environment, infrastructure, urban form, etc. (GOPP, 2011).

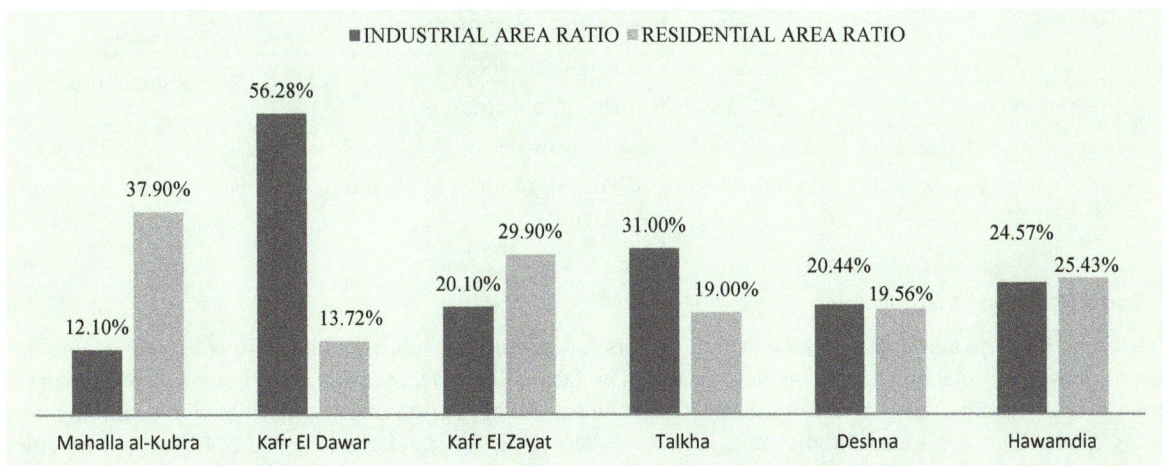

Figure1. Ratio industrial area to residential area in case studies

2.3 The Role of Governmental Organizations in Redrawing the Egyptian Industry Map (Administrative Bodies Responsibilities and Interrelations for Egyptian Industrial Map)

The placement of industrial development is considered the most economical factor guiding the city's urban form, which is the real issue to be discussed. In Egypt, the Industrial Development Authority is responsible for drawing the Egyptian industrial map where its role varies between the preparation of plans and mechanisms necessary to make industrial areas work and the establishment of industrial complexes as shown (Fig.1) (Source: author using data from , Environment Law). To support sustainable development plans complexes of the specialized agencies should target, optimizing the use of available resources, and increasing the added value of the industrial product should be the criteria in the proposed.

The General Authority for Urban Planning is the state body responsible for planning policy, urban development, sustainable development, and preparing plans and programs for these developments at the national and regional levels, as well as the preparation of the strategic plans for Egyptian cities. (GOPP, 2011)

EEAA refers to the Egyptian Environmental Affairs Agency, which is responsible for any environmental impact for any project. It has a law that states that the environmental impact of certain establishments or projects must be evaluated before any construction works are initiated or a license is issued by the competent administrative authority or licensing authority.

The numbers of projects subject to this provision are many and will form a heavy burden to administrative authorities and the EEAA. A flexible system for the management of EIA projects has therefore been developed in order to use limited economic and technical resources in the best possible way.

The system encompasses a flexible screening system and projects are classified into three groups, or classes, reflecting the different levels of environmental impact assessment according to the severity of possible environmental impacts (Egyptian environmental affairs agency (EEAA) – Environment Law).

The "A" list projects for establishments/projects with minor environmental impact.

The "B" list projects for establishments/projects which may result in substantial environmental impact.

The "C" list projects for establishments/projects which require complete EIA due to their potential.

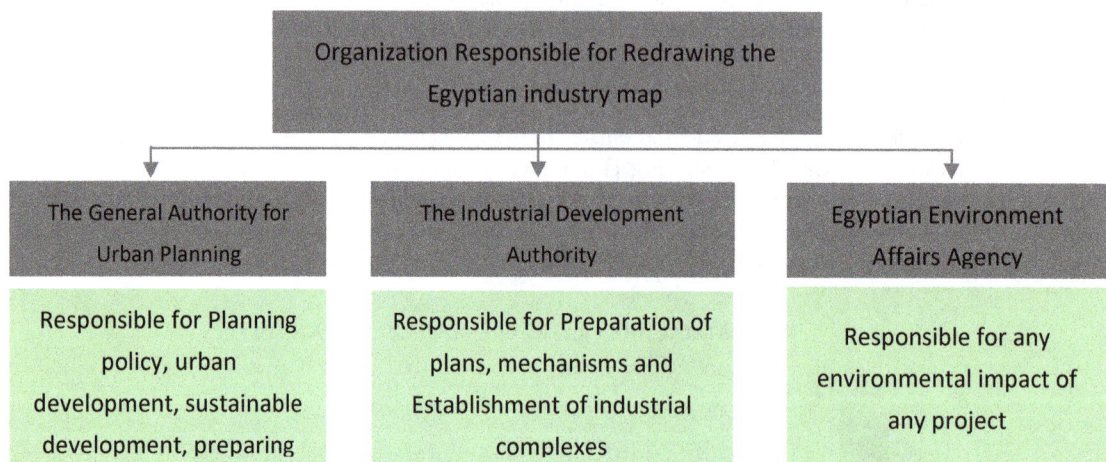

Figure 2. Diagrams shows the Administrative bodies responsibilities and interrelations for Egyptian industrial map

3. Empirical Study Methodology

The research's main aim is to conclude the key factors for sustainable industrial cities. To achieve this aim, this part of the study is met by concluding these factors. The factors that affect the urban environment were finalized from different sources such as the Industrial Development Authority standards, the Millennium Development Goals of the United Nations, Studies and indicators theory, EEAA standards (EEAA, 2014), Egyptian cities indicator, analysis of a number of international cities such as (Joshua city in China – Skikda city in Algeria - the city of Glasgow in UK), and the analysis of the policies of some countries, such as France, England and China, this part of the study consists of three steps. Step 1: after reviewing different sources the paper identifies these factors. Step 2: categorizing the factors (setting main factors' checklist). Finally, step 3; a designed questionnaire for experts and executives to arrange these factors, then the use of statistical analysis program (SPSS) to determine the most and the least influential factor. (Yigitcanlar, T. Dur& F. Dizdaroglu, D, 2015) Next, arranging them in relation to the environment as a result of the presence of the industry, followed by stating the most influential factor in each group (environmental - urban - economic and social), and the positive or negative effects of one factor on the rest.

The statistical analysis was carried out using the SPSS (Statistical Package for Social Sciences) program, which can be precisely illustrated within three main steps. Step 1: This part forms a general view of the elements that affect the urban environment in industrial cities to achieve this quick review for reports, manuals, international and national projects. In addition, to identify the most influential factor in the urban environment based on the number of standards and foundations that are derived from the opinions of specialized and theoretical studies and analysis of local and global experiences.

Step 2: Categorizing the factors (setting main factors' checklist). Factors are classified into three sources; environment, urban, economic and social factors. After identifying the main sources, we extract the factors affecting the physical environment as a result of the establishment of industrial activity. We then start to identify and prioritize these factors strictly in accordance with origin as shown (Tab.3) (Source author using data from, EEAA, 2013). In spite of having a lot of standards for establishing industrial activity, the negative impact of these activities on the physical environment is still present, in particular, in cities with an agricultural background.

Table 3. Final elements affecting the urban environment categorized by source

Studies and other indicators	Industrial Development Authority standards	EEAA standards	The Millennium Development Goals	Analysis of local and global cities	Theoretical framework	Source
Slums area of the total mass	Increase the distance between the industrial zone (inside the city) and the nearest road about 7 km regional	The high proportion of contaminated output of these industries within the city	Eradicating extreme poverty (relying on employment)	The lack of Residential units allows Urban growth without encroaching on farmland	The absence of a regional industrial zone in the distance, not far from the city of more than 40 km	
The presence of a regional industrial zone	The type of industrial activity (in terms of the nature of the activity)	The intensity of the electrical energy consumption	Achieving universal primary education (not to rely on child labor)	The proportion of buildings related to network infrastructure	Factory Site according to sewage water	
The average annual increase for the Urban Mass	The type of industrial activity (in terms of employment)	The type of industrial activity (labour-intensive - capital-intensive)	Promote gender equality and empower women (women allowed to run)	The high proportion of industrial use in the city of about 10% of the total area of the block	Increase the distance between the industrial zone (inside the city) and the areas of marketing for 25 km	Element
High rates of immigration to the city		Environmental effects of the plants (a type of pollution)	The presence of flat growth without urban encroachment on farmland	Adoption of the economic structure of employment in the city on industrial activity	Increase the distance between the industrial zone (inside the city) and manpower for 25 km	
The rate of overcrowding		Industry Classification, according to EEAA	Ensure that the reasons for the survival of the environment	The proportion of poor and rundown buildings	The presence of agricultural hinterland of the city	
			Reduction of child mortality and disease control		The presence of traffic problems	

Step 3: To determine the extent of the importance of these elements and the relative weights, we design a questionnaire with two experts in the field of urban planning and executives in the governments. Afterwards, we merge them to discover the final results and final elements affecting the urban environment and the environmental situation in the cities as a result of the factory's location. About 50 questionnaires from executives, excluded 6 forms of analysis (final number 44), and filled 30 questionnaire by specialists, excluded 1 form of analysis (final number 29), on the form to reach the final count of the forms of executives and experts. Overall, we collected 73 questionnaires as this is the number of forms needed to validate the statistical tests performed through this research. Subsequently, we analysed them using the SPSS program to get the final result which is:

–   Identifying the factors with the most negative effect on the urban environment.

–   Identifying the most influential element in each group (urban – economic and social – environmental)

–   Determining the positive or negative effect of an element on the rest of the elements.

4. Discussion of the Findings

4.1 Final Results of the Elements Affecting the Urban Environment of Cities According to the Opinion of Experts and Executives

The element with the most negative impact on the urban environment as a result of the factor's location is the presence of the industrial zones on the water source. That has achieved the highest percentage of 13% (Fig.3) (author), followed by an arrangement in the high proportion of contaminated output followed by an absence of urban flat that allows growth. The achievement of the Millennium Development Goals of the United Nations has a less negative influence on the urban environment as a result of the factor's location. This has achieved the lowest percentage of 1%, in order for the component to increase the distance between the industrial area and the areas of marketing for 25K followed by an increase in the distance between the industrial areas and manpower about 25 km.

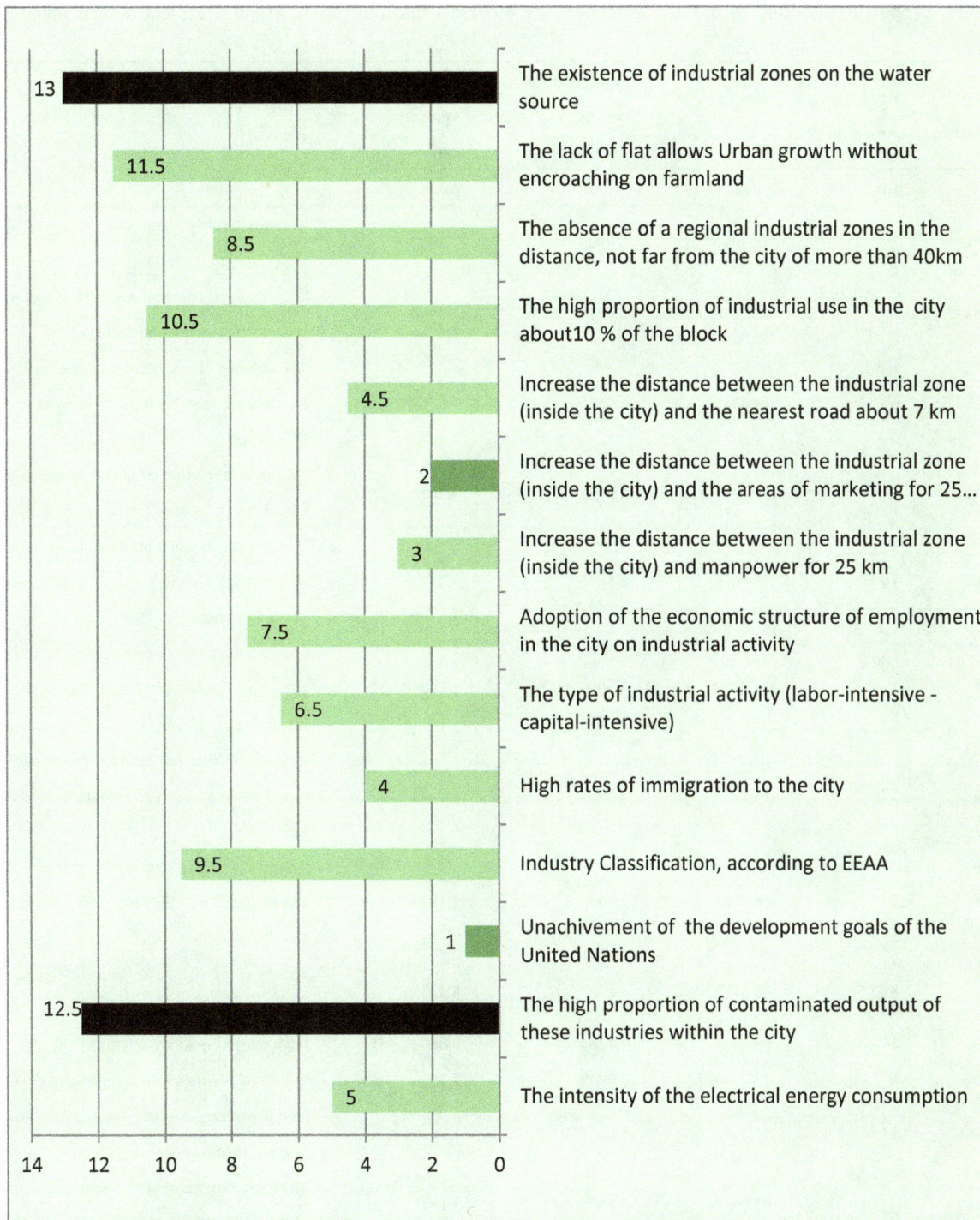

Figure 3. Final results of the elements affecting the urban environment of cities

4.2 The Final Ranking of the Importance of the Elements That Negatively Affect the Urban Environment in Cities According to the Opinion of the Executives and Experts

The existence of the industrial zones on the water source is the most negative element on the urban environment in the city according to the opinion of the executives and experts then the high proportion of the contaminated output of these industries within the city, as shown (Tab.4) (author).

Table 4. The final ranking of the importance of the elements that negatively affect the urban environment in cities

Weight Relative	Final Arrangement	Absolute Questionnaire	According to the opinion of and executives	According to the opinion of experts	Element
%	Arr.	mean	mean	mean	
13%	1	5.0909	2.0909	3.0000	The existence of industrial zones on the water source
11%	3	10.6818	4.6818	6.0000	The lack of flat allows Urban growth without encroaching on farmland
9%	6	14.0315	6.9545	7.0769	The absence of a regional industrial zone in the distance, not far from the city of more than 40 km
10%	4	11.0070	6.5455	4.4615	The high proportion of industrial use in the city about 10 %of the total area of the block
5%	10	17.8636	8.8636	9.0000	Increase the distance between the industrial zone (inside the city) and the nearest road about 7 km regional
2%	13	18.5000	8.5000	10.0000	Increase the distance between the industrial zone (inside the city) and the areas of marketing for 25 km
3%	12	18.1783	8.4091	9.7692	Increase the distance between the industrial zone (inside the city) and manpower for 25 km
8%	7	16.7483	9.3636	7.3846	Adoption of the economic structure of employment in the city on industrial activity
7%	8	16.8112	8.2727	8.5385	The type of industrial activity (labour - intensive – capital - intensive)
4%	11	18.1364	9.1364	9.0000	High rates of immigration to the city
10%	5	13.2448	6.0909	7.1538	Industry Classification, according to EEAA
1%	14	21.8881	11.2727	10.6154	Non-achievement of the development goals of the United Nations
12%	2	7.6993	4.5455	3.1538	The high proportion of contaminated output of these industries within the city
6%	9	17.4161	8.9545	8.4615	The intensity of the electrical energy consumption
100		207.297	103.6818	103.615	sum

4.3 Identifying the Most Influential Element in Each Group (Urban–Economic- Social and Environmental)

The items affecting the physical and environmental conditions are categorized according to the groups to which they belong (environmental - urban - economics –etc.).

The most influential element in the urban characteristics is the presence of the industrial zones on the water source, while the least influential element in the urban characteristics is increasing the distance between the industrial area and the marketing areas for the 25 km^2 as shown (Tab.5) (author).

The most influential element in the economic and social characteristics is an industry classification according to the EEAA and the least influential element is the element of high rates of immigration to the city.

The most influential element in the environmental characteristics is a high proportion of contaminated output and the least influential element is the failure to achieve the Millennium Development Goals of the United Nations.

Table 5. Categorized the items affecting the physical and environmental condition

Arr.	Element	Group
1	The existence of industrial zones on the water source	
2	The lack of flat allows Urban growth without encroaching on farmland	
4	The absence of a regional industrial zone in the distance, not far from the city of more than 40 km	
3	The high proportion of industrial use in the city about 10 % of the total area of the block	Urban
5	Increase the distance between the industrial zone (inside the city) and the nearest road about 7 km regional	
7	Increase the distance between the industrial zone (inside the city) and the areas of marketing for 25 km	
6	Increase the distance between the industrial zone (inside the city) and manpower for 25 km	
4	High rates of immigration to the city	
2	Adoption of the economic structure of employment in the city on industrial activity	Economic & Social
3	The type of industrial activity (labor-intensive - capital-intensive)	
1	Industry Classification, according to EEAA	
3	Non achievement of the development goals of the United Nations	
1	The high proportion of contaminated output of these industries within the city	Environmental
2	The intensity of the electrical energy consumption	

4.4 Determine the Positive and Negative Effects of an Element on the Rest of the Elements

4.4.1 First Study: The Effect of the Presence of an Industrial Zone on the Water Source

Given the effect of the presence of the industrial zones on the water source to extrusive effect where a high proportion of contaminated output increases and the absence of flat growth in most cities are achieved whenever there are factories on the water source.

On the contrary, the city doesn't achieve the development goals whenever there are factories on the water source.

The influnce of existence of industrial zones on the water source

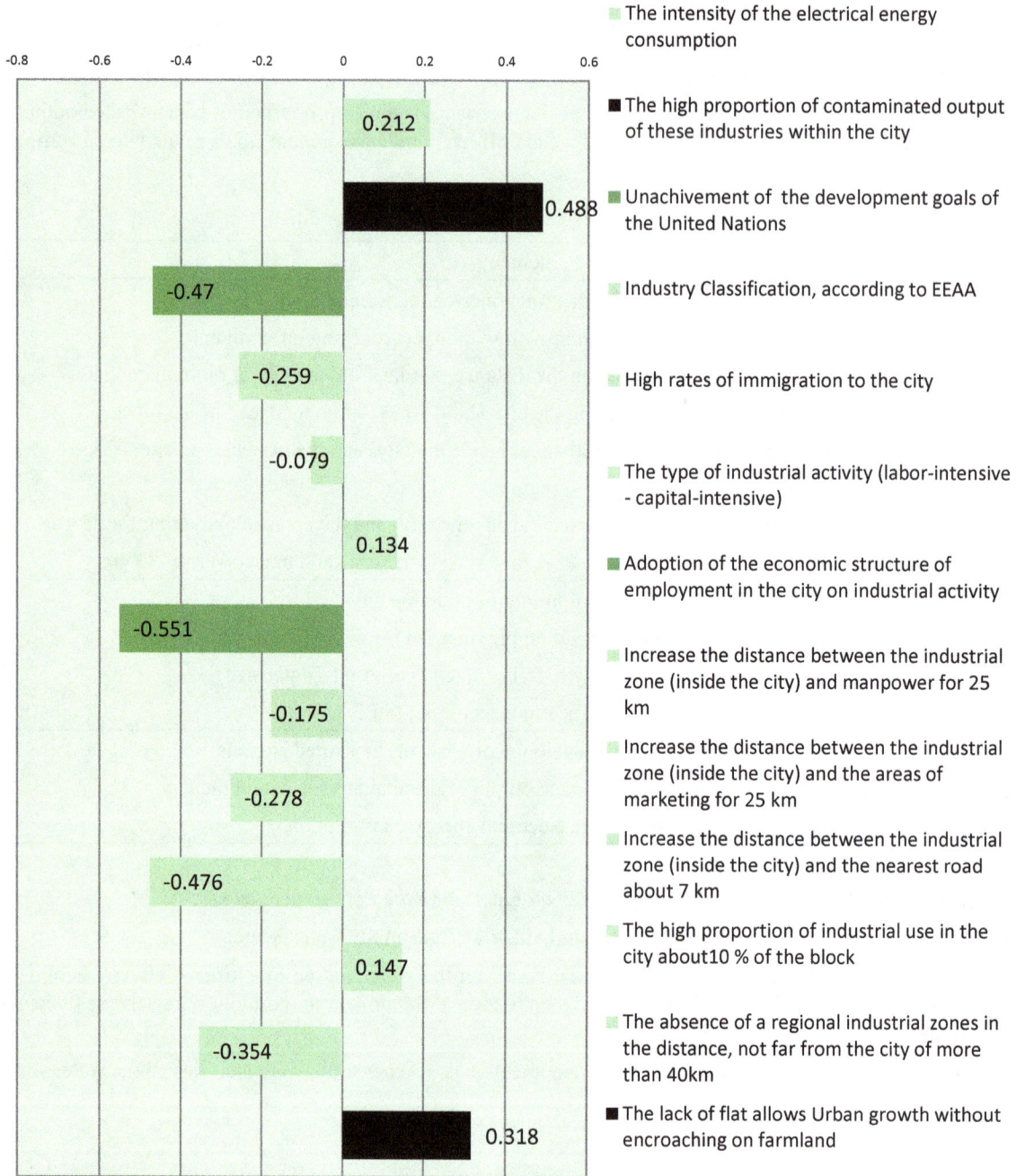

Figure 4. Ratios of the effective elements in the industrial zones

4.4.2 Second Study: The Effect of the High Proportion of Contaminated Outputs

Divided the effect of the high proportion of output to the impact of contaminated extrusive were high-output ratio industries always in the category of polluted black list classified according to the head of Environmental Affairs Industries also be energy-intensive. On the contrary, the city's economic structure doesn't depend on the industry or industrial activity; they are capital-intensive industries.

The influnce of existence of the high proportion of contaminated outputs

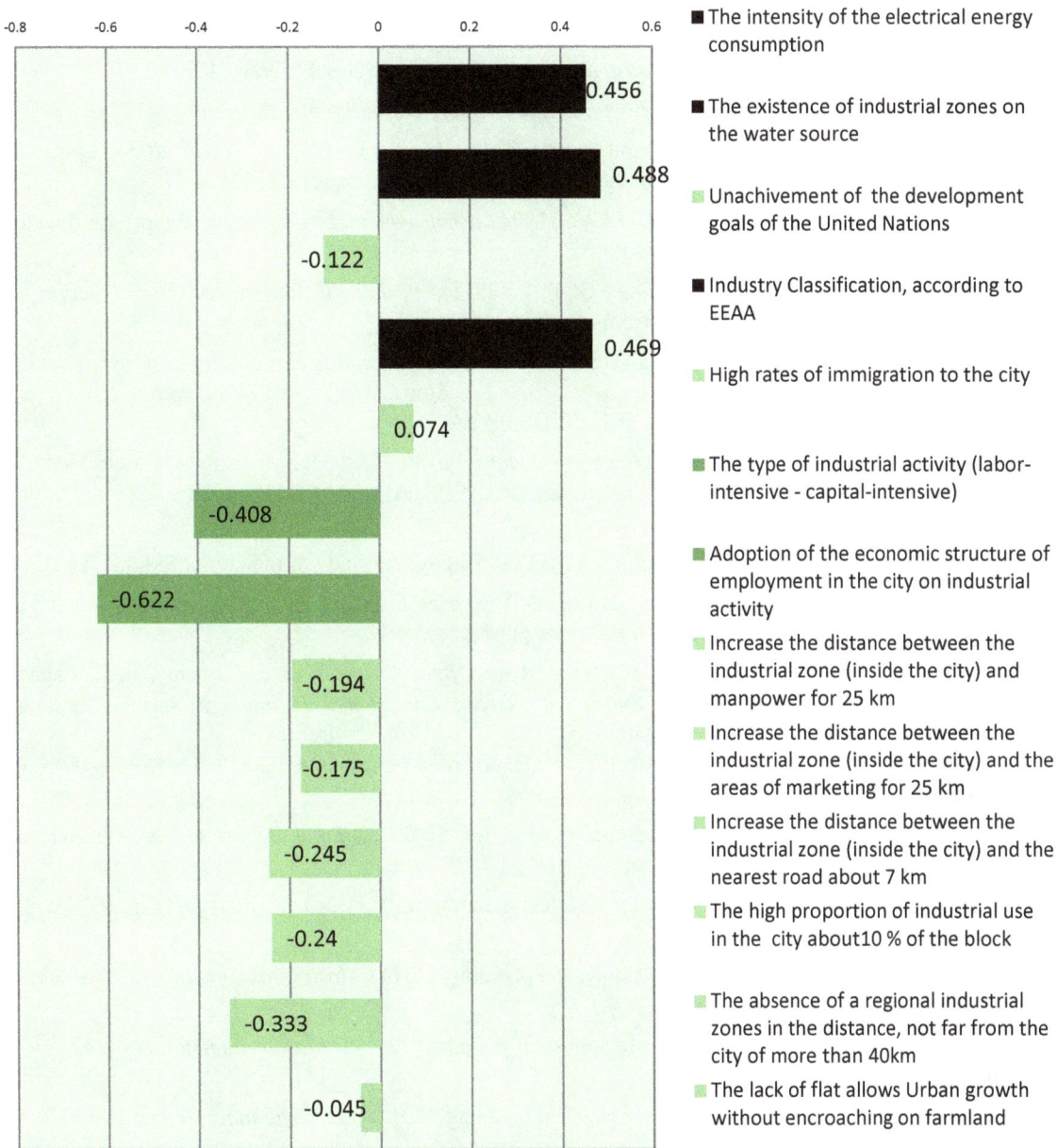

Figure 5. Ratio of the impact of rising output of polluting industries

5. Conclusion and Recommendations

From the analytical side, there is a need for entrances and various policies to address the problem of the industries' endemic within urban areas of cities where these industries cause deterioration of the physical environment. The characteristics of these cities are different and we could not apply a consistent policy of all situations, but we must classify cities into categories and make a special policy for each category (Ilaria & Flavio, 2016).

Therefore, we must stand on the most important factors influencing the environmental and physical situation in cities and the impact of each element on the rest of the elements. In addition, we must focus on the most important elements that contribute to the high rates of emigration to these cities and housing on the outskirts of these cities as the composition of random refocusing lacked most of the infrastructure and services, education, and health (Bisello, A, et al. ,2016).

There is a need for regional planning to contribute to the formulation of an industrial map for the Egyptian cities in general and the agricultural hinterland in particular. Additionally, there is an absence of a mechanism to provide urban extensions resulting from endemic industries (Aly, M Sayed, 2014).

References

Al Omar, H. (2002). *Introduction to The Industrial Economics.* Zat Al Salasel Library, Kuwait.

Allam, A. K. (1995). *Regional planning in Egypt.* Anglo Library, Cairo, Egypt.

Central Agency for Public Mobilization and Statistics. (2016). *Egypt Census Report 2016.* Cairo, Egypt. Retrieved from http://www.capmas.gov.eg/Pages/StaticPages.aspx?page_id=5035

Egyptian Environmental Affairs Agency (EEAA). (1994). *Environment Law* - Cairo, Egypt. Retrieved from http://eeaa.gov.eg/en-us/laws/envlaw.aspx

General Organization for Physical Planning (GOPP). (2011). *Periodical Reports of The Egyptian Urban Observatory.* Cairo, Egypt. Retrieved from http://gopp.gov.eg/

General Organization for Physical Planning (GOPP). (2014). *the national urban development framework in the Arab Republic of Egypt.* Retrieved from http://gopp.gov.eg/wp-content/uploads/2015/01/GOPP-PA_2014.pdf

Glasgow City Council. (2011). *Glasgow city vision 2061.* Glasgow. Retrieved from https://www.glasgowconsult.co.uk/UploadedFiles/GCC%202061%20A4%20Summary%20Final%20onli.pdf

Institute for Urban and Management Techniques. (2011). An analytical study of the city of Skikda, Algeria.

Institute of Design and Urban Planning (Tongji). (2014). *Master planning for the city of Joshua.* Shanghai, China. Retrieved from http://www.tjupdi.com/new/index.php?classid=9127

Maltese, I., Mariotti, I., & Boscacci, F. (2016). Smart City Urban Performance and Energy. In R. Papa, & R. Fistola (Eds.), *Smart Energy in the Smart City, Green Energy and Technology.* Springer International Publishing Switzerland. Retrieved from http://www.springer.com/cda/content/document/cda_downloaddocument/9783319311555-c2.pdf?SGWID=0-0-45-1564088-p179890956

Ministry of Agriculture land protection and land reclamation. (2015). *Annual Report on the Agriculture land status in Egypt.* Cairo, Egypt.

Ministry of Housing, General organization for physical planning. (2013). *Master plan for Kafr El Dawar city 2030.* Cairo, Egypt.

Ministry of Housing, General organization for physical planning. (2014). *Master plan for Kafr El Zayat city 2030.* Cairo, Egypt.

Ministry of Housing, General organization for physical planning. (2014). *Master plan for Hawamdeya 2030.* Cairo, Egypt.

Ministry of Housing, General organization for physical planning. (2014). *Master plan for Talkha city 2030.* Cairo, Egypt.

Ministry of Housing, General organization for physical planning. (2015). *Master plan for Deshna city 2030.* Cairo, Egypt.

Papa, R., & Fistola, R. (2016). *Smart Energy in the Smart City Urban Planning for a Sustainable Future.* Cham: Springer International Publishing. https://doi.org/10.1007/978-3-319-31157-9

Pugh, C. (2013). *Sustainable Cities in Developing Countries.* New York, Routledge.

Sayed, M. A. (2014). *The Impact of Industry on The Urban growth in The Cities With Agriculture Background* (Unpublished master thesis), Cairo University, Cairo, Egypt.

United Nations. (2015). *The Millennium Development Goals Report.* Retrieved from http://www.un.org/millenniumgoals/2015_MDG_Report/pdf/MDG%202015%20rev%20(July%201).pdf

World Bank, Annual Conference on Development Economics. (2014). Interaction between Regional and Industrial Policies, Washington USA.

Yigitcanlar, T., & Kamruzaman, M. (2015). *Planning Development and Management of Sustainable Cities,* Australia, Queensland University of Technology (QUT), School of Civil Engineering and Built Environment,

Science and Engineering Faculty, Queensland University of Technology (QUT), 2 George Street, Brisbane QLD 4001, Australia. https://doi.org/10.1016/j.habitatint.2014.06.033

Yigitcanlar, T., Dur, F., & Dizdaroglu, D. (2015). Towards prosperous sustainable cities: a multiscalar urban sustainability assessment approach. *Habitat International*, *45*(1), 36-46.

8

Effect Relationships on Sustainable Development of Palm Oil Production for Independent Smallholder Farmers toward Sustainable Certification System

Nurliza[1] & Eva Dolorosa[1]

[1] Agribusiness Department, Faculty of Agriculture, University of Tanjungpura, Pontianak, Indonesia

Correspondence: Nurliza, Agribusiness Department, Faculty of Agriculture, University of Tanjungpura, JL. Prof. Dr. H. Hadari Nawawi, Pontianak 78124, Indonesia. E-mail: nurliza.spmm@gmail.com

Abstract

Palm oil is currently the most widely used vegetable oil in the world and its usage is also expected to double by 2020. However, there are some social and environmental impacts of palm oil plantation. Some complications resulted from the plantation may go as far as mass objections to the production of palm oil. On the contrary, demand for palm oil is still vast and constantly rising. In Indonesia, independent small farmers are the most important stakeholders since they are 43% of the whole Indonesian palm oil producers and have become the biggest spotlight of Indonesian palm oil development, including challenges and problems in which they will have to face to substantially increase their role in the global market as well as maintaining sustainability. Challenges today need to be engaged with innovation and inventions in a more productive and effective way. Enhancing independent small farmers will not only enlarge their contribution to sustainability practices, but also ensuring the sustainable products supplied to the market. Thus, supporting sustainable palm oil production is the way forward. Based on this current issue, this research identifies key point relationships (direct and indirect) on sustainable development factors which are based on Indonesian Sustainable Palm Oil Certification System (ISPO), these identified key points will be the primary target to be improved and government support in fostering the sustainability of palm oil industry will be profoundly necessary.

Keywords: sustainability, legality, farmer's organization, environment, certification system, structural equation model, Maximum Likelihood Estimate (MLE)

1. Introduction

Palm oil is the most widely used vegetable oil in the world and is found in half of all packaged products on supermarket, from shampoo to detergents have been using palm oil as one of the ingredients. Palm oil accounts for 35% of the global vegetable oil market. Some empirical research on palm oil showed that there are several reasons to support the following argument. First, its efficiency, Palm requires 10 times less land than the other three major oil producing crops (soya, rapeseed and sunflower). Palm kernel expeller is used extensively in the energy and animal food sector. Palm kernel oil is a widely used ingredient in the personal care market (Sheil *et al.*, 2009; Popp *et.al.*, 2016). Second, its versality. Palm kernel oil can be processed to form a wide range of products with different melting points, consistencies and characteristics (Koushki *et.al.*, 2015). Third, feeding a global population. The importance of palm oil becomes clear when we consider that many people in the developing world rely on it as a cheap and available cooking medium (Sayer *et.al.*, 2012; N. *et.al.*, 2012). Fourth, providing livelihoods. Many people rely on palm oil for their livelihoods (Wilms-Posen *et.al.*, 2014). Palm oil small farmers are some of the poorest farmers in the world. The money they earn from growing oil palm trees is crucial to feed and care for their families. Fifth, supporting economies. Farming and producing palm oil form the backbone for many communities and, indeed, countries (Acheampong & Campion, 2013). Palm oil accounts for 11% of Indonesia's export earnings, with one third of this production attributed to small farmers.

Palm oil usage is also expected to double by 2020, yet the negative impacts of its unsustainable production can be devastating. The social and environmental impacts of palm oil plantation are: First, deforestation. Palm oil has been linked with the destruction of the world's precious rainforests. Development of new palm oil plantations, coupled with small farmers expanding their farms to meet the rising demand for palm oil, has resulted in significant deforestation. Second, threat to species for their survival. The removal of acres of rainforest threatens the rich biodiversity in these finely balanced ecosystems, along with the habitat of species. While palm oil is not the only cause of deforestation, it does play its part. Third, environmental damage. The removal of forest releases carbon into the atmosphere, speeding up global warming. In the tropics, tree roots anchor the soil. Deforestation removes this important structure, allowing heavy rains to wash away nutrient-rich soil. Crop yields begin to decline and farmers then have to use expensive fertilizers, which eat into their profits and further damage the environment. Dealing with adverse environmental and ecological impacts of palm oil cultivation, in particular, avoiding further deforestation must be a priority in a shift towards sustainable palm oil (Obidzinski *et.al.,* 2012; Petrenko *et.al.*, 2016). Fourth, social consequences (WWF & AFGC, 2013). While the global palm oil market creates an opportunity to bring many communities out of poverty, the race for land rights has left many locals on the losing team. Reports of displaced communities and illegal land grabs are not uncommon. The resulting conflicts, loss of income and dependence on large plantations have had a significant impact of the social welfare of many (Jensen *et.al.*, 2009; Obidzinski *et.al.*, 2012).

Some complications resulted from the plantation may go as far as mass objections to the production of palm oil. On the contrary, demand for palm oil is still vast and constantly rising. In Indonesia, independent small farmers are the most important stakeholders since they are 43% of the whole Indonesian palm oil producers and have become the biggest spotlight of Indonesian palm oil development, including challenges and problems in which they will have to face to substantially increase their role in the global market as well as maintaining sustainability. Independent smallholders' according to Vermeulen & Goad (2006) are growers who cultivate palm oil without direct assistance from government or private companies. They sell their crop to local mills either directly or through traders.

It is therefore very strategic to empower independent small farmers in order to effectively respond to the market requirements of Sustainable Palm Oil. Challenges today need to be confronted with innovation and inventions in more productive and effective ways. Enhancing independent smallholder farmer's will not only enlarge their contribution to sustainability practices, but ensure the sustainable products supplied to the market. Thus, supporting the production of sustainable palm oil is the way forward (Padfield *et.al.,* 2012).

In March 2011, the Indonesian government officially launched the Indonesian Sustainable Palm Oil (ISPO) standard in the Ministry of Agriculture's decree No. 19/Permentan/OT.140/3/2011. The standard is designed to make a sustainable palm oil production and complies with Indonesian laws and regulations. The standard has been implemented in 2011 as a mandatory proceeding but however is still on a trial basis, contrary to the Roundtable on Sustainable Palm Oil (RSPO) standard for all oil palm plantation companies operating in Indonesia by 2014 which is a voluntary proceeding. ISPO standard is still in preparation to comprise 7 principles, 39 criteria and 128 indicators covering licensing and plantation management, cultivation and processing, environmental monitoring and management, labor, social and economy empowerment, and business (Dirjenbun, 2011). Recently, the Indonesian Government also issued law No. 19/2013 concerning protection and empowerment to improve palm oil small farmers' livelihood in the future, since it has been quite long that they are not properly taken care of. The inauguration of this act allows relevant parties to be committed to the protection and empowerment to the farmers in Indonesia. It also supported by the recently issued law No. 11/Permentan/OT.140/3/2015 concerning Indonesian Sustainable Palm Oil Certification System (ISPO) for the demand in organizing sustainable palm oil development.

It is also crucial that the independent small farmers can trace back their palm oil usage to meet the standards set. Because of the complexity of the supply chains, therefore companies and farmers organizations need to work together to build a sustainable palm oil industry across the value chain. Sharing practices and developing innovative solutions is extremely important to move forward on this issue. Independent farmers shouldn't take a silo mentality, collaboration is the key. WWF *et.al.* (2012) stated that a responsible and transparent production according to a credible international standards does not generate a net cost for companies and our economic development. In fact, it is quite the opposite; it-compliant operations are simply more profitable. Hand-in-hand, these standards generate broader social benefits to our people and preserve the key natural resources that underpin the wealth of our nations. The key for strengthening the palm oil governance and optimizing development outcomes is to cooperate on making the ISPO as a certification system more valuable and internationally recognized as a part of Indonesia's green development strategy; also to strengthen and improve local government systems for the palm oil sector management (Paoli *et.al.*, 2011).

Thus, this research identifies key point relationships (direct and indirect) on sustainable development factors which are based on Indonesian Sustainable Palm Oil Certification System (ISPO), these identified key points will be the primary target to be improved and government support in fostering the sustainability of palm oil industry will be profoundly necessary. To achieve this objective we developed a framework applied to the independent small farmers. Although access to markets vulnerability is not improved through certification, indirect effects through organizational changes increase productivity. If certification schemes are weakly institutionalized, independent small farmers will easily shift to a more profitable way of production. Here, the relationship among them will be analyzed to stress the necessity effort for government support in strengthening the sustainability of palm oil industry. Government could be a catalyst or an obstacle and we look forward to make sure the social and economic developments will not cause irreversible deforestation. In future years businesses will be accounted for their contribution in a sustainable way.

2. Method

The research was conducted since March – May 2016 in the three districts (Landak, Kubu Raya, and Sambas) which was been the largest population of palm oil small farmers in West Kalimantan as figure 1. In-depth interviews were conducted from independent small farmers by purposive sampling. The number of participants who were selected from the sampling were 150 independent small farmers. This size is the minimum sample in multivariate analysis to estimate the indicators as the research property (Hair *et.al.*, 1992; Fraenkel & Wallen, 1993; Sugiyono, 2003).

In-Depth interviews are one-to-one encounters in which the interviewer uses an unstructured or semi-structured set of questions related to particular issues/topics to guide the discussion. The object of the exercise is to explore and uncover deep emotions, motivations and attitudes. There are some advantages of in-depth interviews. The interviewer can devote complete attention to each research participant, listen actively and establish good result; better sampling because recruiting is easier when scheduling in-depth interviews and researchers need fewer respondents to attain the same results; useful with difficult recruiting because its only need to accommodate one individual; it's also elicit candid responses in a private setting regarding personal and/or professional topics of discussion; fewer distractions; faster and cheaper; more productive; deeper Insights; more flexible, and faster adaptation (Turner, 2010; Mack *et.al.,* 2011; Alshenqeeti, 2014).

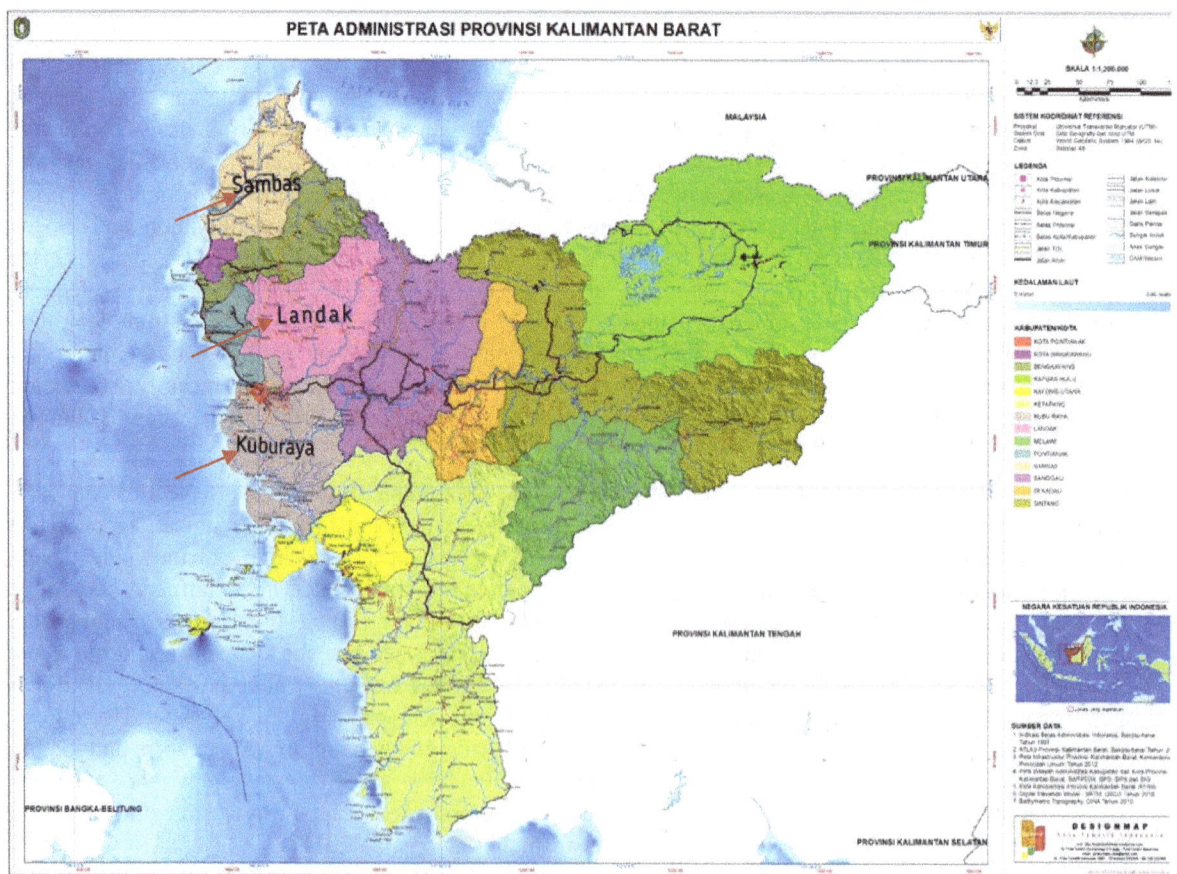

Figure 1. ⟶ Research location in West Kalimantan

The findings of this study are structured base on Indonesian Sustainable Palm Oil (ISPO) which is considered as government policy to improve the competiveness in global market and to reduce greenhouse gas emissions as well as paying attention to environmental issues by Ministry of Agriculture. The sustainability aspects were formulated base on environment, legality, and farmers organizations, to answer these following questions (i) what are the relationship aspects which affected on sustainable development of palm oil production? (ii) What are the direct and indirect effects on the sustainable development of palm oil production? In addition, a review of an extant literature from previous studies contributed to the main findings of this paper. This enables the chain of evidence to be established and to maintain clear linkages between the aspect of ISPO by allowing conclusion to be formulated through the research questions, relevant literature, data collection tools used and the evidence.

Data was analyzed using simultaneous equations (structural equation model) with Lisrel software in the following steps. First, model specs were constructed by defining the latent variables and observed indicators variables of latent variables, to connect the latent variables with each indicator, and to connect each latent variable. Second, identifying models those are qualified for further estimation with over-identified category. Third, estimate the value of the parameters in the model with ML (Maximum Likelihood) because it is commonly used and have some advantages that although estimator is bias against small sample but the asymtotic is not biased; consistent; the asymtotic is efficient so that consistent estimator has smaller asymtotic variant than the others. Afterwards, to evaluate the estimate results to see the possibility of offending estimates in acceptable limits. If the value of the error variance is negative, it can be improved by setting the error value variants with very small positive value (0.005 or 0.01), if the estimated coefficient exceeds the standard or very close to 1, it can be improved by eliminating one of the variables, and to ensure that the value of the standard error coefficient should not be of greater value. Fourth, validity and reliability test of the model with the criteria that the value of t-loading factor must be greater than the critical value (1.96). The data becomes invalid if the value of standard loading factor is more than 0.50 or equal to 0.5 and its better if the value could reach 0.70. Reliability test is shown by CR test. The data become unreliable if CR value is more than 0.7 or equal to 0.7 or VE value is more than 0.50 or equal to 0.5. Fifth, test on goodness of fit. Sixth, model respecification by using the modification indices information was done to get good model. Seventh, goodness of fit test was used again after model respecification. Eight, the last steps is to interpret the results (Wijanto, 2008; Riadi, 2013).

3. Results

The results were analyzed using simultaneous equation with Maximum Likelihood (ML) based on the research question (i) what is the relationship aspects which affected on sustainable development of palm oil production? (ii) What are the direct and indirect effects on the sustainable development of palm oil production? However, it was necessary to test the initial validity and reliability on the data before answering those questions. It has shown that the some t-loading factor value was not valid even though all of the data were reliable. Thus, the next validity and reliability test on the data should be conducted again to qualify all the data until they become valid and reliable. Furthemore, Furthemore, goodness of fit test were done after model respecification and proved in Table 1.

Tabel 1. Validity and reliability test

No.	Criteria	Standard Value	Initial		Final	
			Estimate Value	Conclusion	Estimate Value	Conclusion
1	Chi Square / χ^2	Small value	2.806,7900	Poorly	1.401,8418	Poorly
2	χ^2/DF	$1,0 \geq x \leq 5,0$	9,51	Poorly	5,31	Poorly
3	NCP	Small value with narrow interval	2511,79 (2345,7719; 2685,1809)	Poorly	1137,8418 (1024,3173; 1258,8533)	Poorly
4	SNCP (NCP/n)	Small value	16,75	Poorly	7,58	Good
5	RMSEA	$\leq 0,08$	0,239	Poorly	0,1701	Poorly
6	ECVI	Small value and close to saturated ECVI	I=96,8945 M=19,5892 S=4,7114	Poorly	I=96,8945 M=10,5761 S=4,7114	Good
7	AIC	Small value and close to saturated AIC	I=14437,2735 M=2918,7900 S=702,0000	Poorly	I=14437,2735 M=1575,8418 S=702,0000	Good
8	CAIC	Small value and	I=14541,5500	Good	I=14541,5500	Good

		close to saturated CAIC	M=3143,3855 S=2109,7330		M=1924,7671 S=2109,7330	
9	NFI	≥ 0,90	0,7903	Poorly	0,8673	Good
10	NNFI	≥ 0,90	0,7868	Poorly	0,8559	Good
11	PNFI	Higher value	0,7174	Poorly	0,7045	Poorly
12	CFI	≥ 0,90	0,8065	Poorly	0,883	Good
13	IFI	≥ 0,90	0,8069	Poorly	0,8835	Good
14	RFI	≥ 0,90	0,769	Poorly	0,8366	Good
15	GFI	≥ 0,90	0,4083	Poorly	0,5801	Poorly
16	AGFI	≥ 0,90	0,296	Poorly	0,4418	Poorly
17	PGFI	0-1	0,3432	Poorly	0,4363	Poorly
18	RMR	≤ 0,05	0,1078	Poorly	0,1038	Poorly
19	CN	≥ 200	18,5079	Poorly	26,0011	Poorly

Source: author's analysis (2016)

In Table 1, the respective model was using modification indicates information to improve the model and there were nine criterias that became good. The result of the estimated coefficient model for sustainable development of independent small farmer on palm oil production was presented in Figure 2.

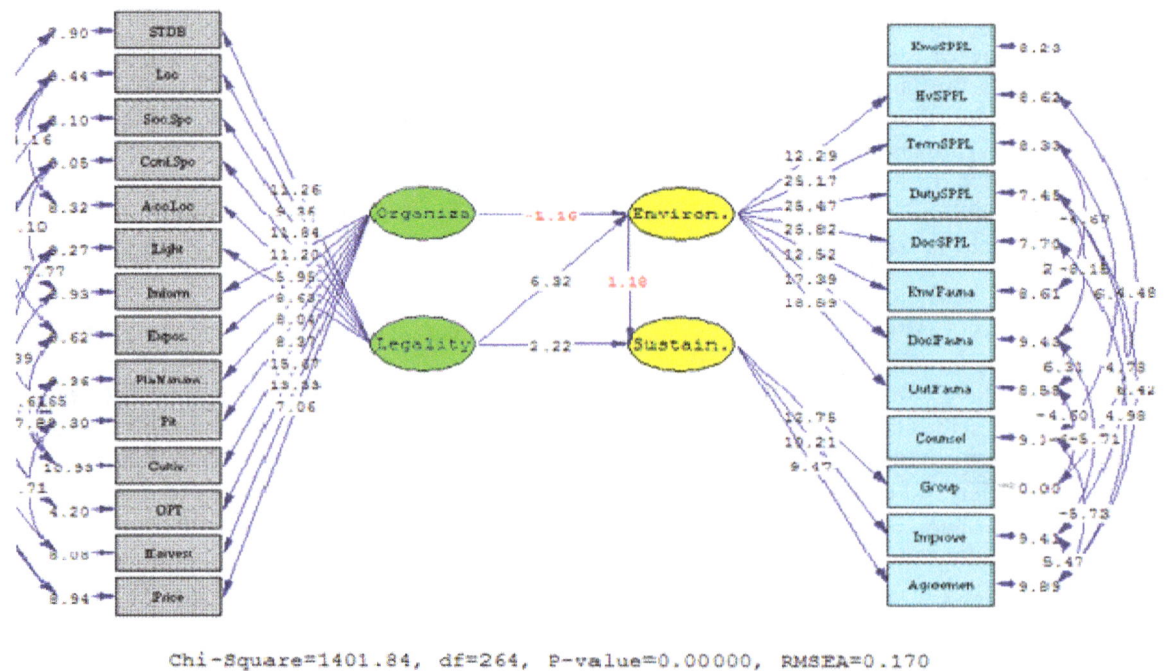

Chi-Square=1401.84, df=264, P-value=0.00000, RMSEA=0.170

Figure 2. Model of sustainable development

In Figure 2, we saw the path of variable latent which affected sustainable development. Illustration summary of the estimate coefficient, the corelation among the laten variables, and the effect that influenced significant or not significant to sustainable development was presented in Table 2.

Table 2. The path of variable latent which effect to sustainable development

Path	Estimate Coeff.	t Value	Conclusion	Correlation	Total Effect	Indirect Effect
Organization → Environment	-0.1906	-1.1578	Not significant	0.8828	-0.1906	-
Legality → Environment	**1.1300**	**6.3224**	**Positive Significant**	**0.9490**	**1.1300**	-
Environment → Sustainability	0.2448	1.1805	Not significant	0.6841	0.2448	-

Organization → Sustainability	-0.0467	-0.8427	Not significant	0.6558	-0.0467	-0.0467
Legality → Sustainability	**0.4629**	**2.2211**	**Positive significant**	**0.6952**	**0.7395**	**0.2766**
Organization → Legality	-	-	-	0.9499	-	-

Source: author's analysis (2016)

In Table 2, there were some paths to describe the effect of latent variable on sustainable development. Yet, there was only one path of variable latent which affected sustainable development. That aspect was legality which proved had positive significant effect with total effect value as 0.7395% and indirect effect value as 0.2766%. The legality had positive effect on the environment as 1.13%. The legality and environment had very strong relationship, which means legality and sustainability also had strong relationship. Thus, if legality was increased 1%, it would increase the environment 1.13% and if the legality was to be increased by 1%, it would also increase the sustainability value to 0.46%. Therefore, the direct effect of legality (62.59%) was stronger than the indirect effect (37.41%). It has shown that legality became the most important aspect to influence the sustainability. However, the environment and the farmer's organization had not significant effect on sustainable development. Farmer's organization also had not proven significant on environment aspect and there wasn't correlation between farmer's organization and legality. Next, we had some results of indicator that has significant effect on latent variable and futher effects on sustainable development as previously described.

Tabel 3. Indicator of latent variable effect on sustainable development

Indicators of Latent Variables	Estimate Coefficient/ Total Effect	Indirect Effect Indicator on Environment	Total/Direct Effect Indicator on Legality	Determinant Coefficient
Sustainability:				
Group	0.7448	0.1823	0.5508	1.0000
Improve	0.6305	0.1543	0.4663	0.6182
Agreement	0.6277	0.1536	0.4642	0.7161
Legality:	0.8281			0.6096
STDB				
Loc	0.7319			0.4522
Soc.Spc	0.9229			0.6467
Conf.Spc	0.8542			0.4115
Acc.Loc	0.6263			0.4964
Light	0.6869			0.3475

Source: author's analysis (2016)

In Table 3, there were three indicators of sustainability that had indirect and direct effect respectively on environment and legality aspects. The biggest effect value was farmer's group then followed by improvement on farming and price agreement of fresh fruit bunches. However, there wasn't a single indicator of legality that had indirect and direct effect on environment.

4. Discussion

The results were to answer the research questions (i) what is the relationship aspects which affected on sustainable development of palm oil production? (ii) What are the direct and indirect effects on the sustainable development of palm oil production? The results confirm that there was only one path of variable latent which affected on sustainable development. That aspect was legality which proved positive significant effect with total effect more than indirect effect on sustainable development. This finding was in line with Colbran & Eide (2008) which established legally binding certification schemes and a reliable monitoring system to ensure that the international certification is effective and enforced and maximizing the possible benefits. Caroko et.al. (2011) even stated that the government should also enable fair community engagement, whereas the government has an active role in supervising and enforcing contract agreements, benefit sharing and dispute settlements. However, Teoh (2010) recognize that small farmers will face financial constraints in preparation for certification so they must be supported. Thus, the proposed fund be required to provide the opportunity for various players in the supply chain to support the overall goal to add value to the oil palm supply chain. This is supported by Diop, et.al. (2013) that local people often lacking knowledge of the formal legal system or how to seek support in the event of contested rights. For some small farmers, the certificate has not been fully understood as a tool to create more sustainable agriculture. Rather, certification is seen as an economic tool in the pursuit of a better livelihood particularly for independent small farmers. Small farmers participate because they have to (scheme small farmers), or because certification is introduced by trustful people who open opportunities for higher incomes (Hidayat *et. al.*, 2015).

Therefore, legal certification alone cannot expect to develop a sustainable development automatically, unless these policy measures are backed by government and the private sector to strengthen the policy frameworks with sustainable development and food security policies in synergy. This translates into the need for supporting policy development in countries with a weak policy framework, building upon the (positive and negative) experiences while enforcing existing policies especially in relation to land tenure, economic and social policies as well as management of natural resources. Those must be put into high consideration, where the local population cannot provide sufficient labour and there is an influx of migrant workers, social and land conflict become inevitable. In cases where labour import is necessary, tenure safeguards must be implemented for local communities. That's why the palm oil plantation development for sustainable development is also reconsidered and planned carefully. With certification, independenet small farmers need to change production processes within their existing resources and power asymmetries. Their relative vulnerable position may influence the farmers' ability to cope with uncertainties related to participation in a certification scheme. This necessarily implies the provision of training, sharing good (environmental and socioeconomic) agricultural practices and facilitating the transfer of adequate technologies and methodologies. Consequently, bureaucratic regulation and government approach, including the application of market instruments, need to be mutually supportive. By mixing policy tools and providing for continual improvement as the state seeks to respond to outside criticism, incremental improvements that involve integrating these approaches may move the state to enhance its capacity for improving policy and implementation over time.

This research also proved that legality has positive effect and very strong relationship with the environment, so legality and sustainability also have positive effect and strong relationship. The direct effect of legality was stronger than indirect effect. It has shown that legality became the most important aspect to influence the sustainability. This finding was also supported by Wisena *et. al.* (2014) that the strategy is carried out on the development of a sustainable oil palm industry should increase the attention to the environmental and law enforcement by the government. The guidelines in legality should complement, not contradict, each other and should not impose unnecessary burden on those who produce biofuel in a socially and environmentally satisfactory way. Those could be the key lessons standing out from the environmental harm which are having a devastating impact on vulnerable people (Colbran & Eide, 2008). McCarthy & Zahari (2010) even concluded that policymakers need to avoid the temptation to raise prices on either market or state regulatory approaches to environmental problems. The solutions need to be crafted in combining different policy instruments to face the specific problems associated with certain sectors of the industry. Legal instruments need to be strengthened to ensure that companies who operate in the shadow of the law are required to improve pollution mitigation systems.

However, the environment and the farmer's organization respectively didn't have significant effect on sustainable development. This finding was different from prior research (Colbran & Eide, 2008; McCarthy & Zahari, 2010; Padfield *et. al.*, 2012) ; Ador *et. al.*, 2014; Montefrio, 2015). Yet, Hidayat *et. al.* (2015) found that non-economic benefits such as social and environmental improvements are less valued by the small farmers unless they lead to economic benefits. This finding also raised the question: is developing economy and protecting environment a pair of contradiction? The answer according to Guo & Ma (2008) depends on the economy development stages. In the developed country the developing economy and protecting environment is not a pair of contradiction because of environment improved along with the economic structure changed. But in the developing country it is a pair of contradiction because of environment worsens with a high speed economy increase (Ador *et. al.,* 2014). So, it is argued that environmental change (or the perception of change) is not an important factor for independent small farmers as the case earlier empirical researches which concern about palm oil contract farming in the exclusion process. Hall (2011) stated that there are four powers those are not exhaustive and that there could be other powers operating in certain contexts, such as environmental change, knowledge and technologies, and political relationships and alliances. Thus, in this analysis, environmental change can only be considered as another form of power or just an intermediary condition that connects the powers from the actual exclusion if it is together with the three of the other factors.

The development of economy and the protection of environment, sometimes, are contradictory in particular in the short-term. But this kind of contradiction has a condition, in the majority situation, this kind of contradictory performance is, protected the environment request to reduce the development path choice space. In the reality, this kind of space nearly always exists, sometimes possibly needs us to develop. However, in the long term, the expense and effect of protecting environment is an important factor affecting environment (Guo & Ma, 2008). Thus, the choice of the industrial structure is a factor affecting the environmental pollution. Technology also is an important indicator to affect the environment. In generally, the country which use low technology may consume more resources and more pollution.

Still, farmer's organization also has no significant effect on sustainable development, eventhough farmer's organization hopefully changes to induce a higher production quality that may benefit small farmers indirectly and financially. Existing research has identified a variety of external drivers, or contextual factors, that affect an organization's tendency to become active with regard to sustainability, including the ecological, organizational field context and individual contexts. This was confirmed by the results that farmer's group had the biggest indirect and direct effect respectively on environment and legality aspects then followed by improvement on farming, price agreement of fresh fruit bunches. The improvement on farming had indirect and direct effect on sustainable development of course has strengthened a consensus is emerging that addressing the new challenges requires a sustainable agricultural intensification in small and large farms throughout the world (SDSN, 2013). Thus, there was an urgent matter to improve performance which mean any or all of increased profitability and productivity, high efficiency and returns from external inputs, improved yield stability, reduced greenhouse gas emissions, enhanced ecological resilience, and environmental service provision.

The improvement on price agreement of fresh fruit bunches has indirect and direct effect on sustainable development. Paoli et al. (2011) in Summary for Policy Makers & Practitioners stated that processing fresh fruit bunches (FFB) from oil palm creates numerous solid and liquid by-products, including empty fruit bunches (EFB), residual fruit material, palm kernel shells and liquid palm oil mill effluent (POME). In the past, these by-products were seen as waste materials requiring disposal, and carried significant pollution risk.

Recently, advanced technologies to contain EFB and POME co-composting (discussed below) generate nutrient-rich organic fertilizers, capture biogas by-products, and eliminate GHG emissions arising from conventional waste treatment.

The factors which specifically promote the formation of partner membership for sustainability were social perceptions, expectations and preferences; technological developments; concerns about globalization; regulatory environment; and decline in government efficacy (Bansal & Roth, 2000). However, in this reseach context, those factors can not be met by farmer's organization. Thus, the role farmer's organization had not significant effect on sustainable development, parrticularly for independent small farmers. Understanding differences in partners' motivations is important because these differences can produce a mismatch within the partner membership and lead to difficulties in working together if motivations are not aligned or complementary. Additionally, when partners have different motivations, different types of partnerships may be needed (Gray & Stites, 2013).

Furthermore, farmer's organization also didn't prove significant on environment aspect and there wasn't correlation between farmer's organization and legality. These findings reinforced the fact that relationship between the absence of farmer's organization role on sustainable development. This could be generate to the absence of effect on environment aspect. However, legality should be legitimate because it refers to social acceptance of an organization based on its conformity to societal norms and expectations. Nikoloyuk et.al., 2010 stated that legitimacy is important for organizations because without it, organizations will have difficulty acquiring critical resources needed for a long-term sustainable success. In this research, independent small farmers was not actively motivated by farmer's organization and caused it wasn't legitimacy. This have led to the absence of the link between farmer's organization and legality. Still, farmer's organization which actively motivated by legitimacy will encourage the sustainability of their farm business to build a reputation, image and brand for social and environmental responsibility; attract and retain employees; and build the social license to operate (Gray & Stites, 2013).

Acknowledgments

We would like to express the sincere gratitude to Directorate of Research and Community Service, Directorate General for Strengthening Research and Development of the Ministry of Research, Technology (RISTEK) and Higher Education (DIKTI) for funding this research through the National Competitive Grant Research Priorities in the National Master Plan for the Acceleration and Expansion of Indonesian Economic Development 2011-2025 (PENPIRINAS MP3EI 2011-2025). We are grateful to Agriculture Faculty of Tanjungpura University for all support that enabled us to do this research. We would also like to thank students and villagers who assisted in data collection in the field. Special thanks for Agribusiness Department of Agriculture Faculty for all support that enabled us to do this research.

References

Acheampong, E., & Campion, B. B. (2013). Socio-economic impact of biofuel feedstock production on local livelihoods in Ghana. *Ghana Journal of Geography, 5*, 1-16.

Ador, S. F., Siwar, C., & Ghazali, R. (2014). A Review of palm oil impact on sustainability dimension: SPOC initiative for independent smallholders. *International Journal of Agriculture, Forestry and Plantation, 2*, 104-110.

Alshenqeeti, H. (2014). Interviewing as a data collection method: A critical review. *English Linguistics Research, 3*(1), 39-45. https://doi.org/10.5430/elr.v3n1p39

Bansal, P., & Roth, K. (2000). Why companies go green: A model of ecological responsiveness. *The Academy of Management Journal, 43*(4), 717-736. https://doi.org/10.2307/1556363

Caroko, W., Komarudin, H., Obidzinski, K., & Gunarso, P. (2011). *Policy and institutional frameworks for the development of palm oil–based biodiesel in Indonesia.* Working Paper 62. Bogor, Indonesia: CIFOR.

Colbran, N., & Eide, A. (2008). Biofuel, the environment, and food security: A global problem explored through a case study of Indonesia. *Sustainable Development Law & Policy, 9*(1), 4-11.

Diop, D., Blanco, M., Flammini, A., Schlaifer, M., Kropiwnicka, M. A., & Markhof, M. M. (2013). *Assessing the impact of biofuels production on developing countries from the point of view of Policy Coherence for Development.* European Union: AETS Consortium.

Dirjenbun. (2011). Policy of government of Indonesia on Sustainable Palm Oil (ISPO). *The International Conference and Exhibition of Palm Oil.* Jakarta 11-13 May: Directorate General of Estate Crops, The Ministry of Agriculture.

Fraenkel, J., & Wallen, N. (1993). *How to Design and Evaluate Research in Education* (2nd ed.). New York: McGraw.

Gray, B., & Stites, J. P. (2013). *Sustainability through Partnerships: Capitalizing on Collaboration.* Pennslyvania, USA: Network for Business Sustainability.

Guo, L., & Ma, H. (2008). Conflict between developing economic and protecting environment. *Journal of Sustainable Development, 1*(3), 91-97.

Hair, J. F., Anderson, R. E., Tatham, R. L., & Black, W. C. (1992). *Multivarate Data Analysis with Readings* (3rd ed.). New York: Macmillan.

Hall, D. (2011). Land grabs, land control, and Southeast Asian crop booms. *The Journal of Peasant Studies, 38*(4), 837-857. https://doi.org/10.1080/03066150.2011.607706

Hidayat, N. K., Glasbergen, P., & Offermans, A. (2015). Sustainability certification and palm oil smallholders' livelihood: A comparison between scheme small farmersand independent small farmersin Indonesia. *International Food and Agribusiness Management Review, 18*(3), 25-48.

Jensen, T. M., Degn, L. T., & Bertule, M. (2009). *Sustainability of Smallholder Palm Oil Production in Indonesia.* Roskilde University, Department of Society and Globalization, International Development Studies.

Koushki, M., Nahidi, M., & Cheraghali, F. (2015). Physico-chemical properties, fatty acid profile and nutrition in palm oil. *Journal of Paramedical Sciences (JPS), 6*(3), 117-134.

Mack, N., Woodsong, C., MacQueen, K. M., Quest, G., & Namey, E. (2011). *Qualitative Research Methods: A Data Collector's Field Guide.* North Carolina, USA: Family Health International (FHI).

McCarthy, J., & Zahari, Z. (2010). Regulating the oil palm boom: Assessing the effectiveness of environmental governance approaches to agro-industrial pollution in Indonesia. *Law & Policy, 32*(1), 153-178.

Montefrio, M. J. (2015). Green Economy, Oil Palm Development and the Exclusion of Indigenous Swidden Cultivators in the Philippines. Land grabbing, conflict and agrarian-environmental transformations: perspectives from East and Southeast Asia. An international academic conference. *Conference Paper No. 22.* 5-6 June (pp. 1-19). Chiang Mai, China: BRICS Initiatives for Critical Agrarian Studies (BICAS), MOSAIC Research Project, Land Deal Politics Initiative (LDPI), RCSD Chiang Mai University, Transnational Institute.

N., A. E., A., O. E., & U., N. H. (2012). Assessment of oil palm production and processing among rural women in Enugu North agricultural zone of Enugu State, Nigeria. *International Journal of Agricultural Sciences,*

2(12), 322-329.

Nikoloyuk, J., Burns, T. R., & Man, R. D. (2010). The promise and limitations of partnered governance: the case of sustainable palm oil. *Corporate Gorvenance, 10*(1), 59-72. https://doi.org/10.1108/14720701011021111

Obidzinski, K., Andriani, R., Komarudin, H., & Andrianto, A. (2012). Environmental and social impacts of oil palm plantations and their implications for biofuel production in Indonesia. *Ecology and Society, 17*(1). http://dx.doi.org/10.5751/ES-04775-170125.

Padfield, R., Papargyropoulou, E., & Preece, C. (2012). A preliminary assessment of greenhouse gas emission trends in the production and consumption of food in Malaysia. *International Journal of Technology, 1,* 56-66.

Paoli, G. D., Gillespie, P., Wells, P. L., Sileuw, A., Franklin, N., & Schweithel, J. (2011). *Oil Palm in Indonesia: Governance, Decision making, and Implications for Sustainable Development, Summary for Policy Makers & Practitioners.* Jakarta: The Nature Conservancy Indonesia Program, Jakarta.

Petrenko, C., Paltseva, J., & Searle, S. (2016). *Ecological Impacts of Palm Oil Expansion in Indonesia.* Washington, USA: International Council on Clean Transportation.

Popp, J., Harangi-Rákos, M., Gabnai, Z., Balogh, P., Antal, G., & Bai, A. (2016). Biofuels and their co-products as livestock feed: global economic and environmental implications. *Molecules, 21*(285), 1-26. https://doi.org/10.3390/molecules21030285

Riadi, E. (2013). *Aplikasi Lisrel untuk Penelitian Analisis Jalur.* Yogyakarta: CV. Andi Offset.

Sayer, J., Ghazoul, J., Nelson, P., & Boedhihartono, A. K. (2012). Oil palm expansion transforms tropical landscapes and livelihoods. *Global Food Security, 1,* 114–119. https://doi.org/10.1016/j.gfs.2012.10.003

SDSN-The Sustainable Development Solutions Network. (2013). *Solutions for Sustainable Agriculture and Food Systems.* New York: United Nation (UN).

Sheil, D. et al. (2009). *The impacts and opportunities of oil palm in Southeast Asia: What do we know and what do we need to know?* Occasional Paper No. 51. Bogor, Indonesia: CIFOR.

Sugiyono. (2003). *Metode Penelitian Bisnis.* Bandung: Alfabeta.

Teoh, C. H. (2010). *Key Sustainability Issues in the Palm Oil Sector: A Discussion Paper for Multi-Stakeholders Consultations (commissioned by the World Bank Group.* Washington, D.C., United States: International Finance Corporation (IFC), World Bank.

Turner, D. W. (2010). Qualitative interview design: A practical guide for novice investigators. *The Qualitative Report, 15*(3), 754-760.

Vermeulen, S., & Goad, N. (2006). *Towards better practice in smallholder palm oil production.* London, UK: IIED (Internal Institute for Environment and Development).

Wijanto, S. H. (2008). *Structural Equation Modeling dengan Lisrel 8.8.* Cetakan Pertama. Yogyakarta: Graha Ilmu.

Wilms-Posen, N., Boomkens, M., d'Apollonia, S., Klarer, A., Kraus, E. M., & Tynell, L. L. (2014). Land-use and livelihoods – A Malaysian oil palm scheme and its social and ecological impacts. *The Journal of Transdisciplinary Environmental Studies, 13*(2), 1-11.

Wisena, B. A., Daryanto, A., Arifin, B., & Oktaviani, R. (2014). Sustainable development strategy for improving the competitiveness of oil palm Industry. *International Research Journal of Business Studies, 7*(1), 13-37.

WWF, & AFGC. (2013). *Palm Oil in Australia: Facts, Issues, and Challenges.* Autralia: Net Balance Foundation.

WWF, F. A. (2012). *Profitability and Sustainability in Palm Oil Production: Analysis of Incremental Financial Costs and Benefits of RSPO Compliance.* Retrieved from http://www.rspo.org/publications/download/47ddf731d851469: WWF

9

Living Shoreline Treatment Suitability Analysis: A Study on Coastal Protection Opportunities for Sarasota County

Briana N. Dobbs[1], Michael I. Volk[2] & Nawari O. Nawari[3]

[1] Graduate Student, College of Design, Construction, and Planning, University of Florida, USA

[2] Assistant Research Professor, School of Landscape Architecture and Planning, College of Design, Construction, and Planning, University of Florida, USA

[3] Associate Professor, School of Architecture, College of Design, Construction, and Planning, University of Florida, USA

Correspondence: Briana N. Dobbs, Graduate Student, College of Design, Construction, and Planning, University of Florida, USA. E-mail: bdobbs@ufl.edu

Abstract

Increases in the world population, sea level rise, and urbanization of coastal areas have put tremendous pressures on coastlines around the world. As a result, natural shoreline habitats are being replaced by seawalls and other hardened forms of coastal protection. Evidence shows that hardened shorelines can have a negative impact on the environment and surrounding habitat, leading to a loss of biodiversity and ecosystem services. This research aims to increase the different forms of coastal protection used throughout Sarasota County, Florida by conducting a geographic information system (GIS) suitability analysis for living shoreline treatment. Living shorelines or hybrid solutions are a more ecologically sustainable alternative to traditional forms of coastal protection, which use natural ecosystems or alternatively- structural organic and natural materials such as plantings, rocks, and oyster beds to stabilize shorelines and enhance shoreline habitat. The GIS model identifies coastlines that are 1) most suitable for living shoreline treatment, 2) most suitable for a hybrid solution, or 3) not suitable for living shorelines by analyzing the bathymetry, land use, land value, tree canopy, population, wave energy, shoreline sensitivity, and shoreline habitat. The suitability for living shoreline treatments was assessed independently for each parameter and assigned a value ranging from 0, areas that should consider using traditional methods of coastal protection to 3, shoreline segments most suitable for living shoreline treatment. The results from the individual analyses for each parameter were combined using a weighted overlay approach to determine general suitability for living shorelines within the study area. The result found that over 95% of the shoreline segments are potentially suitable for hybrid shoreline stabilization solutions.

Keywords: living shoreline, GIS, suitability analysis, coastal protection

1. Introduction

1.1 General

Over one-third of the human population lives within 100 km of the coastline (Moschella, Abbiati, Aberg, Anderson, Bacchiocchi, Bulleri, Dinesen, Frost, Gacia, Granhag, Jonsson, Satta, Sundelof, Thompson, & Hawkins, 2005). This trend is projected to increase significantly, leading to a high demand for infrastructure to protect coastal lands from erosion and fight sea level rise (Hartig, Zarull, & Cook, 2011). Today, the standard form of coastal protection is known as armoring and commonly occurs in the form of engineered structures such as seawalls, bulkheads, and pier pilings. These forms of protection are likely to become more common as coastal populations continue to increase (Chapman & Underwood, 2006). Traditionally, seawalls have been installed when an area must be dredged or deepened to accommodate commercial industries (Hartig et al.,2011), or to protect residents from storm- surge and erosion (Chapman & Underwood, 2006). Other times, shorelines are hardened after habitats like dunes, which naturally protect against coastal hazards, have been removed. The current rate of armoring is approximately 200 km of shoreline a year (Gittman, Fodrie, Popowich, Keller, Bruno, Currin, Peterson & Piehler, 2015). If this rate continues, the percentage of hardened shorelines, factoring in the current bans on armoring within the United States, will increase from its current state of 14% to approximately

33% by 2100.

In 2015, the National Oceanic and Atmospheric Administration (NOAA) established that nearly 14% of the United States' shoreline was coated in concrete (Popkins, 2015). The south Atlantic and Gulf Coasts are among the fastest growing and largest contributors to hardened seawalls. Today, approximately 50% of the Gulf of Mexico coast is lined with tidal wetlands (Gittman et al., 2015). The southern portion of the Gulf Coast, where Sarasota County is located, is expected to have one of the largest increases in density, as well as seawall construction in upcoming years (Popkins, 2015). This endangers the large percentage of remaining tidal salt marshes and mangrove forests left in the area.

Statewide, approximately 20% of Florida's shorelines are artificially hardened while specific coastal communities such as Sarasota, Florida are dominated by an even larger percentage (Hauserman, 2007). A 2005 Gulf Shoreline Stabilization Inventory showed that approximately 32% of Gulf coast shorelines in Sarasota County were stabilized or hardened while a newer GIS study showed about 25% of the total shorelines as stabilized (Sarasota County, 2014). The County has lost over 1,600 acres of wetlands since 1950 and lost more than 100 miles of natural shoreline along Sarasota Bay to hardened structures (Sarasota Bay Estuary Program, 2015). These hardened shorelines may have damaged intertidal habitats vital to sustaining a healthy ecosystem and lead to a loss of biodiversity and ecosystem services (Chapman & Bulleri, 2003). As a result, the County has set a goal to restore one percent of their shoreline a year, a total of 18 acres a year (Sarasota Bay Estuary Program, 2015).

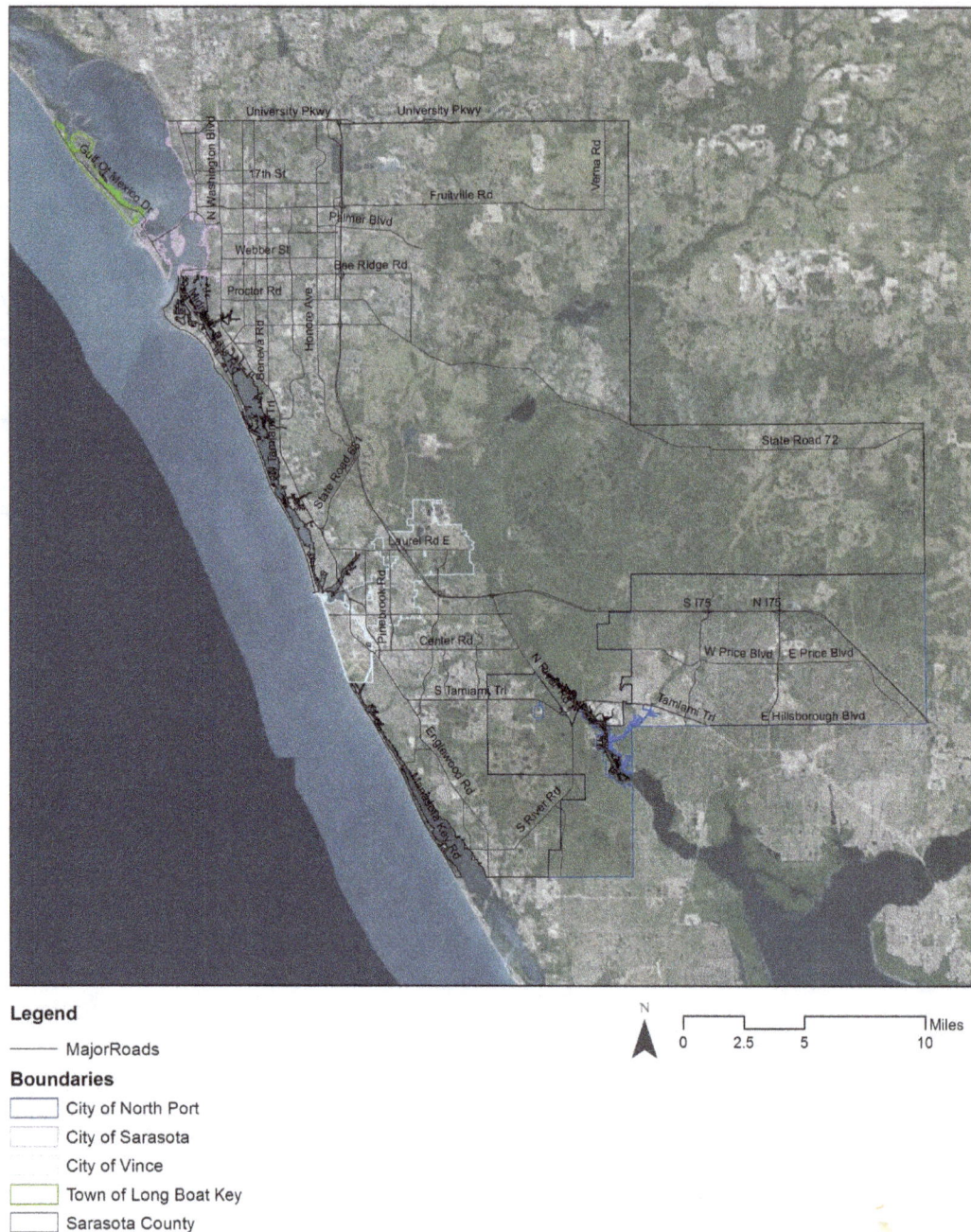

Legend

——— MajorRoads

Boundaries

☐ City of North Port

☐ City of Sarasota

☐ City of Vince

☐ Town of Long Boat Key

☐ Sarasota County

Figure 1. Sarasota county study area

The installation of seawalls results in displacement, degradation, and fragmentation of wetlands and beaches, two of the most valuable global natural resources (Gittman et al., 2015). In a study on impacts from seawalls on mangrove habitat, Heatherington & Bishop (2012) found that mangrove forests near seawalls were in some cases $1/3^{rd}$ in width and had two times the pneumatophore density compared to mangroves away from seawalls. In 1997, Costanza conducted a study to determine the monetary value of marine and terrestrial habitats throughout the United States. Wetlands (tidal salt marshes / mangroves and swamps / flood plains) were calculated to be the largest contributors providing $4.9 trillion dollars each year out of the total $33 trillion dollars all ecosystem services provide annually (Costanza, 1997). Doubtless, these numbers have increased. Wetlands also provide numerous ecosystem services such as water purification, storm protection, recreational opportunities, and carbon sequestration (Gittman, Peterson, Currin, Fodrie, Piehler, & Bruno, 2016). The continued loss of these coastal systems will clearly have monetary impacts, reduce the level of ecosystem services they are able to provide, and change the character of the coastal areas we inhabit.

1.2 Solutions

Coastlines throughout the United States are rapidly changing due to coastal development, climate change, and efforts to prevent erosion, flooding, and the negative effects caused by coastal hazards and by sea level rise. Living shoreline treatment (LST) is one type of sustainable coastal protection that can be a substitute for hardened seawalls. This technique uses natural barriers such as mangroves and fringe marshes to defend the shoreline (Davis, Takas, & Schnabel, 2007). LST and hybrid solitons known as ecological engineering and soft engineering techniques have become increasingly popular as the value of wetlands and marshes are being recognized. Not only do these solutions provide habitat for estuarine species, they can slow erosion rates, reduce runoff, and create a natural buffer that lessens storm surge (Sarasota Bay Estuary Program, 2015).

The technique of combining hard material and plantings while also enhancing ecosystem services is considered a hybrid living shoreline (Gittman et al., 2016). Research shows that different species colonize on artificial structures when compared to the colonization that occurs in natural coastal habitats. One reason is the lack of structural complexity in seawalls (Chapman & Underwood, 2011). Seawalls are typically made of sheet metal or concrete. Both materials are smooth and lack crevasses or spaces to support habitat and species. One option is for the design and construction of seawalls to be carefully engineered to achieve these goals by designing seawalls that can increase biodiversity. Hybrid solutions attempt to merge structural engineering and ecological needs in order to reduce the environmental impacts of a seawall. Ng, Lim, Ong, Teo, Chou, Chua, & Tan (2015), investigated transplanting reef biota onto newly constructed seawalls. Results identified species that were able to survive in intertidal habitats of seawalls in the tropics (Ng et al.,2015). These conclusions could lessen negative impacts of already fabricated walls by enhancing the ecological value and accelerating the colonization of organisms.

Another alternative is to remove and replace seawalls and other traditional coastal armoring structures with more ecologically sustainable solutions. In a study by Brown and McLachland (2002), a seawall located in San Francisco Bay was removed and replaced with marsh plantings. The case was successful and found that the number of species at the seawall was less than the number of species present at the marsh even during seasonal declines. Gitman et al., (2016), researched how living shorelines can potentially enhance ecosystem services. In the study, marsh plantings and breakwaters were used for LST and were compared to non-vegetated bulkheads. The living shoreline supported a higher number of species and greater biodiversity, showing that living shorelines do in fact have the opportunity to increase certain ecosystem services. In Puget Sound, a seawall was completely removed resulting in a successful application of LST (Chapman & Underwood, 2006). Removing the wall created habitat for juvenile salmon to grow. While this method is the most sustainable it does not work in all areas because it requires land to be inundated by the sea and specific environmental conditions for success.

Living shoreline treatment and hybrid solutions perform differently depending on surrounding environmental conditions and locations. By identifying shorelines that are most suitable for LST or alternative hybrid techniques, coastal municipalities have an additional tool to increase the success and palate of alternative nature-based approaches that restore vital wetlands while defending the coast. Similar studies that have been done to-date, such as the Living Shoreline Suitability Model for Worcester County, Maryland completed by the Center for Coastal Resource Management Virginia Institute of Marine Science (VIMS), and College of William and Mary (2008), approached this problem by examining fetch, bathymetry, marsh presence, beach presence, bank condition and tree canopy. These studies have primarily focused on identifying suitability based on environmental variables and have not considered anthropocentric variables such as land value, land use, and population, which are likely to be a critical part of the decision making process when choosing locations for LST construction

This project assessed shorelines in Sarasota County to quantify and determine the locations where LST and hybrid solutions would be most effective and resilient using a combination of anthropocentric and environmental variables. The study sought to answer what segments are 1) most suitable for living shoreline treatment, 2) most suitable for a hybrid solution, or 3) not suitable for LST by analyzing the bathymetry, land use, land value, tree canopy, population, wave energy, shoreline sensitivity and shoreline habitat across shorelines that occur within the County. It was hypothesized that living shoreline treatment would perform best in sheltered areas and therefore useful as a tool to for Sarasota County that may help the county achieve their goal of restoring at least one percent of their wetlands a year (Sarasota Bay Estuary Program, 2015). In addition, though modifications or improvements may be appropriate to the methodology described here, this project provides a basis for conducting future living shoreline assessments that include both anthropocentric and environmental considerations.

1.3 Types of Hybrid Living Shoreline Treatment (LST)

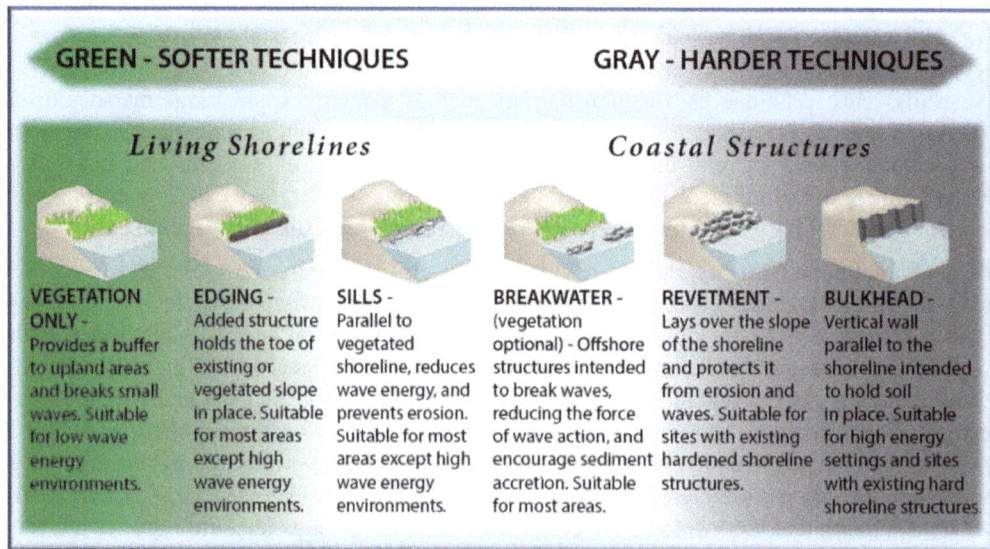

Figure 2. A range of shoreline stabilization techniques (NOAA, 2015).

1.3.1 Marsh with Structures

Marsh with structures include marsh sills, marsh toe revetments and marsh with groins and may be implemented where marshes already exist but are beginning to erode along with the shoreline. Structured marshes are placed in shallow water to allow for the growth of new seagrasses (Zylberman, 2016), which can eventually create a perched beach (Miller et al., 2015). The beach furthermore stabilizes the shoreline and replaces the eroded land. At times sand is placed between the shoreline and the sill to accelerate the development of the beach. Marsh with structures may be used on shorelines that experience higher levels of wave energy and require additional structural enhancement on top of marsh planting.

1.3.2 Breakwaters

Breakwaters run parallel with the coast to prevent coastal erosion and to reduce the wave energy behind the breakwaters. The lower water energy facilitates the growth of marsh vegetation and / or assists in beach establishment which increases habitat complexity (Miller et al., 2015). Often times they are placed outside of marinas, harbors or along open coasts. Breakwaters can be differentiated from sills as they are constructed in deeper water, farther away from the shore, can handle higher wave energies and are usually larger.

1.3.3 Revetments

Revetments, similar to rip-rap, are placed along the shoreline or bank to prevent erosion (Miller et al., 2015). Their locations reduce wave energy on the slope facing the water. Rock and concrete are the most common materials used to build revetments but they can also be constructed of fallen trees or debris. Contrary to rip-rap, revetments are strategically sized and placed to incorporate vegetation, increase the biodiversity, and further stabilize the structure. One disadvantage, however, is revetments do not allow for a connection between people and water that is strongly encouraged in LST. The property owners can observe the water from a distance but are unable to actually touch the water (Hardaway, Milligan, & Duhring, 2010). Additionally, if the revetment is installed incorrectly it could cause more damage and increase erosion (Zylberman, 2014).

1.3.4 Living Reefs

Living reefs, also known as offshore living reef breakwaters and living reef sills, have become increasingly popular in recent years (Miller et al., 2015). They work best in sheltered areas with lower fetch and wave energies and assist in reducing erosion rates in a manner similar to sills and constructed breakwaters. Unfortunately, naturally occurring reefs that assist in decreasing erosion rates have lessened due to anthropocentric forces and natural causes. As a result, reefs are commonly being built off-site and then brought to the site to grow. In some cases, recruitment and growth of the species occurs naturally, and the larvae settles upon the supplied substrate. As time passes, the species used in the living reef colonize and continue to grow as large species and develop as a form of natural breakwater protection as well as aquatic habitat (Miller et al.,

2015). Similar to sills, living reefs assist in facilitating the growth of vegetation behind the reefs by dissipating the tidal and wave energy. Living reefs are most commonly found in the southern parts of the United States.

1.3.5 Reef Balls

Reef balls provide an opportunity for reef development in intense wave conditions (Miller et al., 2015). In the ideal conditions, the reef structures grow independently, forming a large reef over time as the larvae finds a hard surface to naturally grow upon. Reef balls are most commonly found in the southern regions of the U.S and in the Caribbean. Not only do reef balls lessen wave energy, they also create habitats that provide numerous ecosystem services. Reef Balls work best at reducing erosion in areas with low to moderate wave energy.

1.4 Environmental Parameters for LST and Hybrid Solutions

1.4.1 Bathymetry

The bathymetry adjacent to a shoreline determines where a wave breaks, the amount of sediment it transports and the size of the wave that will hit the shore (Miller et al., 2015). Generally, the less intense the slope, the less energy a wave has when hitting the shore. Steeper slopes usually result in high wave energies that reflect off of the shoreline. When the contours produce a milder slope, the energy is more likely to be absorbed into the shoreline, dissipating the energy. Gentle slopes with vegetation indicate a stable slope and bank condition that should be considered for non-structural LST (Zylberman, 2016) and many hybrid approaches (Miller et al., 2015).

Previous GIS-based studies have been completed to assess site suitability of LST. The Living Shoreline Suitability Model for Worcester County, Maryland completed by the the Center for Coastal Resource Management (CCRM), Virginia Institute of Marine Science (VIMS), and College of William and Mary (2008), looked at bathymetry as a suitability variable. Their research showed that in order for marsh grasses to grow successfully, near-shore water depth must be shallow (Center for Coastal Resource Management [CCRM], Virginia Institute of Marine Science, & College of William and Mary, 2008). In the VIMS research, the bathymetry contour was combined with the distance away from the shoreline to determine shallow areas and not shallow areas. If the 1M contour was less than 10M away from the shoreline, then it was considered not shallow, and unsuitable for LST. If the 1M contour was greater than 10M from the shoreline, it was determined shallow and a suitable condition. This can also be considered in terms of slope. If the nearshore slope is greater than 10% the shoreline is less suitable for LST.

1.4.2 Shoreline Habitat

LST and many hybrid stabilization methods require new growth of grasses and plantings. Shorelines that are already vegetated are likely to be more suitable for LST and may have greater potential for supporting new growth (CCRM et al., 2008). An additional factor to consider is the width of existing marshes or other coastal ecosystems. Wider wetlands are likely to be healthier and more established. However, fringing marshes, or marshes less than or equal to 20M wide, deliver some equivalent functions when compared to broader marsh meadows (Bilkovic & Mitchell, 2013). Marshes that are less than 10M wide have been found to be effective at reducing wave energy and sediment deposits.

1.4.3 Shoreline Sensitivity

In 1976, the idea of ranking and mapping shoreline sensitivity originated for the Lower Cook Inlet (Petersen Michel, Zengel, White, Lord, & Plank, 2002). Since then, the idea has developed and been expanded to include North American, Central America, and portions of the Middle East. Today, the Environmental Sensitivity Index (ESI) is typically used to determine the sensitivity to oil spills but also parallels with living shoreline suitability. The shoreline slope, shoreline type, shoreline biological productivity, and overall sensitivity to waves and tidal energies determine a shoreline's ESI score. When examining these four factors, high ESI rankings are more likely to be suitable for LST while low ESI rankings are not suitable for LST. Shoreline type is classified based on the grain size, the tidal elevation, and the substrate. Within the data set, shoreline areas that are exposed to high wave energies and with low biological productivity or sensitivity, generally rank lowest on the scale. In contrast, areas with high biological productivity and low wave energy, rank on the higher end of the scale. Both high wave energy and low biological production are unsuitable situations for LST. As a result, the higher the ESI ranking, the more suitable LST would be on that specific site (Petersen et al, 2002).

1.4.4 Tree Canopy

The amount of sunlight a shoreline receives influences terrestrial and aquatic species (Miller et al., 2015). The tree canopy determines whether there is enough sunlight to grow the necessary vegetation below and the growth rate. Marsh plantings require a minimum of six hours of sunlight a day. Additionally, sunlight is a necessary component for photosynthesis to occur (Miller et al., 2015). Oxygen cannot be created without photosynthesis, which directly alters the water quality and biological productivity of the organism. Marsh plantings should not be planted around large shade trees or structures that prevent the sunlight from reaching the ground. Hybrid solutions that do not require marsh plantings, such as revetments, breakwaters, and reef balls are not influenced by sunlight exposure and can work under low or high sunlight conditions.

1.4.5 Wave Energy

Hydrodynamic parameters are one of the most important factors when evaluating the different variables for living shoreline suitability. Hydrodynamic characters include fetch, wakes from boat traffic, currents, ice, and storm surge (Miller et al., 2015). Wide expanses of open waters create a large fetch, while sheltered coasts create a smaller fetch (Zylberman, 2016). The higher the fetch, the greater the wave energy and size. Low wave energy is required to establish the growth of seagrasses, marshes, and mudflat habitats (NRC, 2007). Marsh sills, living reefs, and reef balls work best in environments with low to moderate wave energy. Revetments can withstand moderate to high waves, while breakwaters can withstand high energy (Miller et al., 2015).

1.5 Anthropocentric Considerations for LST and Hybrid Solutions

It has been over ten years since living shorelines started to gain popularity. Despite the available research, the number of living shoreline projects is still limited. In 2014, Virginia Institute of Marine Science held a public workshop with 144 participants and mailed out a questionnaire to identify the gap between the number of projects that have been constructed and the number of projects that should have been constructed according to suitability research that shows where living shorelines are appropriate (Center for Coastal Resource Management [CCRM], 2014). The results found that many people looked to their friends, family, and contractors, who have traditional forms of coastal protection, for advice on whether LST is the best decision for them and often times found they were more comfortable with standard construction practices. Additionally, the results indicated that individuals are also concerned about the effectiveness of LST because of the limited number of public projects. One way to fight this concern is to develop more public projects. Implementing living shoreline projects in high intensity, urban areas with high populations will result in a larger percent of the population having an opportunity to see living shorelines working in action and an opportunity for education. Areas that have a high number of visitors are great places to present the benefits of living shoreline projects and educate the public, allowing the project to act as a passive learning platform (Swann, 2008). Additionally, these areas offer an opportunity for monitoring compared to private, less populated shorelines. Monitoring the results will contribute to future research and collaborations with surrounding colleges to further education. Integrative projects can also assist in competitive grant opportunities that could help with funding for future projects.

A third concern of citizens was the cost of constructing living shoreline projects (CCRM, 2014). Maintenance costs, in addition to construction costs, must be included in the price for standard types of protection and LST. Living shorelines tend to be more resilient to storms, resulting in lower long-term maintenance costs. Standard practices must include the additional costs of mitigating wetlands if riparian vegetation or wetlands are removed, whereas living shoreline projects either include the vegetation in the design or mitigate any potential impacts on site.

Since the cost of living shoreline projects ranks highly on the list of concerns by interested individuals, it should be considered when determining the correct location for LST. Assessing the land value per acre is a coarse way to evaluate whether a shoreline is an appropriate place to implement LST with easily available data, though there are different ways of assessing suitability. Despite the evidence that shows that LST is comparable in cost to shoreline hardening, there is still a financial fear associated with living shorelines that discourages people from constructing new LST projects.

2. Methodology and Analysis

2.1 The Suitability Index

The suitability index used for this research was adapted from the Living Shoreline Suitability Model developed by the Virginia Institute of Marine Science (VIMS) for Worcester County, Maryland (2008). The research is considered a non-experimental study and was completed using secondary GIS data layers pulled from existing online databases. The variables used in this experiment were altered from the VIMS study due to the availability or lack of GIS layers and to include anthropocentric attributes such as property value, land use, and population which were not included in past studies (CCRM et al., 2008). Suitability for LST was mapped, assessed and modeled independently for each parameter. The results from the individual analyses for each parameter were combined using a weighted overlay approach to determine general suitability for LST within the study area. The weighted overlay tool calculates the final suitability score by multiplying the value of each parameter by the weight of its importance and summing the results together.

The shoreline was represented using the Florida Fish and Wildlife Conservation Commission polyline data (2004). The Sarasota County boundary line was buffered by 400M to clip the shoreline data, ultimately creating the Sarasota County Shoreline layer. The Sarasota County Shoreline layer was buffered 400M on both sides and used as the mask for the model to identify all shorelines within the county and an upland buffer. All data features were converted to raster datasets and assigned a cell size of 10. Each raster was then reclassified and assigned a value of 0-3 based on suitability for living shoreline construction. Three identifies areas that are most suitable for living shoreline with no structural components, while 0 represents locations that were entirely unsuitable. Areas that receive a 2 are segments suitable for hybrid solutions with minimal structural components while shorelines that received 1 are candidates for hybrid solutions that incorporate vegetation and structural components. Areas of no data were also assigned a value of 0 to permit a complete overlay among the 8 variables. The 0 value was chosen to be conservative about the potential suitability.

2.2 Bathymetry

Gentle slopes with vegetation indicate a stable bank condition that should be considered for non-structural LST (Zylberman, 2016) and many hybrid approaches (Miller et al., 2015). Bathymetry data from National Oceanic and Atmospheric Administration (2002) was used to calculate the slope of the seafloor in terms of percent (percent rise) adjacent to the shorelines in the study area. Nearshore slopes that ranged from 0-3% received a value "3" and were considered shorelines that were most suitable for LST. Shorelines with a slope from 3-6% received a value of "2", areas with a slope ranging from 6-10% received a value of "1", and areas with a slope greater than 10% received a value of "0" because it was an unfit condition.

2.3 Land Use

The Florida Fish and Wildlife Conservation Commission's Cooperative Land Cover (2015) raster dataset was used to identify land use. Land uses that were defined as high intensity urban areas, were considered the most suitable for LST and received the value of a "3". Areas defined as low intense urban areas, received a value of a "2", while rural areas were assigned a value of "1". The Florida Fish and Wildlife Conservation Commission's Cooperative Land Cover (2015) raster dataset was used to identify land cover in addition to land use. As a result, the data referring to land cover was reclassified as unsuitable and assigned a value of "0".

2.4 Land Values

The surrounding land values can assist in determining whether installing a living shoreline treatment is the best option. Information was retrieved from the U.S. Census Bureau (2010) dataset. Within the attribute table, a new column was created to calculate the value of land per acre. The new values were reclassified on a scale from 0 – 3. Areas that averaged $0 - $75,000 an acre received a value of "1". Areas that averaged $75,000 - $250,000 an acre received a value of "2". Areas that averaged $250,000- $15,000,000 an acre received a value of "3". The higher the property value, the more suitable the environment for the installation of LST as the individual or business would feel more comfortable investing money to protect their property.

2.5 Population

Locating living shoreline projects in areas with high populations will increase the number of people that see the project. This creates an opportunity for passive educational exhibits where community members can learn about the advantages of LST first hand. The U.S Census Bureau data set (2010), was clipped to the Sarasota County Shoreline Layer. The range of values and the mean were calculated to determine the correct values for reclassification. The highest value was 175 people per acre while the lowest value was 0 people per acre. The mean, 3.6 people per acre, was used to determine the areas that received a value of "2". Populations ranging from 0 -1 people per acre received a value of "0". Areas of the population that ranged from 1 – 3 people per acre were assigned a value of "1". Areas with a population ranging from 3 – 9 people per acre received a value of "2". Shorelines with populations of 9-175 people per acre received a value of "3" and were considered to be segments that are most suitable for LST.

2.6 Sensitive Shorelines

The Florida Fish and Wildlife Conservation Commission's (FWCC) Sensitive Shoreline vector line (2013) dataset was used to determine the sensitivity of the shorelines in the study area. The dataset was classified according to the Environmental Sensitivity Index and was used to determine the sensitivity of coastal and marine environments and species to oil spills. The data classified the shorelines from 1-10 based on their sensitivity to oil spills. The Environmental Sensitivity Index looked at four different aspects: the shoreline type, the ease of clean up, biological productivity and sensitivity, and the wave and tidal energy. For the purpose of this research, the data was reclassified and assigned a value from 0-3. The reclassification of the data was evenly distributed while also paying attention to the descriptions of the NOAA Environmental Sensitivity Index Guidelines for Attribute Domain Value. The most sensitive shorelines received a value of "3", less sensitive shorelines received a value of "2", and the least sensitive shorelines received a "1". Shorelines that are exposed to high wave energies and have low biological productivity, generally rank lowest on the Environmental Sensitivity Index and also rank low on the suitability index. Shorelines with high biological productivity and low wave energy, rank on the higher end of the Environmental Sensitivity Index and are more suitable LST.

2.7 Shoreline Habitat

The Florida Cooperative Land Cover (2015) raster dataset contains information on ground cover and land uses was used to identify shoreline habitat types. In past studies, shorelines with marsh presence were considered suitable and ranked higher than other land cover types (Center for Coastal Resources Management, Virginia Institute of Marine Science, & College of William and Mary, 2008). As a result, this method was adopted and adapted. Using the CLC dataset, the areas defined as isolated freshwater marsh, marshes, and salt marshes were considered most suitable and were assigned a value of "3". All other land cover types that were capable of growing vegetation and near the shoreline were given a value of "2". These included cypress, freshwater forested wetlands, mangrove swamp, tidal flat, wet flatwoods, cultural palustrine, cultural- lacustrine, estuarine, isolated freshwater swamp, marine, other coniferous wetlands, and other hardwood wetlands. The remaining land cover types, such as uplands, and additional land use data received a "0'. Determinations were based on the definition of each type of land cover defined by the CLC.

2.8 Tree Canopy

The tree canopy of an area is an indication of whether enough sunlight underneath the canopy exists to allow vegetation to grow. The 2011 National Land Cover Database's (NLCD) raster dataset, created by the Multi-Resolution Land Characteristics Consortium was used. The dataset depicts the percent of tree coverage for the state of Florida, ranging from 0-100%, with 100 representing full canopy coverage. The data was reclassified in equal intervals. Tree canopy coverage of 0-33% received a value of "3". Shorelines with 33-66% tree canopy were assigned a value of "2". Areas that showed a tree canopy between 66-100% were given a value of "1".

2.9 Wave Energy

High erosion conditions are likely caused by high wave energy. Steep banks and areas with high wave energy are both unfavorable growing conditions for LST. Sarasota County Water Bodies (2016) data was used to determine which types of waterbodies produced the highest wave energy. After examining the location of each on the map, understanding the general waterbody definition and considering the possible boat traffic and waves that could be produced, each type of waterbody was reclassified and given a value from 0-3 based on the wave energy on the shoreline. Bayou, lagoon, slough, tidal creek, and canal were reclassified as low wave energy and received a value of "3". Inlet, pass, waterway, and basin were reclassified as medium wave energy and assigned a value of "2". Gulf, channel, and bay were classified as the highest wave energy and given a value of "1". The remaining waterbody types, such as freshwater lakes and detention ponds were given a "0".

2.10 Suitability Index

The suitability index was calculated based on eight criteria: bathymetry, land use, land value, population, shoreline habitat, shoreline sensitivity, tree canopy and wave energy. All datasets were reclassified and assigned a value between 0-3 based on their suitability. The higher the suitability for LST, the higher the assigned value. After each variable was reclassified, the final suitability score was identified using a weighted overlay approach (see Table 1).

In this experiment, the weighted overlay was calculated in two different ways. The first method weighted all eight variables evenly. This means each variable received a weighted value of 0.125 and was considered equally important (see Figure 3). The second calculation separated the weights by environmental and anthropocentric attributes. Environmental factors that determined whether or not a living shoreline could persist, such as bathymetry, shoreline sensitivity, shoreline habitat, tree canopy and wave energy were assigned a higher weight of 0.155. Land value, land use and population, variables that encourage or discourage LST but do not limit its success based on growth, received a weight of 0.075. (see Figure 4).

Table 1. Suitability results for individual GIS parameters

Parameter	Suitability Score	Miles of Shoreline	% of Total Shoreline
Bathymetry	0	2860.44	51.61
	1	0.26	0.00
	2	26.28	0.47
	3	2655.43	47.91
Land Use	0	471.03	65.56
	1	3.48	0.48
	1	3.48	0.48
	3	214.28	29.83
Land Value	0	0.00	0.00
	1	1520.10	45.22
	2	498.28	14.82
	3	1343.26	39.96
Population	0	681.57	94.30
	1	41.03	5.68
	2	0.19	0.03
	3	0.00	0.00
Sensitive Shoreline	0	0.00	0.00
	1	131.62	7.20
	2	796.91	43.59
	3	899.66	49.21
Shoreline Habitat	0	405.02	56.37
	1	0.00	0.00
	2	278.88	38.82
	3	34.55	4.81
Tree Canopy	0	0.00	0.00
	1	466.62	8.42
	2	1219.32	21.99
	3	3858.99	69.59
Wave Energy	0	481.04	16.97
	1	1408.39	49.69
	2	140.48	4.96
	3	804.36	28.38

3. Results

3.1 Equally Weighted Suitability Scores

In order to analyze the final results, data results were rounded to the nearest integer (see Table 2). For example, shorelines that originally received a 2.6 were rounded up to a 3. Shorelines that received a value of 2.1 were rounded down to a 2. The minimum result was 0 and the maximum value received was 3. The mean score value was 1.27. This shows on average that shorelines are less than moderately suitable for LST. The most common suitability score was 1. There was a standard deviation of .26 between the data results. Over 66% of the total shoreline segments received a value of 1 and will require some kind of structural support. About 30% of the shorelines received a value of 2, meaning these shoreline segments are candidates for a living shoreline treatment with minimal structural support. A minuscule amount of shorelines, less than 1%, received a value 3 for only vegetative LST.

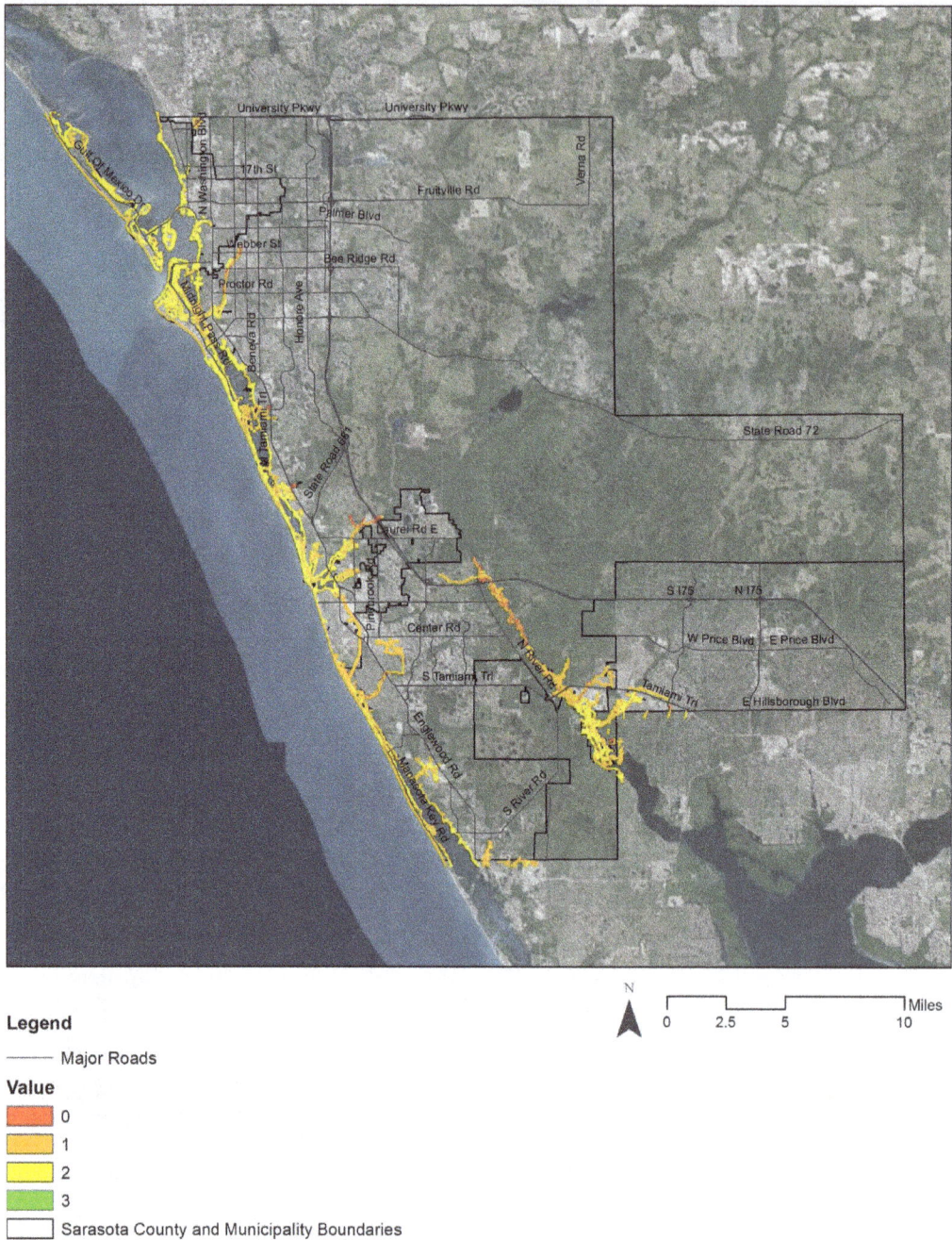

Figure 3. Equally weighted suitability scores

3.2 Unequally Weighted Suitability Scores

After rounding to the closest integer, the maximum value was a 3 and the minimum was a 0 (see Table 2). The mean score was 1.28. This shows on average shorelines are less than moderately suitable for LST. There was a standard deviation of .27 between the data. Approximately 64% of the data received a value of 1 and about 32% of the shorelines received a value 2. When compared to the equally weighted suitability results, more shorelines are received a suitable score of 2. However, more segments received a 0 and fewer shorelines received a 3, making it less suitable for LST with vegetation only. The most common suitability score was 1.

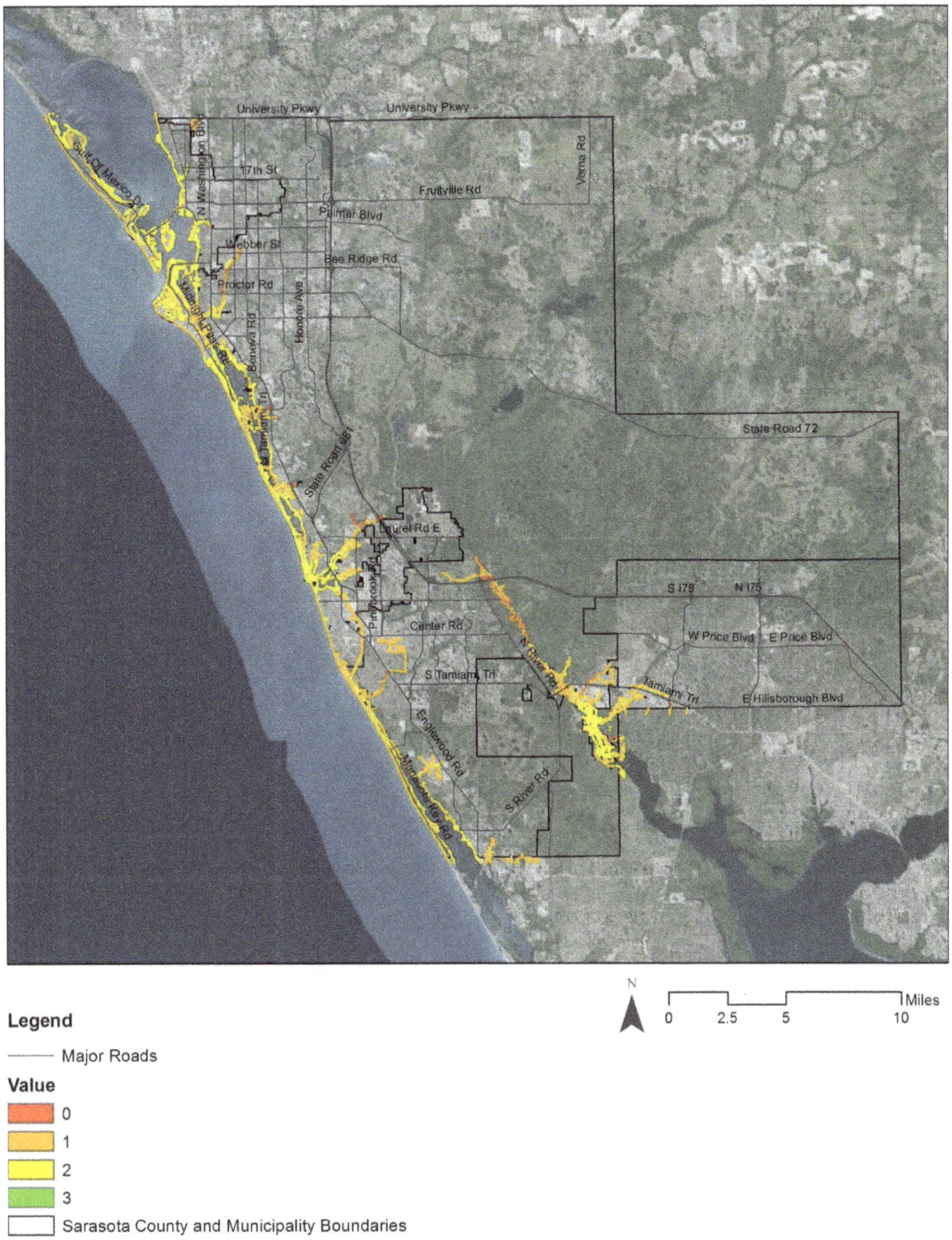

Legend

─── Major Roads

Value

▮	0
▮	1
▮	2
▮	3
☐	Sarasota County and Municipality Boundaries

Figure 4. Unequally weighted suitability scores

Table 2. Final suitability index scores

Parameter	Suitability Score	Miles of Shoreline	%
Equally weighted final suitability scores.	0.00	14.04	3.04
	1.00	308.84	66.84
	2.00	139.18	30.12
	3.00	0.02	0.00
Unequally weighted final suitability scores	0.00	15.95	3.45
	1.00	298.74	64.65
	2.00	147.38	31.89
	3.00	0.00	0.00

4. Discussion

Overall results for the two weighting schemes were very similar. A greater differentiation could have perhaps been achieved by increasing the difference between weights. The results also showed that sheltered shorelines are generally less suitable based on the methods.

It is important to note that there are many factors that can influence the suitability of living shorelines. This study examined the relationships between eight different variables. It is impossible to consider every situation and some information is not available in GIS format, or if it is available it may be too coarse, inaccurate, or otherwise unsuitable. One of the major limitations for this research was the availability of data and/ or its accuracy which can lead to a biased result. This was experienced in the bathymetry and land use data layers. The bathymetry suitability results were unique when compared to the other variables. Approximately 52% received a value of "0" and were unsuitable while about 48% received a value of "3". After further investigation, the bathymetry contours did not expand throughout the whole study area. As a result, many of the shoreline segments received a suitability score of 0 because there was no data available, even though there is a chance that they are suitable.

If this study was completed again, the areas that were originally given a "0" for "no data", should receive an alternative classification of "2". This would allow for areas of "no data" to receive a neutral suitability score and result in fewer shoreline segments receiving a value of "0" when there is a possibility that the area could be suitable. Assigning the value of "0" to areas of "No Data" largely impacted and perhaps skewed the results.

A similar outcome occurred when calculating the suitability index for land use. The GIS layer used for the land use category was the same GIS layer used for shoreline habitat. As a result, over 90% of the attributes were related to shoreline habitat and not land use. The shoreline habitat attributes were given a value of "0" when calculating the suitability scores for land use and were considered not suitable despite their lack of relationship with land use. This in essence resulted in the two different variables cancelling each other out. When the shoreline segment received a value of "1" or higher for land use suitability, it received a score of "0" for the shoreline habitat. When the shoreline segment received a value of "1" or higher for shoreline habitat, it received a "0" in the land use suitability score. An alternative method would be to use one GIS data set for each variable.

Additionally, it would be best to reclassify the Sarasota County Waterbody Layer differently. Streams and rivers should have received a classification of "3" not "0" based on the lower wave energy. This new reclassification value would alter the results and increase the overall suitability score for the sheltered shoreline segments. It would also be beneficial to calculate the fetch in order to achieve more accurate results.

Lastly, as areas of "0" should represent segments that are entirely unsuitable, the ranges of the tree canopy should have been divided in to quarters instead of thirds. Under the current methodology, shoreline segments that were covered by 0-33% tree canopy received a suitability score of "3", areas with 33-66% received a "2" and shorelines with 66-100% tree canopy received a "1". Instead, the tree canopy percentages should have been reclassified into four divisions ranging from 0-25%, 25-50%, 50-75% and 75-100% to accurately represent the data.

In the future, additional variables could be included in the study. Existing structures of coastal protection are not included in this research. When the County completes a GIS map highlighting the location of hardened shorelines, this study could be reexamined to include that data. Additional parameters to include in the future are erosion history, sea level rise, and tidal ranges. The tidal range is a critical factor to consider in terms of the "living" aspects for LST (Miller, Rella, Williams, & Sproule, 2015). For example, when including oysters or similar living reef elements, the placement is dependent upon the water level. Lastly, understanding the quality of water is important for successful habitat development, a key feature in almost every living shoreline project. Different habitats require different conditions for growth including water temperature, salinity, and turbidity.

Finally, any GIS-based or model-based approach to identifying suitability for LST is inherently coarse and needs to be ground-truthed. This analysis is only an early step in the process of identifying suitable locations for LST.

4.1 Challenges of Living Shoreline Implementation

Currently, the regulatory process and permitting is a major challenge when implementing living shoreline treatment, especially hybrid alternatives. There are many specifications that must be met which can turn construction approval into a lengthy process that deters residents from using LST strategies. Research, like this document, can help expedite the process by having a basic understanding of suitable locations for LST. Projects must also follow local, state and federal laws. Within the City of Sarasota, one of the main problems lies in the definition of fill. The regulations allow for nonstructural methods of living shoreline treatment but dismiss hybrid solutions in all areas. Cities around the United States are facing similar issues in regard to permitting hybrid living shoreline treatments.

One solution is to clarify and specify the permitting process. The New Jersey Department of Environmental Protection (2011) released a white paper discussing different situations that would require different types of permitting. This would allow more flexibility in LST construction and replacement (Frizzera, 2011). Others have drafted model ordinances for living shoreline projects that include the definition, purpose, and requirements for a project (Boyd & Pace, 2013). Looking to existing language in surrounding counties is another solution. Brevard County defines living shorelines in Florida Sec. 62-3661 of the code as "erosion management techniques, such as the strategic placement of plants, stone, sand, and other structural and organic materials, that are used primarily in areas with low to moderate wave energy, and are designed to mimic natural coastal processes" (Brevard County, 1994). It furthermore encourages living shorelines as the primary method of coastal protection in areas where bulkheads and rock revetments are prohibited, but stabilization is needed. Additionally, by engaging the residents at the beginning of the research, instead of using public participation data from VIMS, the methodology could have been personalized for Sarasota, helping to build momentum with the citizens to change these restricting regulations.

5. Conclusion

The coastlines of Sarasota are one of the County's most valued resources. As sea level rises and urbanization continues, the demand for coastal protection will increase in order to protect developed shorelines. Standard methods of coastal protection include shoreline armoring which can result in a loss of habitat and ecosystem services. Sarasota has placed restrictions limiting the new construction of hardened shorelines like bulkheads and seawalls. This presents a great opportunity to introduce alternative nature-based approaches. These approaches include but are not limited to, living shorelines, reef balls, and living reefs. Not only do these techniques assist in coastal erosion and sea level rise, but could support the County's goal of restoring 18 acres of wetlands a year while also increasing the health of the water, providing habitat for marine and terrestrial species, delivering an aesthetic benefit and assisting in a variety of ecosystem services.

Suitable conditions for living shoreline treatment (LST) and hybrid solutions depend on the surrounding environment. Nonstructural alternatives, such as beach nourishment and marsh restoration, require low energy shorelines, with existing vegetation, gentle nearshore slopes and adequate sunlight. This research used ArcGIS to create a suitability index to identify shorelines that were suitable for LST and hybrid solutions by examining eight different variables. The suitability for each parameter (bathymetry, land use, land value, population, shoreline habitat, sensitive shorelines, tree canopy and wave energy) was completed independently. In order to determine the overall suitability for LST within the study area, the results from each of the individual suitability analyses were combined using a weighted overlay approach. The weighted overlay approach was calculated using two different methods. The first method assigned an equal value to each of the variables to calculate the final suitability analysis. The second method assigned a higher weight to the environmental parameters and a lesser weight to the anthropocentric attributes.

The results from the study encourage alternative methods of coastal protection in Sarasota County. The equally weighted and unequally weighted suitability analysis showed that over 95% of the shorelines may be candidates for hybrid living shoreline techniques while less than 1% of the shorelines resulted in areas that would likely support LST without structural components. Implementation of LST and hybrid options face many challenges such as costs, lack of awareness and permitting. However, as research continues and alternative methods of coastal protection are demanded, LST will increase. This model can assist in the first step of screening for LST and begin the conversation between residents, homeowners, and officials to take the next steps to becoming a resilient County.

References

Bilkovic, D. M., & Mitchell, M. M. (2013). Ecological Tradeoffs of Stabilized Salt Marshes as a Shoreline Protection Strategy: Effects of Artificial Structures on Microbenthic Assemblages. *Ecological Engineering, 61*(A), 469-481. https://doi.org/10.1016/j.ecoleng.2013.10.011

Boyd, C., & Pace, N. L. (2013). *Coastal Alabama living shorelines policies, rules, and model ordinance manual.* Retrieved from http://floridalivingshorelines.com/wp-content/uploads/2015/05/Boyd-Pace-2013-Coastal-Alabama-Living-Shorelines-Policies- Manual.pdf

Brevard County. (1994). Code of ordinances of Brevard county, Florida volume I. municode.

Center for Coastal Resources Management, Virginia Institute of Marine Science, & College of William and Mary. (2008, May). *Living Shoreline Suitability Model Worcester County, Maryland.* Retrieved from http://ccrm.vims.edu/publications/projreps/worcester_living%20_shoreline_v2.pdf

Center for Coastal Resources Management. (2014). Living Shoreline Implementation: Challenges and Solutions. *Rivers and Coasts, 9*(2), 1-8. Retrieved from http://ccrm.vims.edu/publications/pubs/rivers&coast/RC914.pdf

Chapman, M., & Bulleri, F. (2003, Feb). Intertidal Seawalls—New Features of Landscape in Intertidal Environments. *Landscape and Urban Planning, 62*(3), 159-172. https://doi.org/10.1016/S0169-2046(02)00148-2

Chapman, M., & Underwood, A. (2006). Evaluation of Ecological Engineering of "armored" Shorelines to Improve their Value as Habitat. *Journal of Experimental Marine Biology and Ecology, 1*(2), 302-312.

Costanza, R., d'Arge, R., de Groot, R., Farberk, S., Grasso, M., Hannon, B., Limburg, K., Naeem, S., O'Neill, R.V., Paruelo, J., & Raskin, R.G. (1997). The value of the world's ecosystem services and natural capital. *NATURE, 387,* 253-260. https://doi.org/10.1038/387253a0

Davis, L., Takacs, R., & Schnabel, R. (2007, Jan). *Evaluating Ecological Impacts of Living Shorelines and Shoreline Habitat Elements: An Example From the Upper Western Chesapeake Bay. Research Gate.* Retrieved from https://www.researchgate.net/publication/259673745_Evaluating_Ecological_Impacts_of_Living_Shorelines_and_Shoreline_Habitat_Elements_An_Example_From_the_Upper_Western_Chesapeake_Bay

ESRI. (2016). Retrieved from http://desktop.arcgis.com/en/arcmap/10.3/tools/spatial-analyst-toolbox/how-slope-works.htm

Florida Fish and Wildlife Conservation Commission (FWCC). (2013). *Florida's Environmentally Sensitive Shorelines – 2013.* Retrieved from http://www.fgdl.org/metadataexplorer/explorer.jsp

Florida Fish and Wildlife Conservation Commission (FWCC). (2015). *Cooperative Land Cover Map.* Retrieved from http://myfwc.com/research/gis/applications/articles/Cooperative-Land-Cover

Frizzera, D. (2011, Sept). *Mitigating Shoreline Erosion along New Jersey's Sheltered Coast: Overcoming Regulatory Obstacles to Allow for Living Shorelines.* Retrieved from http://www.nj.gov/dep/cmp/docs/living-shorelines2011.pdf

Gittman, R. K., Fodrie, F. J., Popowich, A. M., Keller, D. A., Bruno, J. F., Currin, C. A., Peterson, C. H., & Piehler, M. F. (2015). Engineering Away Our Natural Defenses: an Analysis of Shoreline Hardening. US. Frontiers in Ecology and the Environment, 13(6), 301–307. https://doi.org/10.1890/150065

Gittman, R., Peterson, C., Currin, C., Fodrie, F., Piehler, M., & Bruno, J. (2016). Living Shorelines can Enhance the Nursery Role of Threatened Estuarine Habitats. *Ecological Applications, 26*(1), 249-263. https://doi.org/10.1890/14-0716

Hardaway, C., Milligan, D., & Duhring, K. (2010). Living Shoreline Design Guidelines for Shore Protection in Virginia's Estuarine Environments, Gloucester Point, VA: Virginia Institute of Marine Science, College of William and Mary.

Hartig, J., Zarull, M., & Cook, A. (2011). Soft Shoreline Engineering Survey of Ecological Effectiveness. *Ecological engineering, 37,* 1231-1238. https://doi.org/10.1016/j.ecoleng.2011.02.006

Hauserman, J. (2007, April). *Florida's Coastal and Ocean Future a Blueprint for Economic and Environmental Leadership*. Retrieved from https://www.nrdc.org/sites/default/files/flfuture.pdf

Heatherington, C., & Bishop, M. (2012). Spatial Variation in the Structure of Mangrove Forest with Respect to Seawalls. *Marine and Freshwater Research, 63*, 926-933. https://doi.org/10.1071/MF12119

Miller, J. K., Rella, A., Williams, A., & Sproule, E. (2015). *Living Shoreline Engineering Guidelines*. Stevens Institute of Technology: Davidson Laboratory, Center for Maritime Systems.

Moschella, P., Abbiati, M., Aberg, P., Anderson, J., Bacchiocchi, F., Bulleri, F., ... Hawkins, S. (2005). Low-crested Coastal Defense Structures as Artificial Habitats for Marine Life: Using Ecological Criteria in Design. *Science Direct, 52*, 1053-1071.

National Land Cover Database. (2011). *NCLD 2011 Land Cover*. Retrieved from http://www.mrlc.gov/nlcd11_data.php

National Oceanic and Atmospheric Administration (NOAA). (2002). *Bathymetric Contours for the State of Florida and Surrounding Areas*. Retrieved from http://www.fgdl.org/metadataexplorer/explorer.jsp

Ng, C., Lim, S., Ong, J., Teo, L., Chou, L., Chua, K., & Tan, K. (2015). Enhancing the Biodiversity of Coastal Defense Structures: Transplantation of Nursery-reared Reef biota onto Intertidal Seawalls. *Ecological Engineering, 82*, 480-486. https://doi.org/10.1016/j.ecoleng.2015.05.016

Petersen, J., Michel, J., Zengel, S., White, M., Lord, C., & Plank, C. (2002, March). Environmental Sensitivity Index Guidelines. Retreived from http://response.restoration.noaa.gov/sites/default/files/ESI_Guidelines.pdf

Popkin, G. (2015, Aug 8). *Fourteen percent of U.S. coastline is covered in oncrete*. Retrieved from http://www.sciencemag.org/news/2015/08/fourteen-percent-us-coastline-covered- concrete

Sarasota Bay Estuary Program. (2015). *Wetlands*. Retrieved from http://sarasotabay.org/habitat-restoration/wetlands/

Sarasota County. (2014, Aug 27). *Sarasota County Comprehensive Plan*. Chapter 2: Environment. Retrieved from https://www.scgov.net/CompPlan/Comp%20Plan%20Amendments/Chapter%202%20-%20 Environment.pdf

Sarasota County. (2016). *Sarasota County Boundary*. Retrieved from https://www.scgov.net/GIS/Pages/DataDownload.aspx

Sarasota County. (2016). *Water Bodies*. Retrieved from https://www.scgov.net/GIS/Pages/DataDownload.aspx

Swann, L. (2008). The use of living shorelines to mitigate the effects of storm events on Dauphin Island, Alabama, USA. American Fisheries Society Symposium 64.

U.S. Census Bureau. (2011). *2010 US Census Blocks in Florida (with Selected Fields from 2010 Redistricting Summary File and Summary File 1)*. Retrieved from http://www.census.gov/

University of Florida's GeoPlan Center. (2015). *Florida County Boundaries with Detailed Shoreline*. Retrieved from http://www.fgdl.org/metadataexplorer/explorer.jsp

Zylberman, J. M. (2016). Modeling Site Suitability of Living Shoreline Design Options in Connecticut. Master's Thesis Paper 875. Retrieved from http://digitialcommons.uconn.edu/gs_theses875

10

Compliance with the Requirements of the Environmental Impact Assessment Guidelines in Zimbabwe

Rowan Kushinga Machaka[1], Lakshmanan Ganesh[1] & James Mapfumo[2]

[1] Centre for Research, Christ University, Bangalore, India

[2] Christ College, Harare, Zimbabwe

Correspondence: Rowan Kushinga Machaka, Centre for Research, Christ University, Bangalore India. E mail: kushinga.machaka@res.christuniversity.in

Abstract

This research set out to find out how well projects are complying with the requirements of the Environmental Impact Assessment Guidelines in Zimbabwe. Data was collected from EIA reports completed between 2007 and 2012. Questionnaires and interviews were used to collect the experiences of practitioners in the EIA sector. The results show that EIA reports contain below 65% of required information for decision making. Critical sections of the EIA report are the most deficient. Compliance varies significantly between consultants, stage of EIA process and size of project. Recommendations are: need to review and expand the existing guidelines and promote their use, building objectivity into the EIA report review process, upholding professional standards of practice for the consultants, improvement of compliance monitoring and enforcement, the use of economic incentives and disincentives other than enforcement to promote compliance, increased awareness raising of EIA in the business sector, and increased political will and transparency.

Keywords: environmental impact assessment (eia), environmental compliance, environmental assessment, environmental policy

1. Introduction

Like most African countries, Zimbabwe initiated the process of environmental impact assessment (EIA) policy formulation in response to the Rio Local Agenda 21 Declaration (1992) resulting in the policy being codified in 1994 (Ministry of Environment and Tourism, 1997). In 1997, EIA Guidelines were published and operationalised to guide EIA practitioners and stakeholders in the process of carrying out EIA studies. Although it was not law then, the EIA policy set out the parameters that needed to be followed by those who opted to subject their development initiatives to the EIA policy. Funding agents, donors as well as local banks played a role in the adoption of EIA practice by demanding compliance with EIA requirements before committing resources to projects. As a result, the EIA policy was widely implemented in all sectors of the society even before it was law.

In 2003, the EIA policy was incorporated into law within the Environmental Management Act (Chapter 20.27) thereby giving the regulating authority more powers to regulate the application of the EIA system. Zimbabwe is one of the countries with well documented step-by-step guidelines for carrying out EIA studies and implementing environmental management plans. The stated purpose of the EIA Guidelines (1997) is "to facilitate compliance with the EIA policy by government departments, project developers and the general public".

In short, the EIA Guidelines require that a study of a project's anticipated impacts be done. Thereafter, mitigation measures are formulated to reduce or avoid the anticipated impacts. The result of the study is an EIA report which is submitted to the relevant authority for approval/acceptance.

The proponent (or developer) has the responsibility to conduct an EIA study. To do so, the proponent engages a consultant who has competency to conduct the EIA study and produce a report.

The EIA process can be split into pre-certification and post-certification stages. The main steps of the EIA Guidelines are:

(1) Screening – to determine if a full EIA study is required;

(2) Prospectus – a document produced by the proponent informing the regulatory authority about the main environmental issues of the project which need to be considered during the EIA Study:

(3) Terms of Reference (TOR) and Scoping – specify how the EIA study is to be conducted;

(4) EIA Study – a scientific process of studying the baseline, the impacts and formulating the mitigation measures which culminates in the EIA report which contains the environmental management plan (EMP);

(5) EIA Report Review – decision point by the regulatory authority whether to allow the project to be implemented or not (acceptance);

(6) Terms and Conditions – if project is given acceptance, additional conditions can be included for implementation by the proponent; and,

(7) EMP implementation and Monitoring and Auditing – execution of the mitigation measure including monitoring to ensure that the EMP is being implemented according plan and that any emergent impacts/ issues are addressed.

Therefore, the EIA Guidelines stipulate the content which is required in each of the sections of the EIA Report.

1.1 The Problem

The EIA system is based on the precautionary principle, which requires action to be taken to avert anticipated environmental problems even without scientific certainty that the problems will occur. It is a system of anticipating negative impacts to the environment and avoiding or mitigating them. For this reason, a series of EIA process steps were designed to ensure that environmental impacts are incorporated into economic projects. Compliance with these steps is crucial to ensure that the EIA system is on track. If compliance is lacking, there is no basis to expect that the environment is being protected as required by law.

In addition, establishing and administering the EIA system demands large quantities of resources for all the stakeholders involved. Hence, it is important to review the level of compliance in order to justify the amount of resources committed to EIA systems.

This study fills in the need to assess compliance issues surrounding implementation of the EIA system in Zimbabwe and draw out lessons to feed into relevant policies. This study will attempt to evaluate how projects are complying with the requirements of the EIA system.

1.2 Literature Review

The importance of compliance to the effectiveness of EIA system as noted by several authors since long ago (Wasserman, 1992; Benson et al, 2006, cited in Kahangiwre (2011)). It is not surprising that much research has been conducted on this theme. Compliance is thus critical to effective EIA systems.

Some studies on EIA compliance have been carried out in Africa. In Uganda, Manyindo (2002) found that compliance and impact management are neglected while there is a misconception that the EIA process is done to obtain a certificate. In Malawi, Mhango (2005, p. 389) found out that 93.75% of the EIA reports had less than 50% compliance grade. This inadequacy was attributed to capacity issues and to lack of mandatory guidelines. Mhango (2005) also found out that other than screening and impact analysis, other requirements of EIA reports are poorly adhered to.

Betey and Godfred (2013:45) reviewed the EIA systems of four African countries (Egypt, Ghana, Mauritius and South Africa) and found out that monitoring of post-certification compliance was generally lacking except where there are complaints or disasters.

Outside Africa, in Pakistan, Nadeem and Hameed (2006) attributed poor EIA practice to proponent's attitudes, lack of consultants' experience, inconsistent EIA review criteria, and inadequate expertise for the review of EIA reports. In a case study of post-certification compliance monitoring using specific parameters in Nepal, Khadka and Khanal (2008, p. 93) found that monitoring was sometimes as low as 23%. They noted that compliance improved to 90% due to the following reasons: appointment of qualified personnel, training of personnel and imposed penalty for non-compliance.

Therefore, it appears that there is a general trend of inadequate compliance in different country, albeit for differing reasons.

EIA evaluation studies tend to measure effectiveness from a procedural point of view (Cashmore et al, 2004; Pope et al, 2013). This focusses on how the EIA system is being implemented other than what has been achieved. This study is no exception.

1.3 Delineation

This Section describes the constructs used in this study.

The EIA system is defined as the sum total of institutions, processes, resources and relationships which form the mechanism for implementing the EIA policy or law.

The EIA process is a series of steps that are stipulated in the EIA policy or law and especially the guidelines through which the EIA policy or law is implemented.

The compliance with the EIA system is defined as the average proportion of the EIA system steps that are complied with by all projects. The other part of the variable is the degree to which the proponents adhere to the steps of the system, i.e. is the compliance being carried out to the extent expected. Therefore, when a proponent produces for example, an environmental management plan, the proponent has complied with the requirement. However, the environmental management plan may not be of the expected quality, which is the quality of compliance. The quality of the compliance is difficult to measure based on the EIA report alone without verification on the location of the project and the implementation. Therefore, this study focused on simply establishing whether the particular step of the EIA system has been complied with regardless of how well.

The goal of this research is to evaluate compliance of new projects with the requirements of the EIA system in Zimbabwe.

1.4 Objectives of the Study

The specific objectives are:

    i.   To measure the proportion of the EIA Guideline requirements that are complied with;

    ii. To find out difference in compliance with respect to consultants, project size, project type (economic sector), annual trend between 2007-2012, and, the stage of the EIA process;

    iii. To understand the factors affecting compliance by proponents; and,

    iv. To find out the major factors affecting compliance with EIA Guideline requirements.

1.5 Hypothesis

The hypotheses of the study are as follows:

    i.   There is no difference in observed and expected compliance of EIA reports at 65%;

    ii. There is no correlation between compliance of EIA reports and years 2007 to 2012;

    iii. There is no difference in compliance of EIA reports between economic sectors;

    iv. There is no difference in compliance between small and large projects;

The first hypothesis relates to the first objective, whilst the second, third and fourth hypotheses relate to the second objectives. There are no hypotheses for objectives three and four which pertain to interview data.

2. Method

2.1 Study Design

The study was designed as a process evaluation approach which measured the process which is being used to attain desired results of the environmental impact assessment legislation and policy against the requirements. Being a mixed method design also entailed the use of both quantitative methods for questionnaires and secondary data and qualitative methods for interview data.

The study was designed as a survey using both qualitative and quantitative techniques. Three tools were used to collect data. Secondary data was collected through enumeration using a checklist. The enumeration involved identifying if the required content was available under the relevant sections of the EIA reports. EIA reports obtained from willing consultants were reviewed and the checklist form was completed for each of the 65 EIA reports made available.

2.2 Sampling

Three sources of data were used. For EIA reports (secondary data), ten willing consultants provided the research with sixty-five EIA reports. Therefore, convenience sampling was used to obtain a sample size of 10 consultants from a population of about 60 consultants.

For questionnaires, the population was made up of all consultants and individuals with knowledge of EIA in Zimbabwe who could be reached through online professional websites or email. A total of 578 questionnaires were conveniently administered by email using online survey software. Seventy-five responses were received, of which 66 were considered valid for analysis.

Fifteen individuals known to have been involved in the EIA system for more than 10 years were specifically targeted in a purposive sampling. The reason was that the interviews were necessary to give deeper insights which explain the current observed status of the EIA system especially the historical trends that have shaped the EIA system in Zimbabwe.

Primary data was collected in two ways. Firstly, a questionnaire was designed with a Likert scale of 1 to 5 (strongly disagree, disagree, average/neutral, agree and strongly agree) with constructs that measure experiences of respondents with respect to EIA compliance.

Secondly, interviews were conducted with experts to delve into the deeper underlying factors of the EIA systems. Interview discussions involved the three research open-ended questions followed by circumstantial follow-up questions where necessary. The split-half method using the Cronbach's alpha was used to assess the saturation point of the sample of questionnaires.

2.3 Limitations

Some limitations of the study are as follows. The study was clearly limited by the feasibility of obtaining data pertaining to certain elements of the EIA system. Only those consultants confident of their work were willing to submit their reports hence the actual compliance may be much worse or better than measured by this study. However, the study provides a reasonable snapshot. Therefore, from a statistical point of view, the generalizability of the findings of this study is limited.

The enumeration process used to collect data from EIA Reports only checked the presence of required content in the sections of the EIA reports. The quality of the content was not assessed due to the verification difficulties.

2.4 Data Analysis

The choice of statistical tools was dictated by the need to compare the observed against the expected. Hence, comparison of means, correlation and test of association were used to analyse the data and to test the different hypotheses. Interview data was grouped according to themes using the thematic analysis approach.

3. Results

This section will present the results according to the research objectives.

3.1 To Measure the Proportion of the EIA Guideline Requirements That Are Complied with

Chi-square analysis of secondary data showed that there is a difference ($p=0.05$) between observed compliance (59%) and expected compliance of 65% with the EIA reports alone.

Chi-square analysis of the questionnaire data showed that there is a significant difference ($p=0.05$) between observed compliance and expected compliance of 50%.

3.2 To Find out Compliance with Respect to Consultants, Project Size, Project Type (Economic Sector), Annual Trend between 2007-2102, and, the Stage of the EIA Process

Analysis of variance showed that there is significant difference ($p=0.05$) between compliance of consultants in EIA reports. Pearson's correlation analysis showed that there is correlation ($p=0.05$) between compliance and year of EIA reports. Analysis shows a moderate positive correlation (0.331) which means that EIA Report compliance has moderately improved from 2007 to 2012;

Analysis of variance also showed that there is no significant difference ($p=0.05$) in compliance between economic sectors of EIA reports. The Mann-Whitney U-test was used to analyse compliance between large and small projects, and compliance between pre and post-acceptance. The results show that there is significant difference ($p=0.05$) in compliance between large and small projects, and in compliance between pre and post-acceptance.

3.3 To Understand the Factors Affect Compliance by Proponents

The questionnaires were used to measure the factors that explain why proponents may fail to comply. Six factors were identified as follows (in order of increasing importance): (i) the EIA system is unclear (3.42), (ii) EIA is unnecessary (3.28), (iii) EIA is too stringent (2.61), (iv) EIA is too expensive (2.60), (v) EIA increases costs (2.16), and (vi) EIA delays projects (2.14).

4. Discussion

Results show that EIA reports comply with less than 65% of EIA Guideline requirements and that the contents of

EIA reports vary widely. It implies that on the average the decision-makers have less than 65% of required information to make their decision. Considering that the EIA system is based on the precautionary principle where uncertainty is high, there is need to have as much information as possible to make the most accurate choices. EMA reported that 31.7% of EIA reports do not meet standard (EMA 2011:8).

The difference in compliance between sections of the EIA reports was also assessed. Some of the reports lack critical information necessary for decision-making, successful implementation of mitigation measures and future evaluation of the impact of the project on the environment. The most lacking sections of the EIA are also the most crucial. These are impact analysis, EMP, participation, alternative analysis, EMP and pre-design. Pre-design refers to those voluntary environmental aspects/measures that the proponents build into the project before an EIA study is done.

Without impact analysis it is not possible to accurately predict the impacts and without an elaborate and costed EMP the implementation of mitigation measures is likely to fail. Without adequate participation of stakeholders, the EIA decision misses some relevant information and even impinges on stakeholders' rights. Without adequate alternative analysis the EIA report lacks information on the most feasible and environmentally suitable options. Lack of pre-design shows a disregard of environmental issues from the beginning of the project.

The wide variance in EIA reports' content also suggests that the methods being used to conduct EIA studies are widely different. Interview results shows that it is not standard practice for consultants to refer to the EIA Guidelines when conducting EIA studies or producing the EIA reports. Instead, consultants use previously approved EIA reports as template on which to base future EIA reports.

The results also show that the consultants do not produce EIA reports of comparable quality (standard deviation = 12.49%; Range = 56.08%; Mean 59%). A number of observations raised by interview respondents can explain this discrepancy. Firstly, this maybe a result of the differences in competency levels between the consultants registered by EMA to provide EIA consultancy services. Some registered consultants simply do not have the competency. This reflects on the appropriateness of the vetting system used to screen consultants before they are registered as well as the capacity building that consultants are exposed to before they can be deemed competent enough to conduct EIAs. The impact analysis, alternative analysis and mitigation are especially technical parts of the EIA reports which demand higher competency levels.

Secondly, it has also been observed that EIA studies are sometimes conducted by non-registered consultants even though the EIA reports are submitted under the names of registered companies. Therefore the consultants who are certified to be competent are not necessarily using their competency to produce the EIA reports.

Thirdly, this is also due to the differing levels of effort that the consultants exert depending on how much they are paid for the job. It has also been observed that some consultants charge fees which are too low for a thorough EIA study, hence the quality of the work reflects how much they are being paid. In some instances, in order to reduce costs, consultants constitute smaller teams than required to conduct a proper EIA.

Lastly, the consultants face difficulties in obtaining necessary information for baseline studies resulting in such information being excluded from the report.

The finding that compliance of EIA reports has improved moderately since 2007 is positive for the EIA system and implies that both EMA and consultants have been addressing some of the compliance gaps in the production of EIA reports.

It is somewhat surprising that there is no difference between compliance of economic sectors in the EIA reports. Respondents to questionnaires and interviews tended to think that some economic sectors are more compliant than others. However, given the fact that EIA study guidelines are almost generic across all projects regardless of sector, it is not surprising that there is no difference between EIA report compliance between sectors. If guidelines demanded more stringent sector specific requirements in carrying out EIA studies, the results would be different.

The findings also show that large projects have a higher level of compliance than the smaller projects. Both questionnaire and interview respondents make the same observations. The reason is that large projects often have adequate budgets to hire internal staff responsible for environmental issues. Large projects are also more conspicuous therefore their environmental impacts will attract more attention. Therefore, it is in the interest of a large project to be compliant with all environmental requirements.

This study included assessing if respondents felt that the different stages of the EIA system are complied with differently i.e. the pre-acceptance versus the post-acceptance stages. Compliance of the pre-acceptance stage was rated higher than the post-acceptance stage comprising of the EMP and the quarterly reporting. Therefore there is a general tendency to comply with the pre-acceptance stage better than the post acceptance stage of the EIA process. EMA "monitoring reveals that most proponents were not implementing the provisions of their own EMPs as specified in EIA reports" (EMA 2011:10). The EMA report further states that due to compliance failure "1475 tickets and 169 closure orders were issued of which 80% of offenders paid fines while 20% are pending court cases". This finding agrees with this EMA report and findings of other writers (Kahangirwe, 2011).

Proponents are primarily worried about the costs associated with delays in project implementation and complying with EIA regulations. The period from the time of contracting the consultant until an EIA certificate is granted can last anything from 3 months to more than 6 months of which stakeholder consultations and report review accounts for most of the time. Wood (2003:7) noted that in developing countries "there is probably a majority view" that EIA benefits are not commensurate with the time and costs. This study confirms that view but project delays are a major concern.

Corruption was decidedly the lowest scored of all the ratings in this study (1.94 out of 5) showing that most of the respondents agree that corruption is a serious issue affecting compliance in the EIA system.

4.4 Factors Affecting Compliance with EIA Guideline Requirements

A decision criteria template (EIA Report review criteria) was obtained from the regulatory authority and interviews were held with key informants to answer this question. The analysis showed that there is limited objectivity in reviewing EIA reports since the EIA Report review template has no objective empirical measure of compliance or adequacy of the EIA report content but is based in general subjective impressions of the reviewer. Whilst other government departments are expected to be involved in reviewing EIA reports which touch on their mandates, in practice this is not easy because they are focused on their own work. Hence the review often reflects the viewpoints of EMA alone.

Site visits are carried out by EMA personnel sometimes including other government departments as part of the review of the EIA report to inform EIA certificate decision. However, there is lack of adequate equipment to measure and monitor some environmental indicators such as ambient air quality. Such measurements are necessary to verify claims made in the EIA reports.

Interviewees noted that there is no clear relationship between the EIA Report review template and the EIA Guidelines. Hence the guidelines used to formulate report are, to some extent, different from the template against which compliance is assessed. The discretion of EMA staff performing the review is a major factor to the decision to grant acceptance or not.

Thematically, the major challenges to compliance that were identified through the interviews are:

- Inadequate enforcement of EIA requirements especially EIA report quality, implementation of mitigation measures and impact monitoring.

- Lack of specialisation in EIA, low entry barriers, unethical practices by consultants and inadequate consultant vetting methods resulting in EIA services being provided by incompetent consultants even though the competent ones are available.

- Lack of up-to-date EIA guidelines to facilitate EIA studies that are scientific which can provide adequate decision-making information.

- Lack of baseline information for EIA studies due to inaccessibility of existing information within the regulatory authority's possession as well as inadequate time and equipment for measuring baseline parameters.

- Delays caused by EIA reports which do not meet the regulator's standards necessitating submission and re-submission(s) of the same EIA reports for review before the quality of the EIA report meets approval requirements.

- Lack of awareness of EIA requirements as well as general low priority given to environmental issues by proponents.

- An EIA report review checklist which depend on reviewer's subjective impression and is not properly synchronized with the EIA report requirements in use by consultants resulting in confusion of exactly what an EIA report should carry, hence continually changing but uncommunicated EIA report requirements.

- Various forms of corrupt practices within the EIA system especially between and within the regulator, consultants and proponents.

5. Conclusion

In conclusion, the findings show that although compliance in the EIA system in Zimbabwe improved slightly between 2007 and 2012, it was still inadequate in EIA reports and is below average expectations of EIA practitioners. The EIA Guidelines are not adequately complied with. The EIA system is therefore not effectively achieving its intended objective of incorporating environmental concerns into project planning. The implementation of mitigation measures stipulated in the EMPs is least complied with.

The implication is that the EIA system is not effectively achieving its intended objectives since strict compliance is necessary to achieve the protection of the environment.

6. Recommendations

Many recommendations can be suggested based on the findings. However, some key recommendations are as follows.

Since only 65% of the guidelines are complied with, there is need for emphasis on the EIA guidelines for carrying out EIA studies so that the EIA report quality can be improved and standardised. In addition, since many adjustments have been made by EMA to the process of carrying of EIA studies, it is prudent to produce an update version of the guidelines which reflect current EMA requirements.

The EIA review checklist/template used by EMA requires objectivity to be built into it. There is need to work out a template which contains empirical and objective measures of EIA report content. In addition, the EIA review checklist must be synchronised with the EIA guidelines to ensure that EIA reports are produced according to the format by which they will also be reviewed against. This will provide a common traceable standard of EIA report quality for both EMA and consultant. Literature contains various EIA Report review templates and models from which the EIA system can build its own.

Consultants play a key role in enhancing compliance in the EIA system. Since most consultants are professionals in their own fields without specific EIA skills and they are not fulltime on EIA studies, there is need to ensure that consulting companies have within their teams the capacity to meet EIA competency requirements. Some respondents recommended that, in addition to providing more regular training, EMA should also give performance feedback to consulting companies as well as promote specific specimen of EIA report best practices within EMA's library for reference.

Related to the foregoing point is the observation that the EIA system has gathered much information since 1997 which is stored but unavailable to experts who need it for use in EIA studies. The main hurdle is the Environmental Management Act (CAP 20.27) Section 108 which seems to forbid personal use of EIA reports outside "civil proceedings ... in a matter relating to the protection and management of the environment". This situation is uniquely Zimbabwean since in many countries EIA reports are publicly available without restriction and many can be downloaded online. This situation casts a cover over transparency in the EIA system.

Awareness building among proponents is an ongoing process which EMA has pursued with some success. However, because proponents usually find out the need for an EIA study only when it has become a hindrance to project implementation, there is need to communicate the need for EIA within the government structures such that from any government office a proponent is informed on the need for EIA compliance at an early stage. Early knowledge of the EIA process allows the proponents to prepare for it so that the EIA process is not treated as an inconvenience to be rushed through for compliance sake. Given that most EIA reports relate to mining the Ministry of Mining and Mining Development could be more actively involved in EIA awareness-raising to proponents. In addition, the EIA process needs to be quickened so that it is not viewed as a bottleneck to development. This can be achieved by setting specific duration for each of the steps of the EIA process until certification.

Apart from awareness raising and enforcement, EMA needs to try other measures of promoting compliance such as use of economic incentives and disincentives. These most likely require fewer resources to implement than awareness raising and especially enforcement.

There is need to raise the entry barriers to EIA practice in order to eliminate unqualified and inexperienced consultants. This can be done in many ways including stricter and more stringent vetting of consultants before registration, enforcing stricter standards of EIA report quality and limiting the number of EIA consulting companies to ensure that consultants are full-time, specialise and hence more committed to professional EIA practice.

EMA needs strong capacity building to ensure that compliance monitoring and enforcement are more effectively practiced. Capacity building should incorporate increasing the number of staff, engaging and retaining well qualified staff, provision of equipment for measuring environmental indicator parameters, transport and managerial competency.

A full-scale national evaluation of the EIA system is long overdue since this is now 20 years since the EIA policy was formulated. There has not been a deliberate forum of multi-stakeholder contribution to the improvement of the EIA system.

References

Ahmad, B., & Wood, C. M. (2002). Environmental impact assessment in Egypt, Turkey and Tunisia. *Environmental Impact Assessment Review, 22*, 213-234.

Badr, E. A. (2009, September). Evaluation of the environmental impact assessment system in Egypt. *Impact Assessment and Project Appraisal, 27*, 193-203.

Betey, C. B., & Godfred, E. (2013). Environmental Impact Assessment and Sustainable Development in Africa : A Critical Review. *Environment and Natural Resources, 3*(2), 37–51.

Cashmore, M., Gwilliam, R., Morgan, R., Cobb, D., & Bond, A. (2004). Effectiveness of EIA - The interminable issue of effectiveness: impact assessment theory. *Impact Assessment and Project Appraisal, 22*(4), 295-310.

El-Fadel, M., & El-Fadel, K. (2004). Comparative Assessment of EIA Systems in MENA countries: Challenges and prospects. *Environmental Impact Assessment Review, 24*, 553-593.

Environmental Management Agency. (2010). *Annual Report 2009*. Harare: Ministry of Environment and Tourism.

Environmental Management Agency. (2011). *Annual Report 2010*. Harare: Ministry of Environment and Tourism.

International Cooperation Bureau. (2009). *ODA Evaluation Guidelines*. Tokyo: ODA Evaluation Division, Ministry of Foreign Affairs of Japan.

Kahangirwe, P. (2011, March). Evaluation of Environmental Impact Assessment (EIA) Practice in Western Uganda. *Impact Assessment and Project Appraisal, 29*, 79-83.

Khadka, R. B., & Khanal, A. B. (2008). Environmental management plan (EMP) for Melamchi water supply project, Nepal. *Environmental Monitoring and Assessment, 146*(1-3), 225–34. http://doi.org/10.1007/s10661-007-0074-8

Lee, N., Colley, R., Bonde J., & Simpson J. (1999). Reviewing the Quality of Environmental Statements and Environmental Appraisals. *Occasional Paper number 55*. Manchester; Department of Planning and Landscape. University of Manchester.

Manyindo, J. (2002). *Monitoring Compliance with EIA Recommendations in Uganda: Opportunities for Progress*. Uganda Wildlife Society, Kampala, Uganda

Mhango, S. D. (2005). The quality of environmental impact assessment in Malawi: a retrospective analysis. *Development Southern Africa, 22*(3), 383–408. http://doi.org/10.1080/14797580500252837

Ministry of Environment and Tourism. (1997). *Environmental Impact Assessment Guidelines*. Harare: Government of Zimbabwe.

Ministry of Environment and Tourism. (1997). *Environmental Impact Assessment Policy*. Harare: Government of Zimbabwe.

Ministry of Environment and Tourism. (2002). *Environmental Management Act (Chapter 20.27)*. Harare: Government of Zimbabwe.

Nadeem, O., & Hameed, R. (2006). A Critical Review of the Adequacy of EIA. *International Journal of Human and Social Sciences, 1*(1), 54–61.

Pope, J., Bond, A., Morrison-saunders, A., & Retief, F. (2013). Advancing the theory and practice of impact assessment : Setting the research agenda. *Environmental Impact Assessment Review, 41*, 1–9. http://doi.org/10.1016/j.eiar.2013.01.008

Ruffeis, D., & Loiskandl, W. (2010, March). Evaluation of the environmental policy and impact assessment system in Ethiopia. *Environmental Protection, 28*, 29-40.

Talime, L. A. (2011). *A Critical Review of the Quality of Environmental Impact Assessment Reports in Lesotho*.

(Unpublished master's thesis) University of Free State, South Africa.

Wood, C. (1999). Comparative Evaluation of Environmental Impact Assessment Systems. In J. Petts (Ed.), *Handbook of Environmental Impact Assessment* (Vol. 2, pp. 10-34). Oxford: Blackwell.

Wood, C. (2003). *Environmental Impact Assessment in Developing Countries: An Overview. Conference on New Directions in Impact Assessment for Development- Methods and Practice.* EIA Centre, School of Planning and Landscape, University of Manchester, UK.

11

Application of Risk Perception Theory to Develop a Measurement Framework for City Resilience: Case Study of Suita, Japan

Maiko Ebisudani[1] & Akihiro Tokai[1]

[1] Laboratory of Environmental Management, Division of Sustainable Energy and Environmental Engineering, Graduate School of Engineering, Osaka University, Osaka, Japan

Correspondence: Maiko Ebisudani, Laboratory of Environmental Management, Division of Sustainable Energy and Environmental Engineering, Graduate School of Engineering, Osaka University, 2-1 Yamada-oka, Suita City, Osaka Prefecture 565-0871, Japan, E-mail: ebisudani@em.see.eng.osaka-u.ac.jp

Abstract

Risk management has developed as an important aspect of sustainability. In order to manage risk more effectively, an overall evaluation of regional resilience needs to be performed. Therefore, this paper develops a framework to measure overall resilience in a community, focusing on risk perceptions of citizens of Suita City, Japan. The framework includes three main phases: (1) identifying multiple risks in the city through discussions with local experts and city workers; (2) prioritizing those risks by applying principal component analysis (PCA); and (3) understanding the relationships among them using decision-making trial and evaluation laboratory (DEMATEL) analysis. As a result, 21 risks were identified, and subsequently, four risks were prioritized: climate change, lack of self-sufficient energy, damage to the ecosystem, and natural disasters. Lastly, the application of DEMATEL analysis revealed that climate change and natural disasters have the greatest cause-effect relationships among the risks. The framework proves that multiple risks can be prioritized and gives overall suggestions on what kinds of risk a community is facing; where to start considering how to manage resilience; and which functions/services a community should improve to boost resilience. The identification, prioritization, and visualization of significant risk relationships completed in this study can support decision-making processes in strengthening community resilience.

Keywords: sustainability, risk perception, resilience, risk management

1. Introduction

Risk management has developed as an important aspect of sustainable development and sustainable resource management. In Japan, especially after the Great East Japan Earthquake, "resilience" has become a keyword in understanding how to manage risks and achieve a sustainable society (Baba, Masuhara, Tanaka, & Shirai, 2013). Several studies have measured resilience in cities, and it has become a key concept for operationalizing sustainability (Pickett, McGrath, Cadenasso, & Felson, 2014). For instance, Chen, Ferng, Y. Wang, Wu and J. Wang (2008) evaluated the resilience capacity of hillslope communities by assessing damage caused by two specific typhoons; Joerin, Shaw, Takeuchi and Krishnamurthy (2012) assessed the resilience of communities facing a higher probability of future disasters due to climate change; and Prashar, Shaw and Takeuchi (2012) assessed the resilience of urban areas to climate-related disasters. Most studies measuring community and urban resilience focused on disaster risks. However, these studies were limited to a specific risk each and failed to provide an overall consideration of resilience in the areas concerned. The difficulty lies in including every risk and addressing their relationships with limited resources and more diversified and complicated risks. To establish resilience in a community, a holistic assessment approach needs to be developed further.

With regard to the definition of resilience, it is generally stated as a system's ability to respond and recover from disturbances (Fisher et al., 2010). Holling (1996) originally defined resilience in two ways, namely, engineering and ecological resilience. While engineering resilience is the more traditional of the two and focuses on recovery and constancy, ecological resilience focuses on system persistence and robustness to disturbance. Essentially, engineering resilience aims to maintain the *efficiency* of a function, whereas ecological resilience focuses on maintaining the *existence* of a function. These two contrasting approaches are fundamental paradigms of resilience (Holling, 1996). As the world faces more and more uncertain and interacting risks, it is desirable to pursue the dynamic challenge of integrating both paradigms to strengthen communities' capacity to manage resilience. Therefore, this study redefines resilience as a community's ability to cope with multiple risks that it might face.

To measure the resilience of a community, this paper aims to develop a framework by demonstrating the application of the theory of risk perception. It utilizes risk characteristics studied by Slovic (1987) to organize multiple risks. The study is customized for Suita City in the Osaka Prefecture of Japan. Thus, the specific objectives are to (1) identify multiple risks to Suita City by reviewing literature and holding discussions with local experts and city workers; (2) prioritize risks using principal component analysis (PCA) in conjunction with the experts' perceptions [these two steps are based on Slovic's theory of risk perception (1987)]; and (3) capture the causal relationships among risks by applying the decision-making trial and evaluation laboratory (DEMATEL) analysis technique. With the development of this framework, decision-makers will be able to understand the procedure for building community resilience in order to improve existing regulations and strategies.

The paper is structured as follows: Section 2 presents an overview of risk perception theory. The methodology, assessment framework, and its application are described in Section 3. Lastly, a discussion of the results from PCA and DEMATEL is presented in Sections 4 and 5.

2. Risk Classification by the Theory of Risk Perception

Previous studies have attempted to classify new technologies and human activities that may generate new and unprecedented risks. Modern risk analysis employs a trio of risk characteristics to evaluate hazards: threat, vulnerability, and consequence (Linkov et al., 2014). The aim of risk management is to reduce the highest risk events by addressing one or all of the risk variables. In recent decades, the mechanisms underlying catastrophic damages have become more complex, along with developments that sometimes created new and unprecedented risks. Slovic (1987) argued that the most harmful consequences are rare and often delayed; therefore, they can be difficult to assess by statistical analysis. As such, he proposed an alternative method for risk assessment analysis, namely the measurement of "risk perceptions." This refers to the instinctive risk judgement that the majority of citizens depend upon. Since Slovic developed this idea, researchers have been attempting to invent techniques for assessing the opinions that people hold about risk.

The original research on risk perceptions was conducted by Starr (1969), who developed a method to weight technological risks against economic benefit. It explained systematic differences in the acceptability of risk and revealed patterns of risk acceptance. Starr's results suggested a classification of risks by applying the dichotomous trait of "voluntary" versus "involuntary" as a risk characteristic (Winterfeldt & Edwards, 1984). The study approach and conclusion have been developed by a number of researchers and have subsequently yielded nine risk characteristics: volition, severity, origin, effect manifestation, exposure pattern, controllability, familiarity, benefit, and necessity. These risk judgments were then evaluated statistically and used to rate various risks according to their risk characteristics. A factor analysis of the ratings and risks could largely be explained by two represented factors.

Slovic extended Starr's result to a broader set of risk characteristics. The extended study by Slovic, Fischoff and Lichtenstein (1980) designed 90 hazards (instead of 30) to cover a wide range of activities, substances, and technologies, and 18 risk characteristics. All risk characteristics were rated on a bipolar 1–7 scale, representing the range for which the characteristics described the hazard (Figure 1).

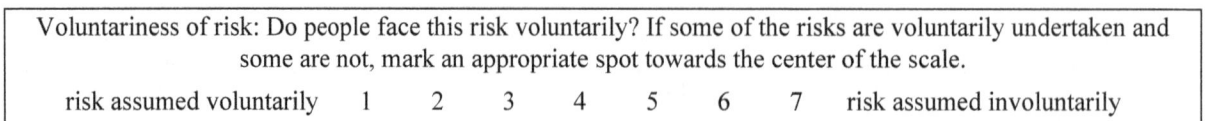

Voluntariness of risk: Do people face this risk voluntarily? If some of the risks are voluntarily undertaken and some are not, mark an appropriate spot towards the center of the scale.
risk assumed voluntarily   1   2   3   4   5   6   7   risk assumed involuntarily

Figure 1. Example of bipolar scale questionnaire

People's judgement ratings were evaluated by a statistical technique known as factor analysis, which indicated two underlying risk factors. As a result, based upon the nature of the characteristics, they labelled the first factor as "Dread" and the second factor as "Familiarity." Within the space of these factors, the designated hazards were plotted on a chart that indicated taxonomic significance. Slovic's study succeeded in applying psychophysical scaling and multivariate analysis to create quantitative representations of risk perceptions. Slovic et al. (1980) concluded that perceived risk was quantifiable and predictable and therefore suggested that the greater the perceived risk, the greater the desired reduction. Furthermore, they stated many of the 18 risk characteristics were presumed by the public to be essential and correlated greatly with perceived risk and preference for risk reduction.

3. Methodology

The fundamental framework approach draws upon the theory of risk perception discussed earlier. Referring to Slovic's original study, this study is composed of three phases.

3.1 Identification of Multiple Risks

This study started by selecting potential risks within a community. A preliminary task entailed reviewing relevant literature to gather designated risks on a global level. Major articles referred to were the following: World Economic Forum (2014), UK Cabinet Office (2010), and Holzmann and Jørgensen (2000). These reports provided information on the environmental, technological, and societal aspects of global risks; World Economic Forum included a set of 14 major risks; the UK Cabinet Office presented 12 major risks; and Holzmann and Jørgensen addressed three significant risks. Referring to these as an initial list of existing risks, several workshops were held, involving local experts and city workers to discuss potential risks. Through this series of workshop discussions, multiple potential risks to Suita City were finally identified.

3.2 Prioritization of Risk Characteristics

This step followed the identification of multiple risks by principal components analysis (PCA). PCA is a statistical technique used to analyze data sets by several inter-correlated quantitative dependent variables. In order to identify patterns of perception and highlight similarities and differentiations, PCA has been deemed an effective tool for analyzing data (Smith, 2002).

In order to prioritize the multiple risks, a questionnaire survey was provided to academic individuals with an understanding of risk management, in order to reveal patterns of risk perception. The survey was conducted with five Master's students, one Lecturer, and two Professors in the Department of Environmental Engineering at Osaka University, Japan.

Both implementation and procedure followed Slovic et al. (1980), and 14 risk characteristics were applied (Table 1). Questionnaire respondents were asked to rate each of the characteristics, and each item was scaled from -2 to 2, with 0 as neutral. An example risk characteristic, "controllability," is shown in Figure 2 below.

Do you think this risk is controllable? (Please mark a score as appropriate).						
Controllable	-2	-1	0	1	2	Not controllable

Figure 2. Questionnaire for "controllability" in a risk characteristic

Average scores were calculated and input to a statistics processing tool, in this case, IBM's SPSS Statistics, Version 22.

Table 1. Risk characteristics rated by academic individuals

Controllability	Reduction ability
Delay effect	Voluntariness
Catastrophic	Observability
Critical	Notice to exposure
Even Exposure	Acute effect
Personal exposure	Familiarity
Future generation	Scientifically unknown

3.3 Visualization of Cause-Effect Relationships

Another questionnaire was distributed to the same eight experts, this time to capture causal relationships among risks. Respondents were asked to evaluate an item's impact among the others using an integer scale from 0 to 8.

(0: no direct influence; 2: moderate direct influence; 4: strong direct influence; 8: very strong direct influence). The data were processed by using DEMATEL analysis.

The DEMATEL analysis offers a method to visualize problems by isolating related variables into cause and effect groups, which can provide greater understanding of any causal relationships among the examined variables (Wang, H. Lin, L. Lin, Chung, & Lee, 2012). Utilizing the survey scores, the first step is to find the average matrix. The assigned scores yield an $n \times n$ answer matrix X^*. The $n \times n$ average matrix A was computed by averaging the h experts' score matrices. The (i, j) elements of the average matrix A is denoted as a_{ij} as follows:

$$a_{ij} = \frac{1}{h} \sum_{k=1}^{h} X_{ij}^{k} \tag{1}$$

The second step calculates the direct influence matrix D, which is obtained from normalizing the average matrix A, where s is a constant:

$$D = sA \tag{2}$$

The third step calculates the indirect influence matrix. The indirect influence of division i on division j declines as the power of the matrix increases. The indirect influence matrix ID is obtained from the values in the direct influence matrix D, where I is the identity matrix:

$$ID = D^2 (I - D)^{-1} \tag{3}$$

The fourth step derives the total influence matrix T:

$$T = D(I - D)^{-1} \tag{4}$$

The (i, j) elements of matrix T is t_{ij}: the sum of the ith row and the sum of the jth column:

$$d_i = \sum_{i=1}^{n} t_{ij} \tag{5}$$

$$r_j = \sum_{j=1}^{n} t_{ij} \tag{6}$$

The last step is to obtain the cause-effect impact-relationships map (IRM). The map can be developed from the values of $d + r$ and $d - r$, represented on the x-axis and y-axis. These values demonstrate the degree of effectiveness among risks. The horizontal axis vector $(d + r)$ was labelled as "Prominence" and the vertical axis $(d - r)$ as "Relation." Prominence indicates the importance of each factor. Relation divides factors into two groups: a cause group and effect group. Using the dataset of the $(d + r, d - r)$, the causal diagram demonstrates the IRM in graph form.

4. Results and Discussion

This section discusses the findings from the application of the developed framework.

4.1 Selecting Multiple Risks

From the literature review, an initial list of 29 risks was developed: 14 risks from World Economic Forum, 12 risks from the UK Cabinet Office, and 3 risks from Holzmann and Jørgensen, as follows:

(1)   World Economic Forum (2014)

In terms of environmental global risks, the following were listed: extreme weather events, natural catastrophes, man-made environmental catastrophes, biodiversity loss and ecosystem collapse, water crises, and climate change mitigation and adaptation. Secondly, the paper listed societal risks, including food crises, pandemic outbreak, chronic disease, severe income disparity, antibiotic-resistant bacteria, mismanaged urbanization, and profound political and social instability. Finally, technological risks were mentioned: the breakdown of critical information infrastructure and networks, cyber-attacks, and data fraud/theft.

(2)   The UK Cabinet Office (2010)

The type of risks summarized by the National Risk Register are catastrophes, human disease, flooding, severe weather, animal disease, major industrial accidents, major transport accidents, attacks on crowded places, attacks on infrastructure, attacks on transport systems, non-conventional attacks, and cyber security.

(3) Holzmann and Jørgensen (2000)

The report discussed social risks that arise from biased social protection and the negative impact of economic development and growth. In view of social risk management, an aging population, rising international competition, and income insecurity were particularly of concern.

These key risks were included in a draft version of the workshop discussion itinerary. Several workshops were held with local experts and city workers to particularize the risk to Suita City. As a result, 21 multiple risks were selected, which were presented in Table 2.

Table 2. The 21 multiple risks identified by workshops

1.	Climate change	12.	Social strain
2.	Lack of self-sufficient energy	13.	Population density
3.	Damage to ecosystem	14.	Obstacles to human security
4.	Natural disaster	15.	Lack of preparation by corporations
5.	Development of city infrastructure	16.	Economic crisis
6.	Daily life inconvenience	17.	Intentional harmful activities by individuals or groups
7.	Lifestyle changes	18.	Dependence on a single energy source
8.	Amount of pollution	19.	Energy supply instability
9.	Change of environmental quality	20.	Over-investing in infrastructure development
10.	Availability of natural resources	21.	Disruption to essential utilities
11.	Change in availability of natural benefit		

Comparing this with the study by Slovic et al. (1980), the number of risks was lower but more focused on a larger scale. In this study, global-scale risks, such as climate change, ecosystem damage, population density, and economic crisis were on the list, while in 1980, risks were more localized and on an individual basis, such as nuclear power, DDT, herbicides and pesticides, food coloring, and radiation therapy. Slovic et al. (1980) explained their result as reflecting the news media's presentation of these issues at the time. Natural disasters were not mentioned as a risk in the former study. This can be explained by the lack of first-hand exposure to natural disasters at that time as compared to the present day group; disaster and natural catastrophes are more likely to be recognized as risks when more people experience them. The same is true of energy issues, which are now regularly faced. Additionally, in the present study, the onset of urbanization, infrastructure, pollution, and quality of life/lifestyle were identified as risks.

The multiple risks identified cover a wide range of fields and reflect a range of perspectives; however, further discussion is still required for further specification. For instance, with regard to natural disasters, the issues of damage to ecosystems and environmental quality should be more specific based on actual disaster experiences and regional geographic information. For Suita City, earthquakes, damage from torrential rainfall, wind, and floods can be listed as particularly relevant natural disaster risks. Similarly, some of the risks might overlap each other, and further clarification is needed for this. Future research will face the challenge of reflecting regional geological information and clarifying the description of each risk in the list.

4.2 Prioritizing Multiple Risks

The questionnaire survey was provided to the experts, and all of the selected 21 risks were rated in terms of 14 risk characteristics. The average scores were calculated and transferred to the statistical analysis software, SPSS. Table 3 shows the analysis results. Application of the PCA statistical technique showed four primary characteristics in the first factor: acute effect (risk decreasing or increasing), delay effect (effect immediate or effect delayed), notice to exposure (known to those exposed or unknown to those exposed) and familiarity (old risk or new risk). By referring to the theory of risk perception, these four factors have been labeled as "Unknown." The second factor primarily reflects six characteristics: reduction ability (easily or not easily reduced), catastrophic (not global catastrophic or global catastrophic), personal exposure (individual or catastrophic), future generation (low risk or high risk to future generations), controllability (controllable or uncontrollable) and criticality (consequences not fatal or fatal). The nature of these characteristics suggests that this factor can be called Dread.

Table 3. Result from principal components analysis (PCA)

Variable	Factor	
	1	2
Acute effect	0.907	0.282
Delay effect	0.883	0.316
Notice to exposure	0.840	0.012
Familiarity	0.763	-0.238
Future generation	0.567	0.741
Observability	0.352	0.007
Catastrophic	0.285	0.817
Personal exposure	0.261	0.773
Even Exposure	-0.240	-0.603
Reduction ability	-0.257	0.854
Scientifically unknown	-0.361	0.201
Critical	-0.507	0.653
Controllability	-0.598	0.663
Voluntariness	-0.765	0.451

Each of the 21 risks has a mean score for each of the 14 characteristics, which also has a score on each factor. These scores give the location of each risk within the factor space, and Figure 3 shows such plots for Factors 1 and 2. The high end of risks on the horizontal dimension (Factor 1) were identified to be unfamiliar, not to be noticeable, or for people to take time to realize their exposure. The risks in this dimension, such as the lack of self-sufficient energy, indicate the necessity to educate people to increase their awareness. Items at the high end of Factor 2 are highly dreaded: climate change and damage to ecosystem. Their characteristics also have unknown risks. Natural disasters are also highly dreaded but are, on the other hand, well known. At the negative end of Factor 2, items are posing risks to individuals and are limited in scale in terms of impact on daily life, corporations, and infrastructure.

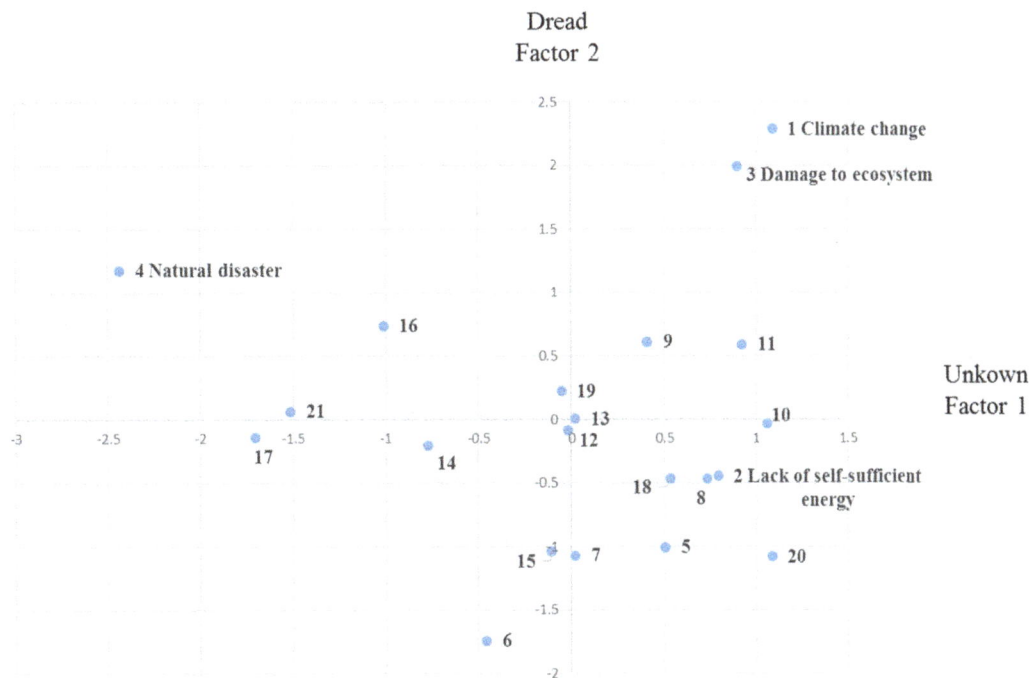

Figure 3. Location of 21 risks on Factor 1 and 2

The results suggest that judgements of multiple risks are related more to characteristics such as Unknown and Dread. These two factors should be considered when thinking about risk management. This intuitive judgement indicates the degree of desire to reduce a risk. The Unknown factor is related to knowledge about a risk, as risks that scored highly on Unknown can be seen as those on which further education is required in order to reduce the risk. The Dread factor was considered as catastrophic, not easily avoided, and posing high risk to future generations. People have stronger desires to reduce occurrence of risks scoring higher in this factor; as such, stronger, more specific regulations related to these may need to be implemented.

Hence, the risk at the higher end of both factors can be identified as most critical, which in turn implies the preference of appropriate countermeasures as part of a community's strategy. As Slovic (1987) suggested, psychometric analysis of questionnaire data in which preferences and perceived risks are identified are useful in quantifying and predicting risk perception. Application of psychometric techniques tends to be well-suited to identifying similarities and differences among groups in terms of risk perceptions. This study identifies four preferential risks among the examined multiple risks: climate change, damage to ecosystems, natural disasters, and lack of self-sufficient energy are thought to be the most significant in the utilized two-dimensional structure.

This study targeted a group of experts as academic individuals and the result came from the perceived risk as expressed by them. To further expand on these results, the same questionnaire could be implemented to undertake comparisons among different groups, which could include public workers, experts, and college students in different areas of the community.

4.3 Demonstrating Cause-Effect Relationships

To understand the relationship among the examined risks, DEMATEL methodology was applied. The values of $d + r$ and $d - r$ were calculated to produce an IRM (Figure 4). This map visualizes the difference between cause groups and effect groups. Climate change and natural disasters are identified as causes, while damage to ecosystems and lack of self-sufficient energy are effects.

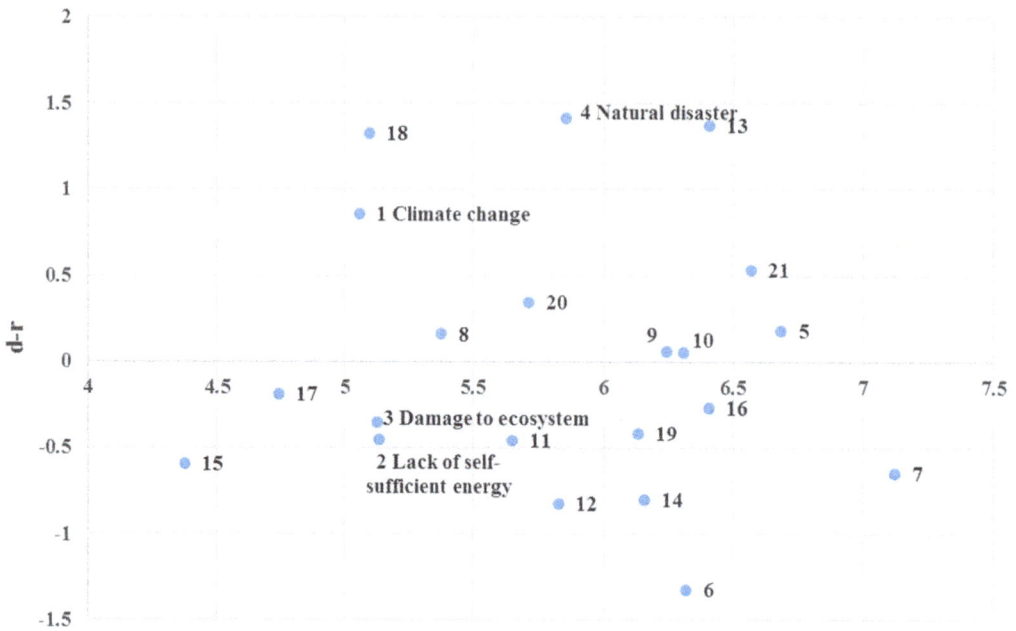

Figure 4. A cause-effect impact-relationships map (IRM)

Furthermore, the net influence matrix N can be used to identify the strength of impact on each item. Using each value calculated for the total influence matrix T (Equation 4), the net influence matrix N is given by the following:

$$N = Net_{ij} = t_{ij} - t_{ji} \qquad (7)$$

For example, Table 4 is the matrix N, which presents the degree of relationship between (1) climate change, (2) lack of self-sufficient energy, (3) damage to ecosystem, and (4) natural disaster. "Climate change" influences all the others, and "natural disaster" highly affects self-sufficient energy (=0.103). Through understanding these relationships, it can be seen that effective countermeasures to climate change and natural disasters will also have a positive impact on reducing other risks.

Table 4. Net influence matrix N of the risk (No. 1–4)

Divisions	1	2	3	4
1 Climate change				
2 Lack of self-sufficient energy	-0.0444			
3 Damage to ecosystem	-0.0666	0.0030		
4 Natural disaster	-0.0406	0.1030	0.0468	

5. Conclusion

By understanding resilience as the ability to cope with comprehensive risks, this study proposed a framework applying Slovic's risk perception theory integrated with the PCA and DEMATEL statistical techniques. The study succeeded in organizing multiple risks for resilience management. As a framework of each assessment, the study was composed of three phases, with the outcome of each summarized as follows:

(1) To identify multiple risks, an initial list of potential risks was provided by reviewing global reports and discussing the topic with local experts and city workers to particularize the risks to Suita City. As a result, 21

multiple risks were identified for Suita City.

(2) In order to prioritize multiple risks, the application of risk perception theory is recognized as a practical approach to identifying and revealing the characteristics of multiple risks facing a community. In the case of Suita City, the 21 multiple risks were characterized by the nature of risk perception: Dread and Unknown. This was demonstrated by applying the PCA statistical technique. The greater the degree of Dread and Unknown, the greater the desire to reduce that risk. Among the 21 multiple risks, four particular risks—climate change, damage to ecosystem, natural disaster and lack of self-sufficient energy—were prioritized as having the greatest degree of Dread and Unknown in Suita City. This information proves effective for helping policy makers redefine local policies.

(3) To understand the internal relationships among the 21 multiple risks, DEMATEL analysis was used to produce a map that visualizes the relevant cause-effect relationships. The result indicates that climate change influences all the others and that natural disasters highly impact the risk of lacking self-sufficient energy. It elucidates the structure of multiple risks, which is useful information for policy makers to draw upon when activating existing management with the aim of increasing community resilience.

The framework proves that multiple risks can be prioritized and gives overall suggestions to decision-makers: what kinds of risk a community is facing; where to start considering how to manage resilience; and which functions/services a community should improve to boost resilience.

For further development of these findings, a follow-up study must engage in further discussion about multiple risks, the group and number of research participants, and include regional characteristics (i.e., natural resources). Through compiling and utilizing a series of datasets, more holistic, accurate information for policy makers will be obtained, and the process will become a suitable tool to integrate into existing decision-making processes.

Acknowledgments

We would like to express our gratitude to the contributors and reviewers of this paper. We also appreciate each participating expert and student, who voluntarily provided the required information in the survey process. The study was completed with the support of the Environment Research and Technology Development Fund, Ministry of the Environment, Government of Japan (1-1304).

References

Ayyub, B. M. (2014). Systems resilience for multihazard environments: Definition, metrics, and valuation for decision making. *Risk Analysis, 34*(2), 340–355. http://dx.doi.org/10.1111/risa.12093

Baba, K., Masuhara, N., Tanaka, M., & Shirai, N. (2013). *A condieration for eshitablishing a concept of "environmental resilience" and providing relevant policy.* Paper presented at the 41th conference on Japan Society of Civil Engineers, Japan. (In Japanese).

Chen, S., Ferng, J., Wang, Y., Wu, T., & Wang, J. (2008). Assessment of disaster resilience capacity of hillslope communities with high risk for geological hazards. *Engineering Geology, 98,* 86-101. http://dx.doi.org/10.1016/j.enggeo.2008.01.008

Fisher, R. E., Bassett, G. W., Buehring, M. J., Collins, M. J., Dickinson, D. C., Eaton, L. K., ... Peerenboom, J. P. (2010). *Constructing a resilience index for the enhanced critical infrastructure protection program. Argonne National Laboratory, 1.* Retrieved from http://www.ipd.anl.gov/anlpubs/2010/09/67823.pdf

Holling, C. S. (1996). Engineering resilience versus ecological resilience. In P. Schulze (Ed.), *Engineering within ecological constraints* (pp. 31-43). National Academy of Engineering. http://dx.doi.org/10.17226/4919

Holzmann, R., & Jørgensen, S. (2000). Social risk management: A new conceptual framework for social protection and beyond. *Social Protection Discussion Papers, The World Bank,* (0006), 1–30. http://doi.org/10.1023/A:1011247814590

Joerin, J., Shaw, R., Takeuchi, Y., & Krishnamurthy, R. (2012). Action-oriented resilience assessment of communities in Chennai, India. *Environmental Hazards, 11,* 226-241. http://dx.doi.org/10.1080/17477891.2012.689248

Linkov, I., Fox-Lent, C., Keisler, J., Sala, S. Della, & Sieweke, J. (2014). Risk and resilience lessons from Venice. *Environment Systems and Decisions, 34*(3), 378–382. http://doi.org/10.1007/s10669-014-9511-8

Litai, D., Lanning, D. D., & Rasmussen, N. C. (1983). The public perception of risk. In V. T. Covello et al. (Eds.), *The analysis of actual versus perceived risks* (pp. 213-224). New York, NY: Plenum Press.

Lowrance, W. W. (1976). Of acceptable risk: Science and the determination of safety. *Journal of The Electrochemical Society*. http://doi.org/10.1149/1.2132690

National Academy of Sciences. (2012). *Disaster resilience: A national imperative*. The National Academies Press.

Okoli, C., Pawlowski, S. D. (2004). The Delphi method as a research tool: An example, design considerations and applications. *Information & Management*, *42*(1), 15–29. http://doi.org/10.1016/j.im.2003.11.002

Park, J., Seager, T. P., Rao, P. S. C., Convertino, M., & Linkov, I. (2013). Integrating risk and resilience approaches to catastrophe management in engineering systems. *Risk Analysis: An Official Publication of the Society for Risk Analysis*, *33*(3), 356–67. http://doi.org/10.1111/j.1539-6924.2012.01885.x

Patton, D., & Johnston, D. (2012). *Disaster Resilience*. http://doi.org/10.17226/13457

Pickett, S. T. a., McGrath, B., Cadenasso, M. L., & Felson, A. J. (2014). Ecological resilience and resilient cities. *Building Research & Information*, *42*(2), 143–157. http://doi.org/10.1080/09613218.2014.850600

Prashar, S., Shaw, R., & Takeuchi, Y. (2012). Assessing the resilience of Delhi to climate-related disasters: A comprehensive approach. *Natural Hazards*, *64*(2), 1609–1624. http://doi.org/10.1007/s11069-012-0320-4

Slovic, P. (1987). Perception of risk. *Science*, *236*, 280-285. Retrieved from http://heatherlench.com/wp-content/uploads/2008/07/slovic.pdf

Slovic, P., Fischhoff, B., & Lichtenstein, S. (1980). Facts and fears: Understanding perceived risk. In R. Schwing, & W. A. Albers, Jr (Eds.), *Social risk assessment: How safe is safe enough?* (pp. 181-216). New York, NY: Plenum Press.

Smith, L. I. (2002). A tutorial on Principal Components Analysis. Retrieved from http://www.cs.otago.ac.nz/cosc453/student_tutorials/principal_components.pdf

Starr, C. (1969). Social benefit versus technological risk: What is our society willing to pay for safety? *Science*, *165*(899), 1232–1238. http://doi.org/10.1126/science.165.3899.1232

UK Cabinet Office. (2010). *National risk register of civil emergencies*. Retrieved from https://www.gov.uk/government/uploads/system/uploads/attachment_data/file/211853/nationalriskregister-2010.pdf

United States Environmental Protection Agency. (2014). *Climate resilience evaluation and awareness tool 2.0 exercise with southern Nevada water authority*. Retrieved from http://nepis.epa.gov/Exe/ZyPDF.cgi/P100KM18.PDF?Dockey=P100KM18.PDF

Wang, W. C., Lin, Y. H., Lin, C. L., Chung, C. H., & Lee, M. T. (2012). DEMATEL-based model to improve the performance in a matrix organization. *Expert Systems with Applications*, *39*(5), 4978–4986. http://doi.org/10.1016/j.eswa.2011.10.016

Winterfeldt, D. v., & Edwards, W. (1984). Patterns of conflict about risky technologies. *Risk Analysis*, *4*, 55–68. http://doi.org/10.1111/j.1539-6924.1984.tb00131.x

World Economic Forum. (2014). *Global Risk 2014 Ninth Edition*. Retrieved from http://www3.weforum.org/docs/WEF_GlobalRisks_Report_2014.pdf

12

Determinants of Corporate Climate Change Mitigation Targets in Major United States Companies

Haoyu Yin[1], Fei Mo[2] & Derek Wang[3]

[1] China Energy Storage Alliance, Beijing, China

[2] Beijing, China

[3] Business School, China University of Political Science and Law, Beijing, China

Correspondence: Derek Wang, Business School, China University of Political Science and Law, Beijing, China.
E-mail: dwang@cupl.edu.cn

Abstract

Setting greenhouse gas emission target is a critical step to meet the challenge of climate change. While the debate on global and national carbon emission targets has dominated every major climate change conference, little is known about how the firms set emission targets. Using a dataset on S&P 500 companies in the United States, we investigate the determinants of firm-level climate change mitigation targets, including target adoption and target metric (intensity target vs. absolute target). We find that companies with larger size, higher growth, better innovation, weaker capital constraint, and higher government pressure are more likely to establish emission targets. Further, firm growth has a negative (positive) and significant association with the use of absolute (intensity) target. This may be due to the fact that intensity target can better accommodate growth than absolute target. Policymakers and corporate managers may resort to those determinant factors in designing climate change policies to induce desirable firm-level target-setting behaviors.

Keywords: climate change, corporate, carbon mitigation, target, target metric

1. Introduction

Climate change has been recognized as one of the greatest challenges for human society, as evidenced by the 2015 United Nations Climate Change Conference in Paris (COP21) and other preceding climate change summits. In order to keep the average global temperature increase to no more than 2°C above pre-industrial levels, we have to limit the emissions of greenhouse gas (GHG). A critical step of coping with GHG emissions is to set appropriate emission targets. Setting emission targets has always been the most important and contentious issue at every major conference on climate change. One of the earliest efforts to limit GHG emissions with binding targets is the 1997 Kyoto Protocol, which sets targets for industrialized countries with the aim to reduce GHG emissions by an average of 5.2% below the 1990 levels by 2012. At COP21, almost all countries present, developed and developing countries alike, have come up with pledges to reduce specific amounts of GHG emissions by specified deadlines. A great amount of theoretical and empirical studies has been conducted to investigate the problems associated with setting country-level and region-level GHG emission targets (Philibert & Pershing, 2001; Pizer, 2006; Lutsey & Sperling, 2008).

While target setting has received a lot of attention at the global, national and subnational levels, the study of setting emission targets at the firm level is much less common. At the same time, as pointed out by Krabbe et al. (2015), striking a global or national agreement is a very slow and difficult process, as shown by the failure to create legally binding targets for major emitting countries at the Copenhagen Summit in 2009. As a result, there is a growing recognition among policymakers and researchers that the private sector may play a more proactive and significant role in tackling climate change. At COP21, John Danilovich, the general secretary of International Chamber of Commerce (ICC), indicates that "One thing is very clear: governments cannot do this alone…business action and engagement will be, without doubt, a central and defining part of the solution to the climate challenge." [1]

In almost all corporate climate mitigation actions, target setting is usually an initial step and precedes the implementation of specific actions (Kolk & Pinkse, 2004; Hoffman, 2007). Target setting can serve multiple important purposes, including as a motivator for carbon abatement actions, an indicator of mitigation commitments, guidelines to choose appropriate actions, and a standard to assess the progress achieved.

The forerunners in the corporate world that voluntarily set internal GHG emission targets are oil and gas companies. BP set up its first reduction target (a 10% reduction from the 1990 emission levels by 2010) and publicly announced it in 2000 (Van den Hove, 2002). Other oil and gas companies have since followed suit. The practice of setting emission targets has then gradually spread to companies in other sectors, especially the energy-intensive ones (Dunn, 2002). As the awareness of climate risk increases, emission targets have proliferated in many industrial sectors. In the United States (U.S.), in the absence of national mandatory GHG emission targets, the U.S. Environmental Protection Agency (EPA) encourages companies to set emission targets through the voluntary Climate Leadership program, which was initially established in 2002 and replaced by the Center for Corporate Climate Leadership in 2011. Companies from various sector, such as the financial, construction and telecommunications, have participated in the program to obtain guidance on setting GHG emission targets and planning GHG mitigation measures.

In this paper, we investigate the drivers behind firm-level target setting among the biggest corporate emitters in the U.S. We examine the targets from two aspects: target adoption and target metric. Target adoption is a binary status indicating whether a company has an established target. Target metric refers to either absolute target or intensity target, both of which have been widely adopted by companies in practice. *Absolute target* sets a cap on the total GHG emissions regardless of the output of a company. *Intensity target* regulates the GHG emissions per unit of output. A company can use absolute and intensity targets simultaneously.

The drivers under study include companies' internal factors and external factors. The internal factors are company size, growth, innovation, capital constraint, and emission intensity. The external factors include pressure from government, market and physical climate change. Through this study, we aim to provide a systematic, holistic and comparative perspective on the strategies for target setting. Our study draws on a sample of 351 S&P 500 companies in the U.S. The sample is extracted and constructed from the Carbon Disclosure Project (CDP: http://www.cdp.net), a non-profit organization based in London that maintains the largest database of firm-level climate change performance and policies. To the best of our knowledge, the CDP provides the most comprehensive and detailed data on companies' internal carbon management activities. The data enable us to look into the black box of emission target setting inside the major companies in the U.S. Given the size and the breadth of our sample and the representativeness of the companies in the sample, we believe that to a significant degree our study has captured the target setting strategies of the corporate world in the U.S. Prior literature has studied the determinant factors behind firms' environmental management practices. However, to the best of our knowledge, nothing has been done on the drivers behind climate change mitigation targets. Our study fills in the gap in the literature.

2. Literature Review and Conceptual Framework

In this section we first present the relevant literature. Then we elaborate on the features of target setting strategies that we aim to study and the potential drivers behind these features.

2.1 Literature

Our research is related to the broad category of literature on the corporate environmental and climate change management strategies. While the response to climate change has traditionally been delegated to national or regional entities, a burgeoning trend of research is exploring various aspects of corporate climate strategies. Prior literature has argued that companies' climate strategies will be a key driving force toward a low-carbon future (Kolk & Pinkse, 2004; Levy, 2005). Further, empirical results point out that there is a growing willingness among business leaders to make bold investments on climate change mitigations (Persson & Rockström, 2011), due to mounting stakeholder (e.g., governments, investors, consumers) pressure and increasing perceptions of opportunities associated with climate change. Indeed, empirical evidences have shown a spectrum of climate strategies employed by companies, including the disclosure of carbon footprints, the introduction of eco-design, the use of low-carbon energy, and the improvement of energy efficiency (Stanny, 2013; Tseng, Tan & Siriban-Manalang, 2013; Rexhäuser and Löschel, 2015; Gerstlberger, Præst Knudsen & Stampe, 2014; Cadez and

[1] http://www.iccwbo.org/News/Articles/2015/Business-rallies-in-support-of-COP21-agreement-at-Le-Bourget/, accessed on September 18, 2016.

Czerny, 2016), etc.

Despite the abundance of studies on corporate climate strategies, target setting, as a critical precondition to mitigation actions, has not received sufficient attention in the past. However, recently, there appears an emerging and rapidly growing stream of research on setting corporate emission targets, including the development of target setting methodology, the characterization of target setting behaviors, and the analysis of motivations behind target setting. Kolk and Pinkse (2004) use the CDP survey data in 2002 to analyze the corporate climate strategies of a broad sample of large multinational companies. They find that the target setting process displays great variations between firms in different sectors, and, overall, 51% of the respondents have targets to reduce or stabilize the direct GHG emissions. Gouldson and Sullivan (2013) investigate the efficacy and reliability of setting corporate emission targets as a response to climate change among the major supermarkets in the United Kingdom (U.K.). They find that the voluntary targets set by the supermarkets in the U.K. not only align with the goals of the U.K. government, but also are high likely to be achieved. Sullivan and Gouldson (2016) compare the target setting behavior of representative U.K. and U.S. retailers, and find U.K. companies tend to set more ambitious reduction targets. Krabbe et al. (2015) suggest that setting corporate emission targets is a critical step in attaining national and global climate goals, and proposes a methodology to derive the corporate targets. Using the CDP data, Ioannou, Li and Serafeim (2015) find that the completion rate of carbon reduction targets is positively affected by both the provision of monetary incentives and the difficulty of the targets.

2.2 Target-Setting Characteristics

In this section we describe the two salient features of targets that we aim to study, i.e., target adoption and target metric.

Target Adoption: This refers to the binary status of whether or not a company has an established target for reducing GHG emissions. Companies may have diverse motives of target setting. Basically, setting an emission target provides a company with a clear criterion to gauge the success of its climate change efforts. Targets can also serve to motivate and pressure a company to take actions. In addition, the adoption of targets may serve as a means for companies to demonstrate to the public their climate friendliness.

Target Metric: The target metric refers to the choice between intensity and absolute measures by a company in benchmarking its GHG emission actions. The absolute target sets a cap on the GHG emissions at a specific time point in the future, regardless of the amount of output of a company. The intensity target puts a limit on the emission per unit of output. Both absolute and intensity targets for environmental purposes are common in reality. The Corporate Average Fuel Economy standards in the U.S. can be regarded as an intensity target and cap-and-trade scheme for sulphur dioxide emissions is on an absolute basis. There have been a great amount of discussions on the advantages and drawbacks of both absolute and intensity metrics. Since the intensity target can tie total emission to economic activity, it has been argued that it provides a more flexible governing framework to accommodate growth (Fischer & Springborn, 2011). Indeed, at the country level, fast-growing countries like China and India have resorted to carbon intensity of GDP rather than absolute terms in Copenhagen Accord and COP21. But others have argued that the uncertain total emissions pose a potential problem for climate change adaption and intensity targets may serve to "wrap up a weak environmental policy and make it look better" (Dudek & Golub, 2003). Some companies employ both absolute and intensity targets.

2.3 Drivers for Target Setting

In this section we outline the potential drivers for setting the GHG emission targets. The factors can be either internal or external to the companies. The internal factors include a company's size, growth, innovation, capital constraint, and emission intensity. The external factors include pressure from government, market, and physical climate change. Below we describe the definitions of the factors and the rationale for including them in this study.

2.3.1 Internal Drivers of Targets

Firm Size: Various studies have indicated that larger companies are more likely to implement environmental management practices, such as the certification of ISO14001 (Nakamura, Takahashi & Vertinsky, 2001) and end-of-pipe pollution control technologies (Demirel & Kesidou, 2011), than smaller ones. Different rationales have been proposed to explain the positive association between company size and environmental proactivity. Nakamura et al. (2001) argue that larger companies may benefit from economies of scale and are more capable of devoting resources to environmental practices. Bowen (2002) suggests that larger companies have societal visibility and hence bear greater pressure from regulators and the general public to comply with certain environmental standards. An alternative view argues that the sign (i.e., positive or negative) of an association depends on the specific environmental initiatives being implemented (Sharma & Henriques, 2005). While larger companies may undertake pollution control and eco-efficiency initiatives in a more efficient manner, smaller companies are more flexible in developing disruptive environmental innovations (Sharma & Henriques, 2005). To the best of our knowledge, no paper has touched on the relationship between firm size and emission targets.

Firm Growth: Russo and Fouts (1997) debate that companies in high-growth industries are more likely to benefit from improved environmental performance than those in low-growth industries, because high-growth is associated with both higher prospect of return and better tolerance of risk from the environmental initiatives. Therefore, we conjecture that companies with high growth potentials are more likely to adopt targets. Moreover, there are a lot of debates at the national level on how to set emission targets to accommodate economic growth. Intensity target is proposed as a mechanism which can balance environmental goals and national economic growth and is more flexible than absolute target, since there is no cap in intensity target restraining the total amount of emission as does that in absolute target (Dudek & Golub, 2003). Similar logic can be applied to firm-level targets: intensity target is linked to per unit of output of a company, the latter in turn reflects the growth rate of that company, therefore, intensity target may balance the environmental target of a company and the growth of a company. Hence, firms with high growth may favor intensity target.

Innovation: It has long been argued that companies with stronger research and development (R&D) capabilities are more likely to benefit from environmental initiatives, especially the initiatives aiming to enhance eco-efficiency (Anton, Deltas & Khanna, 2004; Martin, Muûls, de Preux, & Wagner, 2012). This is because more innovative companies are more adept in undertaking process and product innovation to reduce per unit energy and material input. Hence, innovation may be positively related to the use of target.

Capital Constraint: Investments in environmental protection actions, just like any other investments, are invariably constrained by companies' financial status. A company under heavy capital constraint may not be able to allocate sufficient resources to climate strategies, and hence is likely to eschew emission targets in the first place.

Emission Intensity: A company's emission intensity is likely to impact its decision on setting emission targets. Companies with high emission intensities are likely to face more pressure from stakeholders such as regulators and consumers than their counterparts with low emission intensities. Therefore, a company with high emission intensity may be more proactive in adopting emission targets.

2.3.2 External Drivers of Targets

Government Pressure: It has long been argued that government pressure is a major driving force behind the companies' environmental practices (Anton et al., 2004). The pressure usually takes the form of existing and potential environmental regulations. In the U.S., debates on limiting GHG emissions via regulatory measures have been going on for years. Even though no legally binding national targets have been set so far, sub-national programs have emerged and expanded (e.g., Regional Climate Change Initiative, California Cap-and-Trade Program, and Quebec Cap-and-Trade Program).

Market Pressure: Another important determinant of companies' environmental conduct is the attitude of the market (Anton et al., 2004). The market factors include reputation and changing consumer behavior. The environmentally conscious customers are increasingly reluctant to purchase products and services from a company with tainted environmental profile.

Climate Pressure: Climate pressure refers to pressure of physical impacts of climate change, which can affect business operations. Relevant physical impacts include change in mean and extreme temperatures, change in precipitation pattern, rise of sea level, and increased activities of tropical cyclones. Such physical phenomena of climate change may impact the business operations of a company both directly and indirectly. For instance, Coca-Cola acknowledges that climate change, such as droughts, poses great threats to the company's operations and economic bottom-line.[2] In response, the company has started to phase out diesel fleet with renewable trucks, improve energy efficiency of its manufacturing process, and redesign its products to reduce carbon footprints.

3. Data and Variables

This section describes the data source on target setting and the method of constructing the relevant variables from the data.

---

[2]   Coca-Cola   Climate   Protection   Report   (accessed   September   18,   2016): http://assets.coca-colacompany.com/74/9d/4a39e1f6490bad6c44ebc3da4dec/2013-climate-report-pdf.pdf

3.1 Data Source

The data used in this study are extracted from the CDP database (http://www.cdp.net) and COMPUSTAT. The CDP builds the world's largest database regarding corporate performance in relation to climate change. The goal of CDP is to provide relevant tools and information to companies, investors and policymakers to manage climate change risks. The CDP collects the data through an annual survey sent to major companies worldwide at the end of each calendar year. The survey is very detailed, with more than 100 questions covering topics like management attitude toward climate change, incentive policy for carbon abatement, GHG emission level, and abatement activities implemented. For each question, CDP also provides very detailed reporting guidance on the answer, such as the type of information to be included and the format to be used. For instance, when CDP asks the respondent to report the annual monetary saving from GHG abatement initiatives, it states that "enter the amount of monetary savings per year expected from the initiative once it is fully operational. The number ... should be entered in full and without commas... Where savings occur on a non-annual basis, please average out so that an annual figure can be provided." The companies are requested to submit the answers via an online system by the end of May each year. The CDP then analyzes the responses and assigns two scores to each report. The disclosure score measures the reporting quality and the performance score measures the climate change performance. The report itself and the two scores are published at the CDP website. The CDP data set has been used by several previous studies, such as Kolk and Pinkse (2004) and Yu, Wang, Li and Shi (2016). The company characteristic data are obtained from COMPUSTAT.

3.2 Variables

We collect data on target variables from the CDP survey in 2013. The CDP survey asks a company if it has an emission reduction target that was active (ongoing or reaching completion) in the reporting year. The yes/no answer to the question yields a binary variable for *target adoption*. Then the company is asked to specify the metric of the target if there is one. From the answers we can derive the *target metric* variable.

We construct *government pressure, market pressure* and *climate pressure* variables with the answers to an array of questions in the CDP database. First, the CDP asks the respondents to report assessment of a wide array of regulatory tools, including international agreements, air pollution limits, carbon taxes, cap and trade schemes, emission reporting obligations, fuel/energy taxes, and product efficiency standards. For each type of regulatory tool, the companies are requested to report the magnitude of the impact of regulation, which is rated by a 5-level scale from low to high. We let the low magnitude take the value of 1, and the high magnitude take the value of 5, and all other magnitude levels in between. Then we take the average magnitude of all types of regulatory tools as the proxy for government pressure. Market pressure and climate pressure are constructed in a similar manner.

We collect relevant firm characteristic data from COMPUSTAT. We use the natural logarithm of total assets (item 6) to proxy the *firm size*. The *company growth* is measured as the sales (item 12) growth rates for the companies averaged for the past three years at the setup date of the target. A company's *capital constraint* is measured by its leverage, which is the ratio between long-term debt (item 9) and total assets. We use the R&D intensity, measured by the ratio between R&D expense (item 46) and sale, to proxy a company's *innovation* capability. Following the common practice in many previous studies, we replace missing R&D expense in COMPUSTAT with zero. *Emission intensity* is given as the ratio between total emissions and sales.

In running the empirical models, we need to take the time frame of the targets into consideration. Ideally, the potential determinant factors of the target adoption and target metric should be the factors at or prior to the time when the target was set up. Since CDP does not include the date when the target was established, we need to infer the target setup date from the completion time of the target. For instance, McCormick & Company reported in the CDP data for 2013 that the company had a target that aimed to abate 5% of its emissions by 2015 and further reported that the time period for completing the target had elapsed by 50% in 2012. Therefore, we can infer that the target was established in 2009. But for a certain number of companies with their targets 100% completed, it is impossible to derive setup dates of targets using the above-mentioned method. For those companies, we use the variables one year before the reporting year in the regression. Also, we only have the government pressure, market pressure, and climate pressure data of the reporting year. Therefore, we use them in contemporaneous measure in all regressions.

In addition, we control for the potential of being affected by norms and trends of a particular industry through industry dummies and the time through year dummies. The industry dummies are created based on the two-digit Global Industry Classification Standard (GICS) codes. The year dummies are created based on the year when a particular target was set up.

Table 1 presents the summary statistics for the 351 companies. Panel A reports the distribution of the companies and adoption rates of targets by GICS codes. Overall, 74.64% (100%-25.36%) of the companies have established emission targets. Absolute target and intensity target are almost equally common, with each of them being adopted by around 30% of the companies. Our sample covers all 10 sectors under GICS. Among the sectors, information technology sector has the highest ratio of companies without emission targets (41.27%), followed by the energy sector (37.50%). Panel B presents the mean, standard deviation, 25th and 75th quartiles for the key variables used in our study.

Table 1. Targets and summary statistics

Panel A: Distribution of target adoption and metrics by industry

GICS sector	No target		Absolute target		Intensity target		Absolute & intensity targets		Total
	# of firms	Percentage (%)	# of firms	Percentage (%)	# of firms	Percentage (%)	# of firms	Percentage (%)	
Consumer Discretionary	9	16.98	15	28.30	20	37.74	9	16.98	53
Consumer Staples	6	17.14	5	14.29	17	48.57	7	20.00	35
Energy	6	37.50	2	12.50	5	31.25	3	18.75	16
Financials	12	26.09	24	52.17	6	13.04	4	8.70	46
Health Care	7	24.14	13	44.83	7	24.14	2	6.90	29
Industrials	10	20.00	12	24.00	22	44.00	6	12.00	50
Information Technology	26	41.27	17	26.98	15	23.81	5	7.94	63
Materials	6	22.22	5	18.52	9	33.33	7	25.93	27
Telecommunication Services	0	0.00	2	40.00	2	40.00	1	20.00	5
Utilities	7	25.93	13	48.15	3	11.11	4	14.81	27
Overall	89	25.36	108	30.77	106	30.20	48	13.68	351

Panel B: Summary statistics

	Mean	Standard deviation	25th Percentile	75th Percentile
Firm_Size	9.75	1.50	8.82	10.61
Firm_Growth	-0.03	0.12	-0.07	0.02
Innovation	0.06	0.08	0.01	0.09
Capital_Constraint	0.22	0.14	0.11	0.31
Emission_Intensity	3.94	1.95	2.77	5.07
Government_Pressure	0.46	0.54	0.00	0.75
Market_Pressure	0.26	0.38	0.00	0.38
Climate_Pressure	0.46	0.58	0.00	0.64

4. Methods and Results

4.1 Target Adoption

We first investigate the relationship between company characteristics and a company's target adoption decision. We estimate a binary probit regression model in which the dependent variable *Target_Adoption* takes the value of 1 if the firm sets up a target of any metric, and 0 otherwise. The probit model is represented as follows,

$$Prob(Target\_Adoption_i = 1) = \Phi(\alpha + X_i\beta_X + Y_i\beta_Y + C_i\beta_C + \varepsilon_i). \qquad (1)$$

In equation (1), the subscript i denotes company i, Φ is the standard cumulative normal distribution, α is an intercept, X_i is a vector of covariates corresponding to the characteristics of company i (*Firm_Size, Firm_Growth, Innovation, Capital_Constraint, and Emission_Intensity*), Y_i is a vector of covariates corresponding to the external factors of company i (*Government_Pressure, Market_Pressure,* and *Climate_Pressure*), C_i represents a vector of control variables (industry and year dummies), and β_X, β_Y and β_C represent the unknown coefficients we aim to estimate with maximum likelihood.

Table 2 presents the estimated coefficients and marginal effects of the target adoption regression in (1). The purpose of reporting the marginal effects for all the variables is to put the magnitude of likelihood increase in perspective. The marginal effects are calculated as the change in predicted probability when the covariate varies by one standard deviation (from 0.5 standard deviation below the mean value to 0.5 standard deviation above the mean value), with all other covariates fixed at the mean values. For instance, a one standard deviation increase in government pressure is associated with a 14% increase in the probability of adopting emission targets (18.85%

relative to the baseline probability of 74.64% shown at the bottom at the table).

Among the internal factors, the coefficient and marginal effect for *Firm_Size* are positive and significant at the p<0.01 level. The coefficients and marginal effects for *Firm_Growth* and *Innovation* are positive and significant at the p<0.1 level. *Capital_Constraint* is negatively associated with *Target_Adoption* at p<0.1. Among the external factors, we find the coefficient and marginal effect for government pressure are positive and significant at the p<0.05 level. Hence government pressure in the form of regulation or threat of regulation can stimulate the use of emission targets. We find relatively weak evidence of market pressure and climate pressure on the likelihood of target adoption. Overall, the results show that companies with larger size, higher growth, better innovation, weaker capital constraint, and higher government pressure are more likely to establish emission targets.

Table 2. Determinants of target adoption

	Coefficient	Marginal effect
Intercept	-2.6880 ***	
	(0.6204)	
Internal Factors		
Firm_Size	0.1966 ***	0.0548
	(0.0594)	
Firm_Growth	1.1000 *	0.3065
	(0.5514)	
Innovation	1.4710 *	0.4097
	(0.8360)	
Capital_Constraint	-1.2198 *	-0.3337
	(0.6158)	
Emission_Intensity	-0.0001	0.0000
	(0.0001)	
External Factors		
Government_Pressure	0.5050 **	0.1407
	(0.2360)	
Market_Pressure	0.2584	0.0720
	(0.3442)	
Climate_Pressure	0.2189	0.0610
	(0.2264)	
Industry fixed effect	Yes	
Year fixed effect	Yes	
Prob(Target adoption) (%)	74.6439	
Target adoptions	262	
Number of Observations	351	
Pseudo R-squared	0.2352	

Standard errors are in parentheses

* p<0.1; ** p<0.05; *** p<0.01.

4.2 Target Metrics

We examine the impact of internal and external factors on the target metrics using multinomial logit regression. In this case there are four categories, e.g., no target, absolute target only, intensity target only, and both absolute and intensity targets. Note that absolute and intensity targets are not mutually exclusive so the last category is needed to represent the firms that adopt both. The multinomial logit regression allows us to discern the determinants of different target metrics simultaneously. The model is given as follows.

$$Prob(Target\_Metric_i = j) = \frac{\exp(\beta_j X_i)}{\sum_{k=1}^{4} \exp(\beta_k X_i)}, \quad (2)$$

where j is 1 if no target, 2 if absolute target, 3 if intensity target, and 4 if both absolute and intensity targets. The category of no target is omitted and all coefficients are measured relative to this category. Table 3 reports the multinomial logit regression results. We report the results for absolute target and intensity target only.

Table 3. Determinants of target metric

	Absolute Target	Intensity Target
Intercept	-1.5022 **	-0.9155
	(0.6557)	(0.5516)
Internal Factors		
Firm_Size	0.1918 ***	0.1057
	(0.0595)	(0.0632)
Firm_Growth	-1.8810 ***	0.3193 **
	(0.6630)	(0.1567)
Innovation	0.9810	0.0065
	(1.2549)	(0.2318)
Capital_Constraint	0.9975	0.2551
	(0.5506)	(0.1804)
Emission_Intensity	-0.0620	0.0011
	(0.0429)	(0.0406)
External Factors		
Government_Pressure	0.2891	-0.0010
	(0.2108)	(0.0351)
Market_Pressure	-0.0764	0.0026
	(0.1887)	(0.0911)
Climate_Pressure	0.1103	0.0026
	(0.2057)	(0.0930)
Industry fixed effects	Yes	
Year fixed effects	Yes	
Number of Observations	351	
Pseudo R-squared	0.1948	

Standard errors are in parentheses

* p<0.1; ** p<0.05; *** p<0.01.

Firm_Size is positively significantly related to the probability of using absolute target relative to the case of no target at the p<0.01 level. Its association with intensity target is also positive but not significant. Furthermore, *Firm_Growth* is positively (negatively) significantly associated with the probability of employing intensity (absolute) target. The result is in line with the conjecture that high-growth firms are inclined to use intensity target, because it allows the firms to control emission and accommodate economic growth simultaneously. The argument that intensity target balances emission and growth has been made for countries and our result indicates that firms appear to have embraced the same argument for themselves. But at the same time, we would like to note that intensity target leads to greater uncertainty in total emissions as compared to absolute target. So policymakers may need to keep a keen eye on those firms that use intensity target. Also, we note that *Emission_Intensity* negatively affects the use of absolute target, but the effect is not significant. *Innovation* and *Capital_Constraint* are positively associated with both targets, but the results are not significant. The external factors do not display significant associations with either type of targets. Overall, *Firm_Size* and *Firm_Growth* display significant effects on target metrics while the other factors do not show significance.

5. Discussions and Conclusions

In this paper we investigate the determinants behind the use of GHG emission targets among U.S. S&P 500 companies. Target setting is a critical initial step in GHG mitigation. We find that companies with larger size, higher growth, better innovation, weaker capital constraint, and higher government pressure are more likely to set emission targets. Moreover, high-growth firms are more likely to use intensity targets. We do not find any significant impact from market pressure and climate pressure on target adoption. Our study complements the literature on the drivers behind firms' environmental actions (Anton et al., 2004).

Policymakers and firm managers should consider the aforementioned factors and their impacts on target characteristics in order to induce desirable target-setting behaviors in the corporate world. Among the major companies sampled in our study, there are still 25% of companies without emission targets. Going forward, policymakers should try to increase the adoption of emission targets among the firms. For instance, to encourage the use of emission targets, policymakers may seek ways to create or strengthen governmental pressure on the firms in the form of regulatory measures. Hard measures such as carbon tax or emission cap may not be viable in the short term. But there are ways, such as mandatory emission reporting and even simple threat of stringent regulation, that policymakers can leverage to boost the pressure on firms (Anton et al., 2004). In promulgating the emission targets, policymakers may also pay special attention to smaller, low-growth, less innovative firms, which have been found to be less likely to use emission targets. Moreover, the policymakers should be cautious of the emissions from high-growth firms, since those firms are likely to use intensity metric to control GHG emissions and no cap is imposed on the total emissions.

References

Anton, W. R. Q., Deltas, G., & Khanna, M. (2004). Incentives for environmental self-regulation and implications for environmental performance. *Journal of Environmental Economics and Management, 48*(1), 632-654. http://dx.doi.org/10.1016/j.jeem.2003.06.003

Bowen, F. E. (2002). Does size matter?. *Business and Society, 41*(1), 118.

Cadez, S., & Czerny, A. (2016). Climate change mitigation strategies in carbon-intensive firms. *Journal of Cleaner Production, 112*, 4132-4143. http://dx.doi.org/10.1016/j.jclepro.2015.07.099

Demirel, P., & Kesidou, E. (2011). Stimulating different types of eco-innovation in the UK: Government policies and firm motivations. *Ecological Economics, 70*(8), 1546-1557. http://dx.doi.org/10.1016/j.ecolecon.2011.03.019

Dudek, D., & Golub, A. (2003). "Intensity" targets: pathway or roadblock to preventing climate change while enhancing economic growth?. *Climate Policy, 3*, S21-S28. http://dx.doi.org/10.1016/j.clipol.2003.09.010

Dunn, S. (2002). Down to business on climate change. *Greener Management International, 2002*(39), 27-41. http://dx.doi.org/10.9774/GLEAF.3062.2002.au.00005

Fischer, C., & Springborn, M. (2011). Emissions targets and the real business cycle: Intensity targets versus caps or taxes. *Journal of Environmental Economics and Management, 62*(3), 352-366. http://dx.doi.org/10.1016/j.jeem.2011.04.005

Gerstlberger, W., Præst Knudsen, M., & Stampe, I. (2014). Sustainable development strategies for product innovation and energy efficiency. *Business Strategy and the Environment, 23*(2),131-144. http://dx.doi.org/10.1002/bse.1777

Gouldson, A., & Sullivan, R. (2013). Long-term corporate climate change targets: What could they deliver? *Environmental Science & Policy, 27*, 1-10. http://dx.doi.org/10.1016/j.envsci.2012.11.013

Herzog, T., Baumert, K. A., & Pershing, J. (2006). *Target: Intensity. An analysis of greenhouse gas intensity targets*. Washington, DC: World Resources Institute. Retrieved from http://pdf.wri.org/target_intensity.pdf

Hoffman, A. J. (2007). *Carbon strategies: How leading companies are reducing their climate change footprint.* Ann Arbor, MI: University of Michigan Press.

Ioannou, I., Li, S. X., & Serafeim, G. (2016). The Effect of Target Difficulty on Target Completion: The Case of Reducing Carbon Emissions. *The Accounting Review, 91*(5),1467-1492. http://dx.doi.org/10.2308/accr-51307

Kolk, A., & Pinkse, J. (2004). Market strategies for climate change. *European Management Journal, 22*(3), 304-314. http://dx.doi.org/10.1016/j.emj.2004.04.011

Krabbe, O., Linthorst, G., Blok, K., Crijns-Graus, W., van Vuuren, D. P., Höhne, N., ... Pineda, A. C. (2015). Aligning corporate greenhouse-gas emissions targets with climate goals. *Nature Climate Change, 5*(12), 1057-1060. http://dx.doi.org/10.1038/nclimate2770

Levy, D. L. (2005). Business and the evolution of the climate regime: The dynamics of corporate strategies. In D. L. Levy & P. J. Newell (Eds.), *The Business of Global Environmental Governance* (pp. 73-104). Cambridge, MA: MIT Press.

Lutsey, N., & Sperling, D. (2008). America's bottom-up climate change mitigation policy. *Energy Policy, 36*(2), 673-685. http://dx.doi.org/10.1016/j.enpol.2007.10.018

Martin, R., Muûls, M., de Preux, L. B., & Wagner, U. J. (2012). Anatomy of a paradox: Management practices, organizational structure and energy efficiency. *Journal of Environmental Economics and Management, 63*(2),

208-223. http://dx.doi.org/10.1016/j.jeem.2011.08.003

Nakamura, M., Takahashi, T., & Vertinsky, I. (2001). Why Japanese firms choose to certify: a study of managerial responses to environmental issues. *Journal of Environmental Economics and Management, 42*(1), 23-52.

Persson, Å., & Rockström, J. (2011). Business leaders. *Nature Climate Change, 1*(9), 426-427.

Philibert, C., & Pershing, J. (2001). Considering the options: climate targets for all countries. *Climate Policy, 1*(2), 211-227. http://dx.doi.org/10.1016/S1469-3062(01)00003-1

Pizer, W. A. (2006). The evolution of a global climate change agreement. *The American economic review, 96*(2). 26-30. http://dx.doi.org/10.1257/000282806777211793

Rexhäuser, S., & Löschel, A. (2015). Invention in energy technologies: Comparing energy efficiency and renewable energy inventions at the firm level. *Energy Policy, 83,* 206-217. http://dx.doi.org/10.1016/j.enpol.2015.02.003

Russo, M. V., & Fouts, P. A. (1997). A resource-based perspective on corporate environmental performance and profitability. *Academy of Management Journal, 40*(3), 534-559. http://dx.doi.org/10.2307/257052

Sharma, S., & Henriques, I. (2005). Stakeholder influences on sustainability practices in the Canadian forest products industry. *Strategic Management Journal, 26*(2), 159-180. http://dx.doi.org/10.1002/smj.439

Stanny, E. (2013). Voluntary disclosures of emissions by US firms. *Business Strategy and the Environment, 22*(3), 145-158. http://dx.doi.org/10.1002/bse.1732

Sullivan, R., & Gouldson, A. (2016). Comparing the climate change actions, targets and performance of UK and US retailers. *Corporate Social Responsibility and Environmental Management, 23*(3), 129-139. http://dx.doi.org/10.1002/csr.1364

Tseng, M. L., Tan, R. R., & Siriban-Manalang, A. B. (2013). Sustainable consumption and production for Asia: sustainability through green design and practice. *Journal of Cleaner Production, 40,* 1-5. http://dx.doi.org/10.1016/j.jclepro.2012.07.015

Van den Hove, S., Le Menestrel, M., & De Bettignies, H. C. (2002). The oil industry and climate change: strategies and ethical dilemmas. *Climate Policy, 2*(1), 3-18. http://dx.doi.org/10.1016/S1469-3062(02)00008-6

Yu, Y., Wang, D. D., Li, S., & Shi, Q. (2016). Assessment of US firm-level climate change performance and strategy. *Energy Policy, 92,* 432-443. http://dx.doi.org/10.1016/j.enpol.2016.02.004

13

Leveraging Globalization to Revive Traditional Foods

Jena Trolio[1], Molly Eckman[1] & Khanjan Mehta[1]

[1] Humanitarian Engineering & Social Entrepreneurship (HESE) Program, Pennsylvania State University, University Park, PA, USA

Correspondence: Khanjan Mehta, Humanitarian Engineering & Social Entrepreneurship (HESE) Program, Pennsylvania State University, University Park, PA, USA, Canada.
E-mail: khanjan@engr.psu.edu

Abstract

Traditional foods are important to the sustainability of their native regions because they are often keystone assets to food security, economic stability, and quality nutrition. Globalization of agricultural markets, changing lifestyles, and rural-to-urban migration has contributed to the gradual loss of traditional foods in developing countries. The transition from traditional foods to imported refined carbohydrates, sugars, and edible oils has promoted nutrient deficiency, economic instability, and food insecurity. While the effects of globalization have been largely negative for indigenous foods, globalization is inevitable and has potentially useful aspects. Local champions and international supporters can leverage specific technologies and market patterns brought about or influenced by globalization to revive culinary traditions, strengthen local food systems, and bolster indigenous livelihoods. Such approaches include helping farmers benefit from technological advances in efficiency and economy of scale, biotechnology, post-harvest processing, and smart infrastructure combined with ethically-conscious food sourcing. Trends such as human migration, exotic food fads, interest in nutritious and organic foods, the rise of social media, and agricultural extension and education can also support improvements in local agricultural products and their globalizing markets. Collectively, these efforts can help revive sustainable traditional food production and enhance the lives and livelihoods of indigenous communities.

Keywords: agricultural technology, globalization, indigenous foods

1. Introduction

1.1 Roles of Traditional Foods

Traditional foods can be defined as long-established, culturally important staple crops and foodstuffs. These crops are often simple to cultivate and are usually cornerstones of the native diet, providing essential proteins, vitamins, minerals, and amino acids. Their critical nutritional value has made these foods extremely important to communal and societal norms and allowed them to stand the test of time (Trichopoulou, Soukara, & Vasilopoulou, 2007). Additionally, traditional crops are often best adapted to their local environments, exhibiting tolerance to native biotic and abiotic stressors in their respective climates and conditions. Traditional foods are often important to the local economy and nutrition, providing food security, economic stability, and high nutritional quality to those that grow and/or consume them.

Despite their importance, though, traditional foods have been disappearing as a result of globalized phenomena. Different social and political movements, including colonization and the Green Revolution, have drastically changed agricultural practices by placing higher value on the production of export crops. These changes continue today with trade practices, urbanization, and global diffusion of western tastes and foodstuffs. Despite some negative consequences of globalization, the most productive and advantageous response now is to harness global trends for the revival of indigenous foods and livelihoods. Promising opportunities include those focused on improved agricultural production, post-harvest processing, and infrastructure as well as novel biotechnologies and industrial food systems. Modern technologies can improve farming efficiencies and yields, and methods such as drying and canning can protect and add value to food supplies. Global market trends also support the introduction and acceptance of indigenous crops by broader consumer populations. These trends include increasing human migration, popularity of exotic and organic foods, and prevalence of social media—all of which can facilitate the integration of indigenous foods in the global marketplace.

This article presents opportunities for leveraging globalization to revive and sustain traditional foods. It is designed to be of use to innovators and investors interested in market-based sustainable agricultural, environmental, and nutritional possibilities. This paper begins by describing the factors contributing to the disappearance of traditional foods and to the resulting problems. It then takes an example-centric approach to the discussion of how modern agricultural technologies and global markets can be leveraged to revive traditional foodstuffs.

1.2 Why Are Traditional Foods Disappearing?

Colonization and the Green Revolution (c. 1940-1960) are the two major causes of the disappearance of traditional foods. Colonizing powers focused exclusively on export crops, which led to a collapse in crop biodiversity and the loss of indigenous agricultural knowledge (Weis, 2007). Following the colonial era, the Green Revolution's singular focus on favorable environments left many resource-poor regions trapped in poverty even while overall crop production rose exponentially. For instance, innovations like high-yield variety (HYV) crops and advanced fertilizers result in much smaller yield gains in marginal farming environments than the favorable environments they were created to benefit. Fertilizer and HYV wheat can result in 40% yield gains in well-irrigated areas, whereas such investment often only brings 10% gains on more marginal farming lands (Pingali, 2012). This disparity along with high-capital foreign investment helps large commercial farms retain favorable land, displacing small-scale farmers.

From the Green Revolution on, a series of trade agreements and policies continued to foster the replacement of traditional foods with a more Westernized diet. Today, farmers around the world are moving toward a more commercialized, industrialized, and foreign-driven agricultural system. Instead of focusing on the growth and production of native food crops, developing nations around the world have prioritized cash crops intended for export. Additionally, at least 19 countries are using significant land resources to produce biofuels, further limiting indigenous foods production (Motes, 2010). In addition to these land use and economic changes, urbanization has also affected the production of indigenous crops. The movement away from the traditional croplands has disconnected city-dwellers from native agriculture while promoting a more westernized diet and raising the demand for related crops.

1.3 Why Should We Care about the Disappearance of Traditional Foods?

The disappearance of traditional foods has led to nutrition deficiency in some of their origin countries. The current agricultural system makes refined carbohydrates such as refined wheat, rice, sugar, and edible oils more readily available and affordable than traditional food products. Although these carbohydrates and edible oils are allowing people to become more energy secure (Johns & Sthapit, 2004) they are also displacing nutritious beans, legumes, fruits, and vegetables. Additionally, due to various global tariffs and agricultural subsidies, it has become cheaper to purchase imported "junk food" than to purchase nutritious domestic produce. This diet transition is contributing to the rise of type 2 diabetes, cardiovascular disease, obesity, and other chronic noncommunicable diseases (Johns & Sthapit, 2004).

In addition to the health risks, the loss of biodiversity hinders ecological systems and puts farmers in precarious situations. When plants are all of the same variety, there is a much greater risk of losing the entire monoculture to pests, diseases, or drought. Climate change may also cause many agricultural challenges, such as the rise in extreme precipitation events like droughts and hurricanes. Developing countries are predicted to bear the brunt of most of the problems caused by climate change, and it is also speculated that climate change will widen the gap in cereal production between developed and developing countries (Keane, Page, Kergna, & Kennan, 2009). In order to address these challenges, it is important to grow traditional crops and provide crucial biodiversity.

2. Leveraging Modern Technology to Revive Traditional Foods

Globalization has contributed to the disappearance of indigenous foods, but modern agricultural technologies supported by globalization have the potential to bring them back. Technological innovations present a plethora of new opportunities to address nutrition poverty in the developing world. With the multi-faceted problems of malnutrition and environmental degradation, it is necessary to parse through the various agricultural technologies and identify ones that are economically appropriate as well as sustainable and efficient. Strengthening food value chains (FVCs) with agricultural technologies can reduce food spoilage as well as improve the efficiency and yield of farmland. The types of technologies shown in Figure 1 can strengthen each of the six FVC phases: production, processing, storage, marketing, distribution, and consumption (Callan, Sundin, Suffian, & Mehta, 2014).

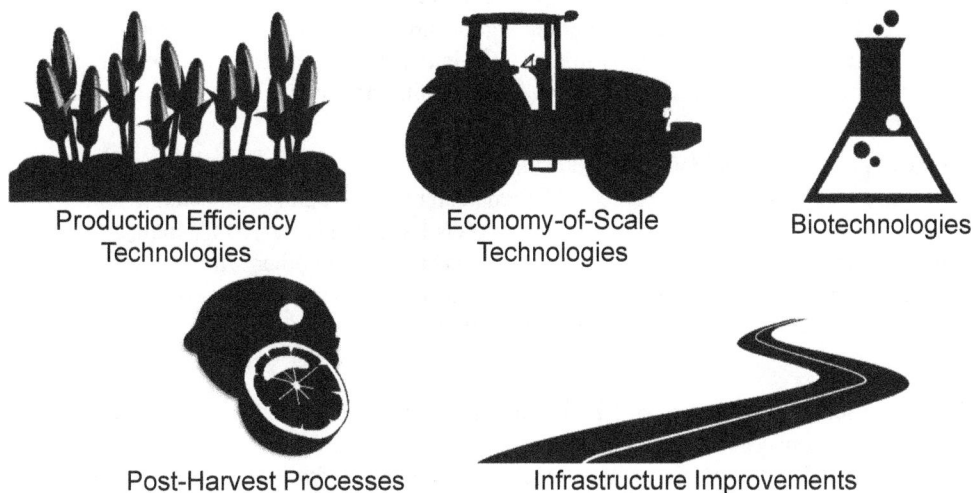

Figure 1. Types of agricultural technologies that can be leveraged to revive traditional foods

2.1 Production Efficiency Technologies

Modern agricultural technologies, including advanced insecticides, improved plant varieties, and more efficient planting tools, can and have improved the efficiencies of small-scale farming operations. For instance, greenhouses have enhanced agricultural production by allowing for year-round crop growth. Use of these structures allows farmers in a variety of climates to increase their revenues by thousands of dollars per year. Until recently, however, greenhouses were not an economically viable option for small-scale farmers in the developing world. Novel affordable greenhouses provide the opportunity to grow traditional foods, thereby improving the local economy (Suffian, De Reus, Eckard, Copley, & Mehta, 2013).

One example of a nutritional benefit is amaranth, a vegetable indigenous to Africa that grows in humid, tropical environments. Amaranth leaves provide a significant number of Africans with as much as 25% of their daily protein as well as significant quantities of vitamins A and C, calcium, and iron (National Research Council, 2006). Greenhouses can extend the growing season by providing the necessary humid environment, making this crop available more frequently.

2.2 Production Economy-of-Scale Technologies

While commercial agricultural technologies are not directly accessible or beneficial to smallholder farmers, they are still key aspects of many agricultural systems. Large economy-of-scale technologies are likely to extend further into developing markets with any increase in crop demand. Though commercial farming can create detrimental competition for smallholders, large-scale technologies can also help underdeveloped economies leverage industrialized efficiencies and develop transportation infrastructure. Economies of scale can also help revive some traditional foods that might not otherwise gain market penetration. The important caveat is that both foreign and domestic markets practice ethical sourcing and consumption.

For example, teff is a grain harvested in Ethiopia and is used in the indigenous dish injera. Injera is a spongy flatbread that is widely consumed across Ethiopia as an inexpensive traditional addition to nearly all meals. Teff is a reliable and resilient crop, often found free of pests and disease and able to survive unfavorable weather conditions (Hrušková, Švec, & Jurinová, 2012). However, it is difficult to sow efficiently and generally has low yields and slow harvests. These difficulties limit its popularity among farmers, and many Ethiopians have replaced it with wheat or other grains. Modern seed dispersal units could make planting and harvesting teff easier and more productive, thus lowering its price and making it available to a larger market. Adopting this large-scale production technology could help revive the traditional Ethiopian staple, benefitting consumers and communities and potentially leading to innovations for smallholders.

2.3 Biotechnologies

Biotechnology, though controversial, has resulted in large gains for both small-scale and large-scale farmers around the world. Bioengineered seeds in particular have made farming more environmentally friendly and economically efficient due to their increased resistance to pests, disease, and unfavorable climate conditions.

Insect resistance has greatly improved the production of soybeans, corn, cotton, and canola seeds worldwide (Motes, 2010). Further research can and is developing similar improvements for other crops, including sorghum (Kumar, et al., 2011) and potentially millet. However, it is important to note that the use of biotechnology has raised many ethical and practical questions, such as if consuming artificially altered foods will lead to health problems in the future.

Other crops are also promising candidates for biotechnology research. Cowpeas are a grain legume with quality amino acid, vitamin, mineral, and protein content. The crop can resist drought, grows well in soils of poor quality, and provides many of the nutrients missing from African diets (National Research Council, 2006). However, an insect pest known as Maruca pod borer can easily destroy this crop and all other African grain legumes. Farmers can face up to a 100% crop failure during serious attacks (Machuka, 2001). Biotechnology research may be able to create pest-resistant varieties of the cowpea and other grain legumes.

2.4 Post-Harvest Processing Technologies

Significant loss of agricultural productivity also occurs after the harvest. Lack of post-harvest protection is a larger problem for fruits and vegetables, as their sale price is largely dependent on their physical appearance. Pests and pathogens degrade produce appearance, forcing down their market price, which encourages farmers to switch to crops with more consistent selling prices. Post-harvest technologies, such as food dehydrators, can improve storage and marketability.

The longan fruit is an example of an indigenous crop with a great market potential that is not currently being realized (Jiang, Zhang, Joyce, & Ketsa, 2002). This fruit is native to areas of northern Burma and southern China. At ambient temperatures, the post-harvest life of the longan fruit is very short, about 3 or 4 days. However, its shelf life can be extended by canning, freezing, or drying the fruit or by utilizing preservation technologies such as fungicide dips and sulfur fumigation (Jiang, Zhang, Joyce, & Ketsa, 2002). There is a strong market in Asia for the dried fruit in particular, and canning retains much of the fruit's flavor (Jiang, Zhang, Joyce, & Ketsa, 2002).

2.5 Infrastructure Improvements

Transportation and infrastructure are generally expensive and inconsistent in rural areas of developing countries, while urban areas take advantage of better infrastructure to facilitate importation into their larger shipping hubs. However, construction of hard infrastructure like paved roads is connecting rural areas of developing countries to the rest of the world at unprecedented speeds. The global spread of communications technologies are also allowing isolated communities to join global markets and networks with greater ease than ever before (Kramer, Urquhart, & Schmitt, 2009). Development initiatives that improve transportation, electricity, and water access can connect farmers to new markets.

Improved infrastructure could directly improve the sale of the butterfruit, a well-known and popular fruit in Central and West Africa (National Research Council, 2008). Its pulp packs high nutritional content including protein, energy, and amino acids. It also combats micronutrient deficiencies by providing phosphorus, potassium, calcium, and magnesium (National Research Council, 2008). In the areas of high heat and humidity where the butterfruit is grown, its shelf life can be only a week or less. Better infrastructure can help get the fruit to market sooner, making it a more viable and attractive crop for consumers and farmers alike.

3. Leveraging Globalized Markets to Revive Traditional Foods

Increasing migration, trade, and communication trends in the global society can be advantageously leveraged to revive traditional foods. Human migration spreads knowledge, customs, and cultures in ways that can also promote the spread of traditional foods. Trends in the global market such as exotic food fads and the growing demand for nutritious and organic foods open niches for traditional crops in entirely new locations. As summarized in Figure 2, unprecedented methods and rates of worldwide communication create many novel opportunities to spread awareness, knowledge, and education pertaining to traditional foods and agricultural systems.

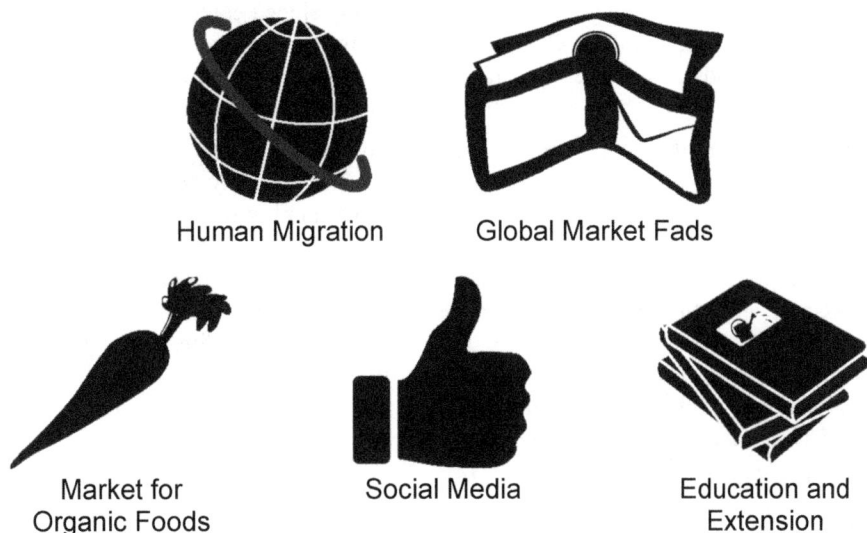

Figure 2. Ways in which globalized interaction can be leveraged to revive traditional foods

3.1 Human Migration

Heavy human migration from rural areas to urban centers as well as from developing to developed countries is occurring globally. As people travel, some traditional foods travel with them. For instance, as migration from African countries has increased, exports from Africa have as well (Willer & Lernoud, 2014). This rise in exports can benefit smallholder farmers if new technologies allow them to price their crops competitively in the global market.

An example of a traditional African crop that traveled out of Africa with modern globalization is sorghum, the second most important cereal crop in Africa. Sorghum is incredibly versatile: indigenous uses include being cooked like rice or made into porridge, beer, flatbreads, or popcorn. It has now been adopted and adapted in Latin America, specifically in Mexico, where it has a new role primarily as animal feed (Stone, et al., 2011). The demand for sorghum as an ingredient in beer, instant porridge, and vegetable oil is also growing. Sorghum's use in adhesives, waxes, and dyes provides other markets for farmers to leverage. Beyond sorghum, countless additional crops have also spread globally and expanded both their indigenous and other uses. Another native African, yam, is already well known beyond Ghana, but has still seen in a modern rise in production and export, from a total production of 877,000 tons in 1990 to 5,960,490 tons in 2010. Smallholder farmers are the main drivers behind this production jump (Anaadumba, 2013).

3.2 Global Market Fads for Exotic Foods

Changes in food consumption usually occur because of lifestyle changes surrounding income, urbanization, demographics, transportation, and quality and safety perceptions (Regmi, 2001). When people generate larger incomes, they desire diets that are more expensive. Globally, this trend results in a booming demand for exotic foods. Developed countries in particular are catalysts of these exotic food fads. Wealthier people living in developed countries are the quickest to join the newest global trends, such as the fad of the paleo and gluten-free diets. These diets use ancient grains like quinoa and chia that were once very popular in their native countries but were overshadowed by modern grains and cereals. These trends spur rapid increases in associated product launches. For instance, launches of foods labeled 'Japanese' grew 230% from just 2009 to 2010 and 'Thai' grew 68% (Mintel, 2011). While this sort of growth is certainly not sustainable long-term, it points to clear untapped niches in the ethnic food markets.

As mentioned, quinoa is an example of an exotic food that has quickly risen in popularity in a new major niche market. Quinoa is native to Bolivia, Chile, and the Peruvian Andes (Oelke, Putnam, Teynor, & Oplinger, 1992). The initial takeover of western crops and foods in these regions led to the marginalization of quinoa, but now western markets offer it the chance of revival. The grain is highly nutritious, with strong protein content and all nine essential amino acids. Consumers have also sparked a new wave of food products releases: up by nearly 50% in 2013 (Schroeder, 2013).

Other traditional foods from around the world have received the "trendy" label. For instance, chia is native to both Mexico and Bolivia, but is becoming a popular product in developed countries for weight loss. It is rich in omega-3 fatty acids, antioxidants, and fiber. Chia, like quinoa, saw an increase in 2013 product launches—again about a 50% increase worldwide. In fact, nearly half of the demand for chia has come from the United States and not South America due largely to weight-loss fads in the United States (Schroeder, 2013).

Perhaps the strongest evidence of demand for these traditional crops, however, is the strength of the negative impact some have on their native locales. The social movement for ethically conscious sourcing has not been able to keep pace with the surges in market demand, leading to a number of problems for smallholder farmers and their environments. For instance, quinoa is now so expensive in its traditional croplands that many locals must turn to cheaper imported junk food (Collyns, 2013). Environmentally, irrigation of asparagus destined for international markets is depleting Peruvian water resources (Lawrence, 2010). In Brazil, increased soya production and displaced ranching operations are leading to increased deforestation (Barona, Ramankutty, Hyman, & Coomes, 2010). Sustainable development efforts must drive interest and investment in ethically-conscious sourcing simultaneous with efforts at market penetration in order to ensure benefit to smallholder farmers, indigenous nutrition, and local environments.

3.3 Growing Market for Nutritious and Organic Foods

The global market for nutritious and organic foods has boomed in recent years. In the United States alone, the market for organic foods grew from $3.6 billion in 1997 to $21.1 billion in 2008 (Dimitri & Oberholtzer, 2009). Even toward the end of that decade, the industry still saw a 50% increase in producers from 2006 to 2008 (Willer & Lernoud, 2014).

Developing countries have become increasingly involved in organic agriculture. As of 2012, Africa has over one million hectares devoted to organic agriculture and Asia has three million, both up from zero in 1999. Latin America saw a similarly strong increase, 6.8 million up from 1.2 million in 1999, and Oceania wen from 3.7 to 11.2 million over the same period (Willer & Lernoud, 2014). In particular, developing countries that are former E.U. colonies with good market ties to Europe can and do benefit from organic exporting (Kim, 2013). This shows that although the demand for nutritious and organic foods is much higher in developed countries, smallholder farmers in developing countries can leverage this market if enabled by efficiency and transport technologies and ethical sourcing. In addition, organic farming tends to improve the biodiversity and sustainability of farming practices in rural communities (Maeder, Dubois, Gunst, Fried, & Niggli, 2002).

One of the first crops domesticated by man, flax, is an example of one whose sales have seen enormous growth due to global demand (Goyal, Sharma, Upadhyay, Gill, & Sihag, 2014). Although it was originally grown in African and Asian locations, the crop has spread worldwide. In the United States today, flax and flaxseed are in high demand. Flaxseed in particular has been promoted for its low calorie and high fiber and omega-3 fatty acid content (Goyal, Sharma, Upadhyay, Gill, & Sihag, 2014). With the right support and technology, sustainable farmers in native flaxseed locations such as Egypt and China can leverage this large global market.

3.4 Social Media

The rise of social media has provided consumers with easy, unfettered access to recipes for traditional meals, advice on where to find ingredients, and tips for food preparation. Searching for 'traditional foods' or 'indigenous foods' on the popular media sharing site Pinterest produces a wide variety of resources. Through this social media website, the common user can discover links leading to websites about foods from places as diverse as the Pacific Islands, Chile, India, and Australia. The ease of cultural knowledge flow over the internet has made traditional foods more accessible to consumers around the globe. This effect is likely to increase through subsequent generations.

The spiky, unusual appearance of the horned melon, or kiwano, has made it the focus of prominent social media coverage. Its distinct look has helped promote worldwide sales even more so than its taste or nutritional qualities (National Research Council, 2008). Many social media users also like to boast in the captions and comments of their pictures that their horned melon was cut and opened in the traditional fashion, a trend that further extends its reach. Performing a search for the fruit on the popular photo-sharing platform Instagram reveals 8,222 posts containing the hashtag "kiwano," and 3,864 posts containing the hashtag "hornedmelon". Through careful understanding of how information spreads through these platforms, increasingly connected smallholder farmers can promote their own crops and thereby increase sales in the global market.

3.5 Improved Education and Extension

'Soft' infrastructure systems include all of the nonphysical, organizational institutions established for a society to function. Some of these systems, such as global communication channels, can be harnessed to educate smallholder farmers in rural areas. Extension programs can enhance and broaden farmers' skills and familiarize them with agricultural technologies that can help revive traditional crops (Elias, Nohmi, Yasunobu, & Ishida, 2013). Knowledge areas include farm management practices, appropriate agricultural technology, and nutrient maintenance (Elias, Nohmi, Yasunobu, & Ishida, 2013). These tools can help small-scale famers handle problems associated with reintroduced indigenous crops. Additionally, enhanced knowledge transfer can help with market timing, allowing farmers to achieve the highest sale price when faced with fluctuating demand.

The bambara bean is an example crop that educated farmers could use to increase the overall quality of their soil. A type of peanut native Africa and exported primarily by Gulf of Guinea countries, the bambara bean is a nutritious, low-growing legume that improves soil fertility and is inexpensive to grow in diverse and difficult environmental conditions (National Research Council, 2006). Education can encourage farmers to diversify their crop rotations to include legumes like the bambara bean in order to revive the crop while simultaneously adding nutrients to their soils.

4. Conclusion

Globalization creates opportunities for both the disappearance and revival of traditional foods. Often the most appropriate and resilient options in their native environments, indigenous crops tie people to their respective societies and cultures. Highly nutritious traditional foods offer many possibilities to promote greater food security and economic advancement for smallholder farmers in developing countries, especially by opening up global niche markets.

Leveraging technology pushes and market pulls along with ethical sourcing can bring about traditional food revival and benefit underserved agricultural economies. Modern agricultural technologies provide the opportunity for more efficient and better quality production that can add value to food products and increase their shelf life. Global market trends have the capability to spread nutritious and organic indigenous foods to a worldwide audience. Government investment in improved infrastructure and outreach programs could give farmers access both to new markets and to knowledge needed to take advantage of global market trends. Additionally, global consumer support of traditional foods can increase the incomes of smallholder farmers in developing countries. The revival and preservation of indigenous crops and foods are extremely important for ecological, food security, health, economic and societal reasons in a globalized world.

References

Anaadumba, P. (2013). *Analysis of Incentives and Disincentives for Yam in Ghana.* Rome: United Nations Food and Agriculture Organization.

Barona, E., Ramankutty, N., Hyman, G., & Coomes, O. (2010). The role of pasture and soybean in deforestation of the Brazilian Amazon. *Environmental Research Letters, 5*(2).

Callan, J., Sundin, P., Suffian, S., & Mehta, K. (2014). Designing Sustainable Revenue Models for CHW-Centric Entrepreneurial Ventures. Global Humanitarian Technology Conference. Seattle: Institute of Electrical and Electronics Engineers. https:/doi.org/10.1109/GHTC.2014.6970357

Collyns, D. (2013, January 14). *Quinoa brings riches to the Andes.* Guardian.

Dimitri, C., & Oberholtzer, L. (2009). *Marketing U.S. Organic Foods Recent Trends From Farms to Consumers.* Washington, D.C.: United State Department of Agriculture.

Elias, A., Nohmi, M., Yasunobu, K., & Ishida, A. (2013). Effect of Agricultural Extension Program on Smallholders' Farm Productivity: Evidence from Three Peasant Associations in the Highlands of Ethiopia. *Journal of Agricultural Science, 5*(8). https:/doi.org/10.5539/jas.v5n8p163

Goyal, A., Sharma, V., Upadhyay, N., Gill, S., & Sihag, M. (2014). Flax and flaxseed oil: an ancient medicine & modern functional food. *Journal of Food Science Technology, 51*(9), 1633-1653. https:/doi.org/10.1007/s13197-013-1247-9

Hrušková, M., Švec, I., & Jurinová, I. (2012). Composite Flours-Characteristics of Wheat/Hemp and Wheat/Teff Models. *Food and Nutrition Sciences, 3*(11), 1484-1490. https:/doi.org/10.4236/fns.2012.311193

Jiang, Y. M., Zhang, Z. Q., Joyce, D. C., & Ketsa, S. (2002). Postharvest biology and handling of longan fruit (Dimocarpus longan Lour.). *Postharvest Biology And Technology, 26*(3), 241-252. https:/doi.org/10.1016/S0925-5214(02)00047-9

Johns, T., & Sthapit, B. (2004). Biocultural Diversity in the Sustainability of Developing-Country Food Systems. *Food and Nutrition Bulletin, 25*(2), 143-155. https:/doi.org/10.1177/156482650402500207

Keane, J., Page, S., Kergna, A., & Kennan, J. (2009). Climate Change and Developing Country Agriculture: An Overview of Expected Impacts, Adaptation and Mitigation Challenges, and Funding Requirements. Geneva: International Centre for Trade & Sustainable Development and the International Food & Agricultural Trade Policy Council.

Kim, M. (2013). Growing Organic around the World: Domestic Regulations, International Relations and Developing Countries' Involvement in Organic Production. Annual Meeting of the International Studies Association. San Francisco: International Studies Association.

Kramer, D. B., Urquhart, G., & Schmitt, K. (2009). Globalization and the connection of remote communities: A review of household effects and their biodiversity implications. *Ecological Economics, 68*(12), 2896-2909. https:/doi.org/10.1016/j.ecolecon.2009.06.026

Kumar, A., Reddy, B., Sharma, H., Hash, C., Rao, P., Ramaiah, B., & Reddy, P. (2011). Recent Advances in Sorghum Genetic Enhancement Research at ICRISAT. *American Journal of Plant Sciences, 2*(4), 589-600. https:/doi.org/10.4236/ajps.2011.24070

Lawrence, F. (2010, September 14). How Peru's wells are being sucked dry by British love of asparagus. Guardian.

Machuka, J. (2001). Agricultural Biotechnology for Africa: African Scientists and Farmers Must Feed Their Own People. *Plant Physiology, 126*(1), 16-19. https:/doi.org/10.1104/pp.126.1.16

Maeder, P. F., Dubois, D., Gunst, L., Fried, P., & Niggli, U. (2002). Soil Fertility and Biodiversity in Organic Farming. *Science, 296*(5573), 1694-1697. https:/doi.org/10.1126/science.1071148

Mintel. (2011). *Ethnic Food-Lovers Developing a Taste for Exotic Flavors.* Chicago: Mintel Research.

Motes, W. (2010). *Modern Agriculture and Its Benefits – Trends, Implications and Outlook.* Washington, D.C.: Global Harvest Initiative.

National Research Council. (2006). *Lost Crops of Africa: Volume II: Vegetables.* Washington, D.C.: National Academies Press.

National Research Council. (2008). *Lost Crops of Africa: Volume III: Fruits.* Washington, D.C.: National Academies Press.

Oelke, E. A., Putnam, D. H., Teynor, T. M., & Oplinger, E. S. (1992). Quinoa. In Alternative Field Crops Manual. Madison, WI: University of Wisconsin and University of Minnesota.

Pingali, P. (2012). Green Revolution: Impacts, limits, and the path ahead. *Proceedings of the National Academy of Sciences of the United States of America, 109*(31), 12302-12308. https:/doi.org/10.1073/pnas.0912953109

Regmi, A. (2001). *Changing Structure of Global Food Consumption and Trade.* Washington, D.C.: United States Department of Agriculture Economic Research Service.

Schroeder, E. (2013, December 16). *Global demand strengthens for ancient grains.* Food Business News.

Stone, A., Massey, A., Theobald, M., Styslinger, M., Kane, D., Kandy, D., ... Davert, E. (2011). *Africa's Indigenous Crops.* Washington, D.C.: Worldwatch Institute.

Suffian, S., De Reus, A., Eckard, C., Copley, A., & Mehta, K. (2013). Agricultural technology commercialisation: stakeholders, business models, and abiotic stressors-Part 1. *Journal of Social Entrepreneurship and Innovation, 2*(5), 415-437. https:/doi.org/10.1504/IJSEI.2013.059314

Trichopoulou, A., Soukara, S., & Vasilopoulou, E. (2007). Traditional foods: a science and society perspective. *Trends in Food Science & Technology, 18*(8), 420-427. https:/doi.org/10.1016/j.tifs.2007.03.007

Weis, T. (2007). *The Global Food Economy: The Battle for the Future of Farming.* London: Zed.

Willer, H., & Lernoud, J. (2014). *Organic Agriculture Worldwide: Current Statistics.* Frick, Switzerland: Research Institute of Organic Agriculture.

14

Success and Success Factors of Domestic Rainwater Harvesting Projects in the Caribbean

Everson James Peters[1]

[1] University of the West Indies, Trinidad and Tobago

Correspondence: Everson James Peters, University of the West Indies, Trinidad and Tobago. E-mail: everson.peters@sta.uwi.edu

Abstract

In the Caribbean, domestic rainwater harvesting (DRWH) projects are being implemented to augment water supplies in water scarce islands and as a no-regret approach to adaptation to climate change. The evaluation of these projects is usually limited to the implementation process i.e. measuring the ability of the project to meet the set deliverables. Factors that are considered are the cost and time specified for the installation of the DRWH systems and the quality of the harvested water. There is seldom a post-project evaluation to determine whether the beneficiaries are able to properly maintain the system and or to improve on it, or whether the project is leading to increased household collection and use of rainwater in the project location and its environs. This paper is based on a survey of key stakeholders actively involved in the promotion of DRWH over a number of years. Active involvement was the basis of accepting the information on their perception as adequate in providing a reliable measure of the level of success of DRWH projects. The metrics for success were based on stakeholders' perspective of the success of DRWH projects as determined by community involvement, rate of uptake of DRWH, increased awareness, impact of training on maintenance of systems, appropriate use of the systems, increased use of rainwater, increased capacity of community leaders to train and improved support by local private sector. It was found that there was willingness to invest in DRWH particularly among the stakeholders who have regularly used rainwater. The stakeholders were also asked to corroborate a set of pre-selected factors that were considered important for the successful development of DRWH projects. A ranking of these factors indicated that although the cost of the DRWH systems was the most important factor for success, technical issues were imperceptibly more important than economic and social issues.

Keywords: success factors, domestic rainwater harvesting, Caribbean

1. Background

"Nothing is more useful than water: but it will purchase scarcely anything; scarcely anything can be had in exchange for it" (Smith, 1776). In 1992, the Dublin Statement of the International Conference on Water and the Environment recognised that water should be considered an economic good, giving credence to Smith's statement. Today, dwindling freshwater resources threatens the availability of conventional water supplies in several parts of the world and as such, greater emphasis is being placed on harnessing alternative sources. Rainwater harvesting (RWH) is a broad term for small-scale, collection, storage and use of rainfall runoff for productive purposes. In this context, RWH is a simple low-cost technique that requires minimum specific expertise or knowledge. Through the ages, RWH, has been an important source of fresh water for agriculture and domestic use.

Harvested rainwater is used for both domestic and commercial purposes. Domestic rainwater harvesting (DRWH), that is, collecting rainwater from roofs and storing it in containers of varying sizes for household uses. Traditionally this has been the main source of potable water, particularly in remote and isolated communities. Recently, DRWH has been given a new lease on life after expansion of conventional public water supply systems was found to be unable to satisfy the demand in all situations. The impact of climate change through the increased and prolonged drought conditions and variations in rainfall patterns have been the driver for growing interest in RWH. As a result, DRWH is finding application in drought mitigation, flood mitigation, poverty reduction, crop irrigation, watershed protection and carbon emission reduction (Ariyananda, 2009). In the Caribbean, the practice of RWH is contributing to resilience in water management. The importance of RWH in sustainable development and in the adaptation to climate change as a no-regret option, has been highlighted (Pandey, *et al.*, 2003). An example is in the case of Grenada where after Hurricane Ivan, a DRWH project was commissioned to build resilience in water communities in terms of access of water in a post-disaster environment and to augment supplies during droughts (CEHI, 2006).

Recognising the potential of DRWH, many regional and international agencies are responding to climate-driven water resource challenges by promoting RWH (Global Water Partnership-Caribbean 2016). In the Caribbean, DRWH projects are being implemented to supplement water supplies in water scarce islands. In many cases, considerable resources are utilized in DRWH projects to enhance potable water supply (GWP-C, 2012) and for commercial application in the tourism industry and agriculture sector (Blake, 2014).

Due to the project approach adopted by developers of DRWH projects, their evaluation usually limited to the construction phase (Bhuiya 2013) and hence focuses on the ability of the project during the construction phase to meet a set of objectives related to time and quantity. In other cases projects are evaluated in terms of water savings (Abdulla and Al-Shareef, 2009; Eroksuz and Rahman, 2010). In the case, of DRWH projects, after the system installation, there is seldom a post-installation evaluation to determine the sustainability of the system. For example, there is no assessment of the beneficiaries' ability to properly maintain the system or to improve on it, or whether the project is leading to greater utilisation of harvested rainwater in the community. Needless, to say, the RWH infrastructure should be built to function efficiently over time, i.e., to provide adequate water to meet the needs of the users in a sustainable way, particularly in relation to climate change.

Although governments, non-government organisations (NGOs) and international development organizations are actively promoting RWH in the Caribbean, there have not been any empirical and coordinated studies to measure the success of such projects by evaluating the efficiency and effectiveness of this technology as an alternative and/or a primary water source. Factors for better policy development, if identified and considered during project design and implementation, can enhance DRWH promotion. CEHI (2006) proposed a programme for promoting RWH in the Caribbean region which recommended that there should be a coordination of monitoring of the success of national DRWH programmes.

In this research, success of DRWH is interpreted as the extent by which RWH technology is adopted for meeting drinking water needs. This paper therefore, evaluates the factors contributing to the success of RWH projects as perceived by principal stakeholders (managers in the international aid agencies and regional agencies, local officials, project team, steering committee members, beneficiaries). It analyses the opinions of these principal stakeholders which were solicited on the following issues: how correctly was the RWH systems installed; how well did water harvested from the systems meet expected quantity and quality; what was the level of consumption of the harvested water; how well were the systems maintained; how adequate was the technical and financial support; and what was the adaptation response as a result of the projects. Further, the opinions were solicited in determining the factors that influence the success of RWH, particularly, with regard to the up-scaling and sustainability of DRWH projects in the Eastern Caribbean. Factors such as the knowledge, attitudes and practices of DRWH, and how favourable are the local cultures, the institutional environments and governance of DRWH are considered.

2. Literature Review

There has been a revival of RWH around the world due to a number of factors (Smet, 2003). These include the failure of centralised water systems due to decreased quality and quantity of the source, safer roof material, lower costs of storage tanks and greater emphasis on community-based approaches and technologies which emphasise participation, ownership and sustainability of water supply (Smet, 2003). The revival is also driven by a number of reported success stories on RWH projects for domestic and commercial uses. For example, in Chennai, India, Krishnan (2003) observed that households were willing to invest in RWH after hearing of other people's experiences. The success of a 2009 RWH project by NetWater and its partners that installed RWH systems at 12 hospitals in Sri Lanka triggered a further 22 RWH applications by 2011 (Women for Water Partnership, 2012). In Central Sudan, Ibrahim (2009) reported on how the success of small RWH dams has encouraged the authorities to expand the RWH initiative as part of a long-term solution for the water supply in the city.

In the Caribbean, the practice DRWH is evolving and expanding (Dempewolf *et al.*, 2015). Generally, DRWH projects fall under a broad heading of not-for-profit developmental projects. As with other developmental projects, the criteria to objectively assess the success of DRWH is project specific. Although, many methodologies developed to assess the sustainability of RWH projects recognise the importance of the socio-economic factors for success (Goyal and Bhushan, 2005), these factors are often not incorporated in the assessment. This non-integration of socio-economic factors was identified as the main cause of failure of RWH projects (Kahindaa *et al.,* 2008). In a study of two RWH projects commissioned by UNICEF (2004), it was found that RWH projects were successful where there was a willingness of beneficiaries to contribute in terms of both labour and funding. The study also suggested that success could be hampered by the involvement of multiple agencies resulting in uncoordinated execution which leads to an increase in overhead expenses.

Typically, the success of RWH projects is measured by some specific objectives like water availability, improvement in household water supply and sanitation (Kahinda *et al.,* 2006) as well as greater household productivity that results in the saving of time and energy (RAIN, 2014). Success of RWH projects should also be measured by levels of adaptation as indicated by increased participation in RWH, growth in the number of systems and improvement in operations of established systems over time. Previous evaluations of RWH systems, however, focused only on quality and quantity issues (Despins *et al.,* 2009; Toronto and Region Conservation Authority, 2010), and economic viability (Roebuck *et al.,* 2011). Therefore, the tendency is to measure the success of DRWH projects by how much of the expected outputs, based on the project design, has been realised over the short-term, that is, during the implementation period of the project

The success of DRWH projects is, of course, not guaranteed in all situations. For example, while domestic rainwater harvesting systems (DRWHS) are being widely promoted as a solution for the growing drinking water crisis in many underdeveloped and developing countries, there are limitations to the use of DRWH. On the other hand, Kumar (2004) argued that RWH systems are not alternatives to public systems in urban and rural areas of regions that receive low rainfall, particularly for urban housing stocks of low and middle-income groups.

Another factor that can influence success is legislation. The literature shows that legislation that enforces rainwater harvesting for new buildings is being introduced in many parts of the world (Rainwater Harvesting, 2002; Rainwater Harvesting, 2010; Government of Jamaica, 2013). However, legislation must be supported by effective institutional and organizational arrangements. In Delhi, despite the legislative requirement for RWH several factors have contributed to failure (Dikshit, 2012). These include inadequate financial assistance, bureaucracy and poor maintenance of structures once they are built (Dikshit, 2012). In another Indian district, the majority of the RWH projects that were set up at various government institutions became defunct after 5 years owing to lack of proper maintenance work as a result of institutional and organizational failures (Kannan, 2014). In the case of the Caribbean, Hutchinson (2010) reported that in Barbados, regulation had a positive effect on the storage and use of rain water for secondary purposes. While success may not be achieved through the passing of legislation (Karnna, 2014), legislation may be necessary but not sufficient as the accompanying innovative policy interventions, incentives and regulations by local urban bodies must be adopted (Kahinda and Taigbenu, 2011).

A high degree of stakeholder participation is necessary to guarantee the acceptance of proposed projects. For many development projects, experience has shown that beneficiaries are often not satisfied with the programmes when these do not meet their expectations (Zimmermann *et al.,* 2012). For example, poor stakeholder participation can cause projects to fail when beneficiaries decide to use the RWH tanks as storage facilities for conventional public water supply rather than for storing rainwater. This could be a case of inertial resulting in an unwillingness to adapt to the DRWHS. However, when such projects are intended for RWH, this is usually an indicator of poor beneficiary targeting and the non-participatory approach adopted in the implementation phase (Ariyananda and Aheeyer, 2011). In this context, Boodram *et al.* (2014) identified targeting and engaging communities, establishing partnerships with private sector organisation, expansion of training to include certification and entrepreneurship training and monitoring of projects as necessary for improving the success in RWH.

Understanding the critical success factors enhances the ability of funders of RWH projects to ensure desired outcomes. However, defining the criteria to measure project success has been recognized as a difficult and controversial task (Baccarini, 1999; Liu and Walker, 1998). Research on success factors for industrial, commercial and general projects is widely available but, research on not-for-profit development projects like RWH is limited. Such projects with humanitarian and social objectives are usually much less tangible, less visible and not easily measurable, compared with projects commonly found in the private sector (Khang and Lin Moe, 2008). Diallo and Thuillier (2004) undertook an important empirical research that focused on the specific success criteria and factors of development projects. They did so by assessing the project success as perceived by the key stakeholders. They found that management dimensions (time, cost, quality) were of high importance while project impacts were rated lowly. Generally, the factors selected for inclusion in evaluating success of projects can vary considerably (Fortune and White, 2006).

In the Caribbean, successful rainwater harvesting projects are generally associated with communities that consider adequate water supply a priority (UNEP, 1997). Consequently, recent DRWH projects in the Caribbean aim to promote and expand the application of RWH such that there is higher rate uptake of the technology (CEHI 2006; GWP-C 2016). Yet there has not been any formal mechanism that evaluated the potential or actual up-scaling of RWH projects (increasing catchment areas, storage capacity, improving quality by installing purification facilities) or increasing investments and sustainability (maintenance of quantity and quality over time). This is partly due to the requirement of long time for monitoring these indicators and the complexities involved. In 2006, participants at a regional workshop (Promoting Rainwater Harvesting in the Caribbean) proposed that the success of RWH projects in the region should be monitored (CEHI, 2006). Boodram and Dempewolf (2015) recognised that while evaluations of Caribbean Integrated Water Resources Management (IWRM) initiatives including RWH, are routinely carried out immediately after project completion, assessments are lacking on the long term viability and sustainability of the initiative. Follow-up is lacking and the long term effects of the project remain unassessed, therefore limiting the ability to replicate projects to be more successful andt effective in the long-term (Boodram and Dempewolf, 2015).

3. Methodology

An evaluation of DRWH projects can be carried out by soliciting the views of key stakeholders who have been or are actively involved in DRWH projects. In this study, the stakeholders were mainly experts in RWH who were identified as suitable for participation in the survey since it required much less resources than that for surveying beneficiaries spread over many islands. Moreover, access to the experts was easier than other beneficiaries. For this study, experts are defined as persons who are or have been actively involved in the development of DRWH through project design, project implementation, training and capacity building, technical support to projects in post construction phase, project monitoring, research, project financing and the promotion in communities. Participants for the research were selected from sources. Firstly, use was made of a data base of Caribbean experts in water and related fields created by Global Water Partnership-Caribbean (Global Water Partnership-Caribbean, 2014). Secondly, the annual conference of the Caribbean Water and Waste Association, the biggest gathering of professionals in water management in the region was used to identify and solicit participation in the survey. Finally, a list experts was supplemented by identifying additional suitable candidates for participation through consultation with regional and international NGOs involved in RWH activities, government departments responsible for water issues and water utility companies in the islands.

A total of 45 persons were identified as experts in RWH meeting the criteria of active involvement in DRWH. Of these, 40 persons, who could have been contacted, were invited to participate in the survey. Participants were selected from among 5 groups namely: government, non-governmental organisations, technical support, water utility companies and researcher.

A pilot survey was administered to three experts to pre-test the questionnaire in order to eliminate inadequacies or ambiguity in the questionnaire. The participants were invited to comment on the format of the questionnaire and make suggestions which were considered for incorporation into the final questionnaire..

The first section of the questionnaire was comprised of seven items to obtain information on the level of involvement of the participants in DRWH. The second section comprised 42 questions that sought the participants' perspective on performance factors that would determine the success of DRWH projects in the Caribbean. Each question in the second section was based on a 7 point Likert scale where 1 = strongly disagree and 7= strongly agreed. There were 4 questions related to community involvement; 6 questions related to training; 9 questions on success; 7 questions on the beneficiaries and users and 16 questions specific to the development and sustainability of RWH projects. The questions on success measured: community involvement, rate uptake of DRWH, increased awareness, impact of training on maintenance of systems, appropriate use of the systems, increased use of rainwater, increase capacity of community leaders to train and improved support by local private sector.

In the third section participants were asked, to use a scale of one to ten, to rate the importance of 14 success factors where one represented low importance and ten was very high importance. The list of success factors was compiled from available literature (UNEP 1999; Pearce-Churchill et al. 2005; and Bhuiya 2013) and feedback from the Pilot survey. The overall ranking was analysed by categories of participants. This was done by taking the arithmetic mean of the responses of each category of participants. The fourth section which was optional required participants to provide any additional comments that were considered relevant to the survey. Follow up telephone interviews with participants were done.

The distribution of the questionnaire was accompanied by a letter requesting participation and advising on the confidentiality of the responses. Participants were advised to by-pass questions which they considered to be sensitive. For the analysis of data, scores of all variables were averaged to obtain the overall indication of the project success. The perception of specific groups in countries was obtained by grouping participants by country and category. The student t-test and correlation coefficients were used to compare the responses between the different categories.

4. Results and Discussions

4.1 Participants

The results of Section I of the questionnaire and the first two items of Section II, on the level of involvement of participants in RWH confirmed that the criteria for the selection of participants were met. There was 80% response from the target group of which 95% completed all sections of the questionnaire.

The main organisations from which participants were selected included the Organisation of Eastern Caribbean States (OECS), Food and Agriculture Organisation (FAO), Global Water Partnership-Caribbean (GWP-C), the St. Vincent de Paul of the Catholic Church, United Nations Environment Programme (UNEP) and University of the West Indies (UWI). The selected participants for the survey were considered to be the best persons to assess the success of DRWH projects if they had more than 3 years working experience in the field. To facilitate the analysis, the participants were categorised into five main groups as shown in Figure 1, were from the following islands: Organisation of Eastern Caribbean States (Antigua and Barbuda, Carriacou, Grenada, St. Vincent and the Grenadines, Saint Lucia, Dominica,), Trinidad and Tobago, the Virgin Islands and the others (Jamaica, Barbados, Guyana and the Bahamas) as shown in Figure 2.

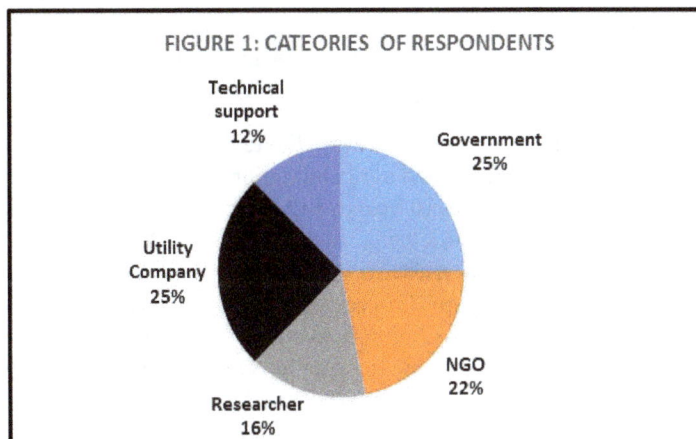

Figure 1. Categories of respondents

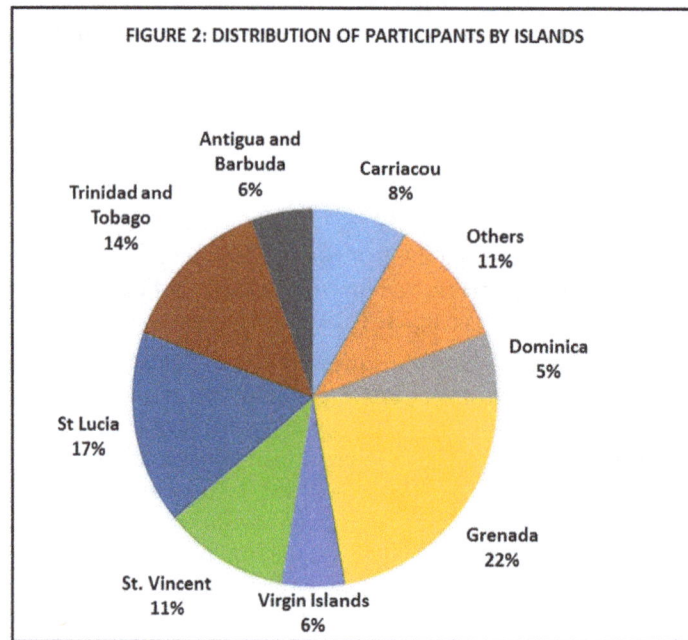

Figure 2. Distribution of participants by Islands

In this study, 53% of the participants' experiences were limited to RWH for domestic purposes, 3% for agriculture purposes and 44% for both domestic and agricultural purposes. Forty eight percent of the participants were involved in the RWH sector for more than 10 years while another 30% were involved for between 3 years and 5 years. More than half (52%) of the participants used RWH on a regular basis and considered it their primary source of freshwater, while 36% used it occasionally, and only 14% considered themselves non-consumers of harvested rainwater. Although 75% of the participants were willing to invest at the personal level in RWH, only about 60% of these were willing to make that investment even if the water from the RWH system was more expensive than that of that public supply. Overall, 80% of the participants considered themselves having an excellent knowledge of RWH with the remainder claiming good knowledge. Based on the responses to the first section of the questionnaire, the overall characteristics of the participants suggest that they are likely to be good promoters for RWH and good judges of whether DRWH has been successful.

4.2 Support for RWH

In order to expand the use of RWH, adequate general support from key stakeholders is required. The study showed that the experts in RWH, as identified for this project, concluded that the support from some key stakeholders is weak and from the responses, the strongest support came from the beneficiaries and users of RWH, particularly in the islands where RWH is well developed (Figure 3). This may be expected as this group has direct benefit. More than half (52%) of the participants were of the view that users were willing to spend their own money to maintain and upgrade their RWH systems while only a minority (12%) were of a contrary view. The other participants were non-committal. The willingness to invest in maintenance and upgrade of RWH systems was highest among participants from Carriacou and St. Lucia. In the case of Carriacou, this may be explained by the high familiarity with DRWH since RWH has been well established on this island for some time, whereas in the case of St. Lucia recent activities in the promotion of DRWH may have enhanced public awareness. The lowest indication of willingness to invest was found to be in St. Vincent, where water supply was generally satisfactory.

The weak support from users (Figure 3) was reported mostly by the participants who were from islands where RWH has not been widely utilised previously. This weak support is reflected in a poor rate uptake of the technology to date, notwithstanding a number of projects undertaken in these islands during the last decade and may be due to satisfactory public water supply in those islands. From the survey it was suggested from the study that where higher level of support was observed from the users of RWH, this can be capitalised such that their involvement and local knowledge could influence costs and allow for innovative improvements of current systems.

The least overall support for DRWH was from the governments. The view of a low level of support from the governments was a view supported by 75% of participants who represented governments. Further, there was consensus (88% strongly agreed) that government support is critical to the success of RWH projects. A minority (12%) of participants observed that little support is given to small communities where the practice of RWH is not well established.

International NGOs and regional and international development agencies have provided mostly moderate to strong support which was acknowledged by all categories of participants. While overall, there is support from international NGOs and regional and international agencies for the development and enhancement of RWH, 28% of participants considered that there was too much dependence on these bodies. At the same time, there is an observation among a minority (23%) of participants that RWH projects failed on the departure of sponsors. Put simply, beneficiaries abandoned the projects or neglected to adequately maintain the RWH systems, particularly when rainwater generally only supplements a more regular supply. This observation is not unique to RWH systems as often in development projects, there is a long standing history of projects failing shortly after the agencies leave due largely to the lack of investment in capacity building (Stergakis, 2011). Failure also occurs as the beneficiaries do not "own" the projects in the first place. Further, there is a larger cultural issue which is based on the thinking that donors will always come in to support these initiatives.

Interestingly, about half (48% of the participants) believed that the growth of RWH depended on the support from the international NGOs. In contrast, support from the local NGOs in the promotion and development of RWH was assessed to be adequate by only 20% of the participants. All told, partnering between the regional and international agencies and local NGOs seems to be weak. Further, the financial resources available for the development of RWH in the region, were assessed to be adequate by only 16% of the participants and these were mainly government representatives.

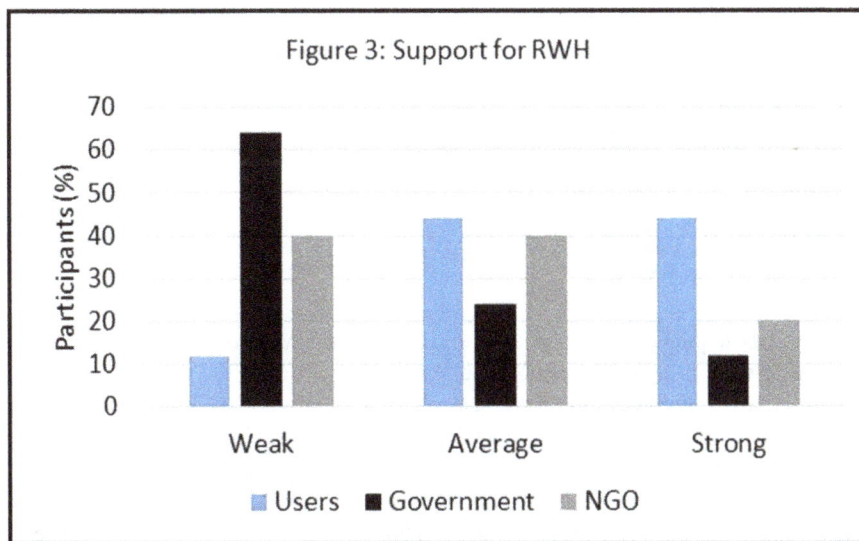

Figure 3. Support for RWH

4.3 Impact of Legislation

The impact of legislation is influenced by both the history of RWH and the motivation for its introduction. Among the islands which have laws requiring RWH are Barbados and the US Virgin Islands. In the U.S. Virgin Islands, the law requires that provision be made in the construction of all new buildings for the capture and storage of rainwater. From interviews with participants from the US Virgin Islands, it was revealed that the stored rainwater is generally fully utilised and is preferable to desalinated water. However, the extent of the influence of legislation on this outcome is not clear. Notwithstanding, the view among leaders in the RWH sector (70% of participants) is that legislation facilitates the growth of RWH and enhances the success of RWH projects.

While the impact of legislation is not conclusive, the results of this study suggest that in the Caribbean, appropriate legislation is likely to enhance RWH development. Nevertheless, there appears to be a hesitation to develop legislation due to the possibility for the use of subsidies for the installation of RWH and a possible reduction in revenue for utility companies (Boodram and Dempewolf, 2015).

4.4 Training

Included in almost all RWH projects are training on systems installation and maintenance, and programmes for public awareness. As shown in Figure 4, most of the participants believed that training was inadequate. This was the case particularly during the post implementation of RWH projects and for the training of community trainers as only 11% and 9% are of the view that training was in the case of post implementation and community training respectively. However, it was generally agreed (60% of participants) that there was increased knowledge of the RWH technology among the communities. Moreover, 82% of participants strongly believed that an increased provision of adequate technical training is important for households to maintain and improve the quality of water from their RWH systems. It was observed by the participants (64%) that households were interested in improving or actually improved their systems through the interaction with neighbours who have been exposed to RWH through formal RWH project.

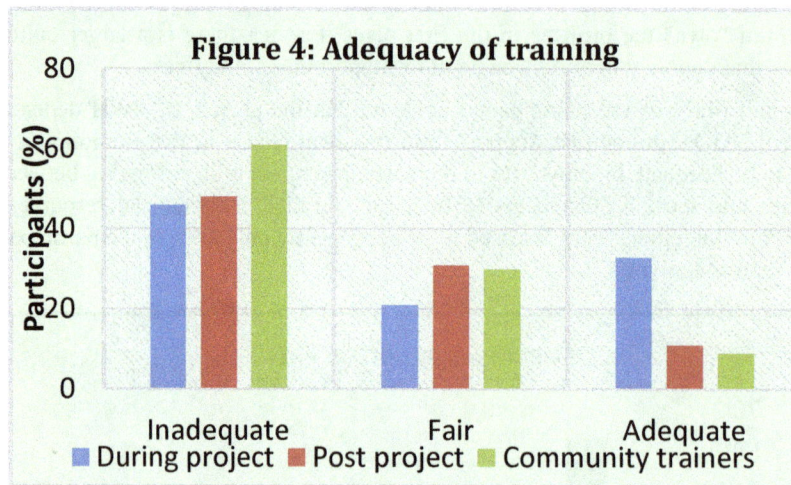

Figure 4. Adequacy of training

4.5 Community Involvement

A significant number of participants (64%) observed that generally RWH projects are intended to be targeted to the communities and individuals that are most vulnerable to water scarcity. However, there was doubt as to whether the best individuals or communities were selected, since only 36% of participants indicated that from observations through active involvement with communities, that individuals and communities were poorly selected. The majority of participants were satisfied with the involvement of the individual beneficiaries at all stages of the implementation of the projects, which suggests that beneficiaries are generally consulted at all stages. At the same time, the majority of these participants believed that the general public should be more involved in the development and promotion of RWH projects. Only a small portion of the participants reported that the majority of the population does not need to be more involved in RWH projects.

Conversely, 76% of participants believe that the cultural practices of communities are not adequately considered in the design of projects. This may partly explain why there is not a strong sense of trust between the sponsors and developers of the projects and the beneficiaries of RWH projects as suggested by the response of 60% of participants.

4.6 The Selection of Beneficiaries

Although the most vulnerable community might be selected, the success of RWH projects can also be influenced by the socio-economic status of the beneficiaries. Only 24% of participants believed that the beneficiaries were in positions to pay for the costs of electricity and other accessories where these were needed in the RWH systems. Consequently, 70% of the participants agreed that simpler systems would have been more affordable to the beneficiaries of the RWH projects. Further, there was consensus (92%) that greater success would have been possible where projects targeted communities with existing RWH systems. Poor selection of beneficiaries means that greater efforts are required to achieve success. In this regard, there was a majority (64% of participants) view, that there is often an over-concentration of efforts on projects where there was evidence of failures and as such there are often signs of project fatigue among beneficiaries as indicated by 56% of the participants.

4.7 Success

There has been a number of interventions to enhance the availability of water to water scarce communities in the region. Generally, in this study, participants have a positive perspective about the success of RWH projects (with an average score of 5.2 on a scale 0 to of 7). For the participants who were regular consumers of rainwater their perspective were slightly less positive (with an average score of 4.94 on a scale of 0 to 7). In the case of participants who never used or occasionally used rainwater there appeared to be a more positive view of the success (with an average score of 5.4 on a scale of 0 to 7). However, statistically there is no significant difference in the perception between the groups as the t-test (at α =0.05) of averages of the groups returned a p-value of 0.47. This suggests that the perception of success was not influenced by the regular consumption of DRWH by participants. On the other hand, there was a significant difference in the view of the level of success by participants who represented the government (average score of 6.12 on the 7 point scale) and non-government representatives (average score of 4.89 on the 7 point scale) as indicated by a p-value of 0.009 for the t-test (at α =0.05). As there is often political interest of governments in reporting success of DRWH projects which impact on vulnerable groups, there is likely to be tendency for government officials to report on the success of the projects. Among the different categories of participants, as shown in Figure 1, representatives from the utility companies showed the least positive view of the success of DRWH projects with an average score of 4.2 on the 7 point scale. Further, less than half (46%) of the participants rated the RWH projects undertaken in the past two decades as highly satisfactory. Although the representatives from the utility companies have a more negative about success of DRWH projects, it is important for DRWH to be fully integrated into the management of the islands' water resources. The utility companies must support the stakeholders of RWH projects as these projects can have positive impacts on the people who need water.

Most of the participants (73%) believe that DRWH has the potential to be sustainable and that the technology would be maintained in its present form for the foreseeable future. In the study, 82% of the participants have endorsed the view that evaluation of the success of projects is generally limited to their designated timeframes with few follow-up after the implementation of physical systems. The greatest success in DRWH was therefore, identified as taking place in the initial stages of the development of the projects, that is, the mobilisation and awareness raising of communities and in the installation of DRWH systems during which enthusiasm and motivation were found to be the highest.

Among the researchers, there was a perception that progress made in the promotion and development of RWH was less than established targets. Sixty two percent of the participants were of the view that greater success of projects was seen where there was a good match between the scope of the project and the available finances. For RWH projects in the region, a minority of participants (20%) believe that success was not achieved even if technology was consistent with the local culture. Two other key factors identified by the majority of participants as critical to success are (a) project design and implementation that is incremental, and (b) functioning indigenous RWH systems.

Some weaknesses were identified in the design and implementation of RWH projects which hamper success. More than half (68%) of the participants were of the view that too much resources were spent on the administration (salaries of consultants and project managers) of projects and too little on actual management of water. This short coming suggests that the financial resources could be better distributed to the different components of projects. This might require a protocol for the proper allocation of funds.

An important challenge for RWH is the availability of funding. While participants acknowledged the government's role, the main source of funding is from regional and international development agencies. The potential role of entrepreneurs seems to be limited. Only a minority (26%) of participants believe that local entrepreneurs can play a greater role in the development of RWH while about half (48%) of the participants did not see RWH as being attractive enough to motivate investment from local entrepreneurs. This position may be influenced by the current state of RWH development and the level of involvement of the private sector and could therefore be temporary. Unlike what is found in this study, Naugle (2011) found that the private sector In Uganda was well positioned to provide RWH products to meet the demand of householders who are prepared to make investment in the construction and up-grading of DRWH systems. This may be explained by the differences in socio-economic conditions between the Caribbean and Uganda

4.8 Success Factors

Each RWH project goes through a project development analysis by the regional or international funding agency and critical success factors are routinely identified. The interview with the experts in RWH from the Caribbean, who participated in this study, confirmed that the 14 factors which were identified were important to the success of projects. These factors were rated according to importance on a scale of 1 to 10 where 10 was most important and 1 least important. Table 1 shows the average rating by participants. For the aforementioned success factors (Table 1), the mean value ranged from 5.89 to 8.52 on a scale to a maximum of 10. The standard deviations show that the spread was less than half the mean, with the least variation in the participant's perception of technical support and the most important success factor being the costs of the systems.

The ranking of success factors by participants varied according to institutional association and place of residence, as show in Table 1. One of the top three success factors was identified in each, of the social, technical and economic groups of factors. It must be noted that all categories of participants identified the initial capital costs of installing RWH systems (particularly storage facilities) as the most important success factor. Overall, education and public awareness through training at the school and community levels, appear to be important factors in ensuring success. The least important factor was the educational levels of the beneficiaries.

The correlation analysis shows that the greatest similarity in the way success factors were ranked was between participants associated with water utility companies and those associated with governments. There was a correlation coefficient of 0.462 between the rankings of these groups (See Table 2). Participants employed in research, the NGOs and project sponsor agencies tended to rank in reverse order as indicated by the Correlation Coefficient of -0.392

An analysis of the results for Trinidad, Grenada, the OECS excluding Grenada and the other countries (Table 3) shows that the highest correlation was between Grenada and the rest of the OECS. Among the OECS, the greatest DRWH intervention has been in Grenada. There was almost no correlation (correlation = -0.012) between the order in which the RWH mature islands (Carriacou, the U. S. Virgin Islands and Antigua and Barbuda) ranked success factors and the ranking by the other countries (Jamaica, Guyana and The Bahamas) where formal DRWH is currently not widely utilised or is an early state of formal development. The above observations on the ranking of the success factors suggest that in planning and developing RWH projects due consideration must be given to the particular islands and the engagement of the different stakeholders should take into consideration what are important. Notwithstanding, the success factors as identified in the islands where RWH has developed to a mature state, should be given the highest priority in planning and implementing RWH projects. Such success factors are quite likely to have been determined by individuals with lessons they have learnt from their experiences over a long period.

Table 1. Success factors: Statistics and ranking in order of importance

| Category | Success factors | Statistics | | | RANKING | | | | | | |
		Mean (N=35)	STD	Overall	Government	Research	NGOs and international agencies	Utilities	RWH mature islands	OECS	Other islands
	Involvement of schools	7.78	2.74	5	8	9	2	6	8	2	5
	Educational level of beneficiaries	5.89	2.53	14	14	3	14	8	12	13	13
Social	Public support	7.70	2.75	13	10	12	8	10	6	11	7
	Training in the community	6.04	2.41	3	5	10	3	2	5	4	2
	Previous experiences with RWH	7.97	2.86	8	5	1	11	2	2	3	14
	Legislation	7.00	3.02	7	13	12	8	7	14	8	9

Technical	Technical support	8.13	1.57	2	3	3	11	1	4	6	3
	Inclusion of a maintenance component in the design	7.78	2.52	4	5	6	4	2	6	4	4
	Follow up activities	6.96	2.71	6	12	3	5	12	12	10	6
	Availability of local tradesmen	7.07	2.23	5	2	11	7	9	3	14	8
Economic	External financial support	6.67	2.45	9	4	12	10	10	8	6	11
	Involvement of the private sector	6.81	2.38	10	11	6	6	14	8	9	8
	Government financial support	6.77	2.90	10	9	12	13	12	8	11	9
	The cost of systems	8.682	1.93	1	1	1	1	2	1	1	1

Table 2. Pair wise Pearson Correlation Coefficient

	Researcher	NGO	Utilities	Government
Researcher	1			
NGO	-0.392	1		
Utilities	0.252	0.40	1	
Government	0.093	0.276	0.462	1

Table 3 Pair wise Pearson Correlation Coefficient

	RWH mature islands	Trinidad	Grenada	OECS
RWH Mature Islands	1			
Trinidad	0.275	1		
Grenada	0.445	0.467	1	
Rest of OECS	0.325	0.528	0.768	1
Other Countries	-0.012	0.242	0.133	0.252

5. Conclusions and Recommendations

The high willingness of representatives of key stakeholder groups to invest in RWH at a personal level has the potential to make them good RWH project champions in their communities. Further, if the participants who are users of RWH are taken as a proxy for the general beneficiaries and users of RWH, then it could be concluded that there is a potential for investment in RWH projects by individuals.

The support from the different stakeholders varied. For example, financial support from government was generally low for ongoing RWH projects. While improved financial support from government could enhance the success of RWH projects, there appears to be no strong desire by governments to do so. Overall, the study found that if general support for DRWH could be maintained and further enhanced, then there is the potential for better growth of DRWH projects.

In DRWH development, the greatest success can be found during the phases of the projects that relate to capacity building and the delivery of pilot projects. Further, the success of projects seemed to be highest in communities where the use of rainwater was already well established. Consequently, selection of beneficiaries' community is important to ensuring success.

The top six success factors for RWH projects were identified in the study as: the cost of affordable systems; involvement of schools; provision of technical support; inclusion of a maintenance component in the design; training in the community; and undertaking follow up activities. Although the cost of DRWH systems is the most important factor for the success of DRWH projects, overall, technical issues appear to be no less important than social and financial issues. Training and public awareness (TPA), has also been identified as an important success factor. While TPA are now included in DRWH projects, it was generally agreed that the greater integration of TPA in the projects increases the chances of success of the overall project. This is important in enhancing confidence in the use of the water and its consumption which can lead to greater care for the systems and ultimately their sustainability. This finding is consistent with research carried out in the Cameroon where the two main challenges of up-scaling RWH was costs of systems and awareness (Folifac *et al.*, 2013).

Notwithstanding the absence of conclusiveness of legislation in the literature as a success factor for RWH, in this study it is a middle ranked success factor. There are no specific policies or legislative provisions to support RWH in some of the islands (Government of Grenada, 2012). If future RWH projects are to be successful and sustainable consideration should be given to enacting appropriate legislation that can help create an enabling environment.

Finally, it is recommended that in the conceptualisation and design of future RWH projects in the region, these success factors identified in the study should be given adequate consideration particularly those related to the long-term performance of the systems.

Acknowledgement

The author wishes to thank members of the water sector in the region who participated in the survey and Professor G Shrivastava, Mrs Patricia Aquing and Ms Laurel Bain for their valuable review contributions.

References

Abdulla, F. A., & Al-Shareef, A. W. (2009). Roof rainwater harvesting systems for household water supply in Jordan. *Desalination, 243*(1-3), 195–207. http://dx.doi.org/10.1016/j.desal.2008.05.013

Ariyananda, T. (2009). Climate Change and Rain Water Harvesting, *The 14th International Rainwater Catchment Systems Conference 2009* "Rainwater Harvesting to Cope With Climate Change"3-6 August 2009, Kuala Lumpur, Malaysia. Retrieved May 20, 2015, from http://www.eng.warwick.ac.uk/ircsa/members/pdf/14th/Slides/k1%20Ariyananda.pdf

Ariyananda, T., & Aheeyer, M. M. M. (2011). *Effectiveness of Rain Water Harvesting (RWH) Systems as a Domestic Water Supply Option.* Retrieved from http://www.lankarainwater.org/projects/docs/effectiveness_rwh_domestic_2011.pdf

Baccarini, D. (1999). The logical framework method for defining project success. *Project Management Journal, 30*(4), 25–32.

Bhuiya, M. R. (2013). Report on Identification of Barriers Against Successful Implantation of Rainwater Harvesting In Urban Areas: A Case Study of Dhaka City, Bangladesh University of Engineering and Technology, Dhaka, Bangladesh.

Blake, G. G. (2014). *Rainwater harvesting in the hospitality sector.* Retrieved July 4, 2015, from www.statiatourism.com/.../Rainwater_Harvesting.ppt

Boobram, N., & Dempewolf, L. (2015). Sustainability of Integrated Water Resources Management Initiatives in the Caribbean: Key Findings from the Sustainability of Integrated Water Resources Management Initiatives in the Caribbean Report.

Boodram, N., Dempewolf, L., Superville, L., & Wells, D. (2014). *Rainwater Harvesting in rural communities in Trinidad: Success stories and lessons learnt.* Retrieved August 21, 2015, from http://www.caribbeanrainwaterharvestingtoolbox.com/Images/Posters/Poster%20-%20RWH%20in%20Rural%20Communities.pdf

Caribbean Environment and Health Institute, CEHI. (2006). *A Programme for Promoting Rainwater Harvesting in the Caribbean Region.* Retrieved from http://www.cehi.org.lc/Rain/Rainwater%20Harvesting%20Toolbox/Media/Print/ProgrammePromoteRWH.pdf

Dempewolf, L., Boodram, N., Cobin, C., Cox, C., Clauzel, S., Vogel, H., & Nacher, E. (2015). *Rainwater harvesting in the Caribbean –State of Play 2015.* Retrieved from http://www.aidis.org.br/PDF/cwwa2015/CWWA%202015%20Paper_GWP-C%20Dempewolf.pdf

Despins, C., Farahbakhsh, K., & Leidl, C. (2009). Assessment of rainwater quality from rainwater harvesting systems in Ontario, Canada, *Journal of Water Supply: Research and Technology—AQUA.* 117-133. IWA Publishing. http://dx.doi.org/10.2166/aqua.2009.013

Diallo, A., & Thuillier, D. (2004). The success dimensions of international development projects: The perceptions of African project coordinators. *International Journal of Project Management, 22*, 19–31. http://dx.doi.org/10.1016/S0263-7863(03)00008-5

Dikshit, S. (2012). *Rainwater harvesting a failure, Times of India, June 8, 2012.* Retrieved September 14, 2015, from http://timesofindia.indiatimes.com/city/delhi/Rainwater-harvesting-a-failure-Sheila-Dikshit/articleshow/13912763.cms

Eroksuz, E., & Rahman, A. (2010). Rainwater tanks in multi-unit buildings: a case study for three Australian cities. *Resources, Conservation and Recycling, 54*(12), 1449–1452. http://dx.doi.org/10.1016/j.resconrec.2010.06.010

Folifac, F., Ndoping, Y., Banseka, H., & Mamba, L. (2013). *Climate change and water supply adaptation: Lessons from domestic rain water harvesting in Sudano Sahelian Cameroon.* Retrieved September 7, 2015, from http://www.gwp.org/Global/ToolBox/Case%20Studies/Africa/CS_460_Cameroon_full%20case.pdf

Fortune, J., & White, D. (2006). Framing of project success critical success factors by a system model. International. *Journal of Project Management, 24*(1), 53–65. http://dx.doi.org/10.1016/j.ijproman.2005.07.004

Global Water Partnership-Caribbean (GWP-C). (2012). Rainwater Harvesting Encourages Sustainability in Rural Communities. *Caribbean Water Insight, 3*(2), 4-7.

Global Water Partnership-Caribbean. (2014). *Caribbean IWRM Expertise.* Retrieved October 27, 2015, from http://www.gwp.org/en/Caribbean-Water-and-Climate-Knowledge-Platform/Databases1/Caribbean-IWRM-Expertise/

Global Water Partnership-Caribbean. (2016). *Mainstreaming rainwater harvesting to build climate resilience to the Caribbean water sector.* Retrieved May 5, 2015, from http://www.gwp.org/Global/ToolBox/Case%20Studies/Americas%20and%20Caribbean/CS_Caribbean_final_final_final.pdf

Government of Grenada. (2012). *Road Map on Building a Green Economy for Sustainable Development in Carriacou and Petite Martinique. Grenada*, Ministry of Environment, St. Georges, Grenada.

Government of Jamaica. (2013). *Government to make Rainwater Harvesting mandatory for housing projects, Jamaica Government Information Services.* Retrieved September 27, 2015, from http://news.caribseek.com/index.php/caribbean-islands-news/jamaica-news/item/66589-govt-to-make-rainwater-harvesting-mandatory-for-housing-projects

Goyal, R. R., & Bhushan, B. (2005). Rainwater harvesting: Impacts on society, Economy and Ecology, In *12th International Rainwater Catchment Systems Conferenc*e' "Mainsteaming Rainwater Harvesting" New Delhi, India - November 2005.

Hutchinson, A. P. (2010). *Rain Water Harvesting – Case Studies from the Barbados Experience, CWWA Conference, Grenada.* Retrieved September 27, 2012, from http://www.cehi.org.lc/Rain/Rainwater%20Harvesting%20Toolbox/Media/Print/A_Hutchinson_RWH_barbados.pdf

Ibrahim, M. B. (2009). Rainwater Harvesting for Urban Areas: a Success Story from Gadarif City in Central Sudan. *Water Resources Management, 23*(13), 2727–2736. http://dx.doi.org/10.1007/s11269-009-9405-6

Kahinda, J. M., & Taigbenu, A. E. (2011). Rainwater harvesting in South Africa: Challenges and opportunities. *Physics and Chemistry of the Earth, Parts A/B/C, 36*(1, 14-15), 968–976. http://dx.doi.org/10.1016/j.pce.2011.08.011

Kahinda, J. M., Lillieb, E. S. B., Taigbenua, A. E., Tauteb, M., & Borotob, R. J. (2008). Developing suitability maps for rainwater harvesting in South Africa. *Physics and Chemistry of the Earth, Parts A/B/C, 33*(8–13), 788–799. http://dx.doi.org/10.1016/j.pce.2008.06.047

Kahinda, J.-M. M., Taigbenu, A. E., Boroto, J. R., & Zere, T. (2006). Domestic Rain Water Harvesting to Improve Water Supply in Rural South Africa. *Physics and Chemistry of the Earth Parts A/B/C, 32*(15), 1050-1057.

Kannan, A. (2014). *Rainwater Plants Go Down the Drain, The New Indian Express, March 18, 2014.* Retrieved July 17, 2015, from http://www.newindianexpress.com/cities/kochi/Rainwater-Plants-Go-Down-the-Drain/2014/03/18/article2114767.ece

Karnna, B. (2014). Rainwater harvesting still not successful in City. *Deccan Herald.* Retrieved September 27, 2015, from http://www.deccanherald.com/content/402889/rainwater-harvesting-still-not-successful.html

Khang, D. B., & Lin Moe, T. (2008). Success criteria and factors for international development projects: A life-cycle-based framework. *Project Management Journal, 39*(1), 72–84. http://dx.doi.org/10.1002/pmj.20034

Krishnan, R. (2003). RWH – Success story from Chennai India. *11th International Rainwater Catchment Systems Conference "Towards a New Green Revolution and Sustainable Development Through an Efficient Use of Rainwater"* Texcoco, Mexico - August 2003. Retrieved September 27, 2015, from http://www.eng.warwick.ac.uk/ircsa/members/pdf/11th/Krishnan.pdf

Kumar, D. M. (2004). Roof Water Harvesting for Domestic Water Security: Who Gains and Who Loses? *Water International, 29*(1), 43-53. http://dx.doi.org/10.1080/02508060408691747

Lee-Look, G., Boodram, N., & Dempewolf, L. (2015) Knowledge Exchange on Integrated Water Resources Management: Successful Strategies for Learning and Sharing at the National, Regional and Global Levels, *The Caribbean Water and Wastewater Association,* 24th Annual CWWA Conference and Exhibition "Improving the Quality of Life with Water & Wastewater Management Solutions" August 24-28th, Intercontinental, Hotel,Miami, 10pp.

Liu, A. N., & Walker, A. (1998). Evaluation of project outcomes. Construction *Management & Economics, 16*, 109–219. http://dx.doi.org/10.1080/014461998372493

Naugle, J., Opio-Oming, T., & Cronin, B. (2011) A market-based approach to facilitate self- Supply for Rainwater Harvesting in Uganda, 6th Rural Water Supply Network Forum 2011 Uganda Rural Water Supply in the 21st Century: Myths of the Past, Visions for the Future. Retrieved September 20, 2013, from http://www.ri.org/files/uploads/RWSN_%20Rainwater%20Self%20Supply%20Final.pdf

Pandey, D. N., Gupta, A. K., & Anderson, D. M. (2003). Rainwater harvesting as an adaptation to climate change. *Current Science, 85*(1), 46-59.

Pearce-Churchill, M., Willis, E., & Jenkin, T. (2005). Barriers To Rainwater Harvesting In An Aboriginal Community In South Australia, *12th International Rainwater Catchment Systems Conference "Mainsteaming Rainwater Harvesting"* New Delhi, India

RAIN. (2014). *Increasing access to water: Upscaling rainwater harvesting through microfinance.* Retrieved from http://www.rainfoundation.org/wp-content/uploads/140307-2-pager-RHW-and-MF_DEF1.pdf

Rainwater Harvesting. (2002). *Legislation on Rainwater Harvesting.* Retrieved September 27, 2015, from http://www.rainwaterharvesting.org/urban/Legislation.htm

Rainwater Harvesting. (2010). *Regulations and Rebates: Rainwater Harvesting Is a Way of the Future.* Retrieved from http://rainharvesting.com.au/knowledge-center/regulations-and-rebates/

Roebuck, R. M., Oltean-Dumbrava, C., & Tait, S. (2010). Whole life cost performance of domestic rainwater harvesting systems in the United Kingdom. *Water and Environment Journal, 3*(25), 356-365.

Smet, J. (2003). WELL FACTSHEET Domestic Rainwater Harvesting, Water, Engineering and Development Centre, Loughborough University. Retrieved June 25, 2015, from http://www.lboro.ac.uk/well/resources/fact-sheets/fact-sheets-htm/drh.htm

Smith, A. (1977). [1776] *An Inquiry into the Nature and Causes of the Wealth of Nations.* University of Chicago Press. http://dx.doi.org/10.7208/chicago/9780226763750.001.0001

Stergakis, A. (2011). *Project Management and Sustainability - Key challenges for donors and recipients.* Retrieved September 27, 2015, from http://www.academia.edu/1511374/Project_Management_and_Sustainability_-_Key_challenges_for_donors_and_recipients

Toronto and Region Conservation Authority. (2010). *Performance Evaluation of Rainwater Harvesting Systems Toronto, Ontario.* Retrieved September 27, 2015, from http://www.sustainabletechnologies.ca/wp/wp-content/uploads/2013/01/FINAL-RWH-2011_EDIT3.pdf

UNEP. (1997). Source Book of Alternative Technologies for Freshwater Augmentation in Latin America and the Caribbean, General Secretariat, Organization of American States, Washington, D.C.

UNICEF. (2004). Evaluation of the Rooftop Rainwater Harvesting Project, Water, Environment and Sanitation Section, UNICEF, New Delhi, India.

United Nations. (1992). *The Dublin Statement on Water and Sustainable Development.* Retrieved May 3, 2015, from http://www.un-documents.net/h2o-dub.htm

Women for Water Partnership. (2012). *NetWwater rainwater harvesting project major success.* Retrieved September 27, 2015, from http://www.womenforwater.org/openbaar/index.php?alineaID=308

Zimmermann, M., Jokisch, A., Deffner, J., Brenda, M., & Urban, W. (2012). Stakeholder participation and capacity development during the implementation of rainwater harvesting plants in central northern Namibia. *Water Science & Technology: Water Supply, 12*(4), 540–548. http://dx.doi.org/10.2166/ws.2012.024

15

A Bottom-Up, Non-Cooperative Approach to Climate Change Control: Assessment and Comparison of Nationally Determined Contributions (NDCs)

Carlo Carraro[1,2]

[1] Ca' Foscari University of Venice and FEEM, Venice, Italy

[2] Vice-Chair, IPCC WG III

Correspondence: Carlo Carraro, Ca' Foscari University of Venice and FEEM, Venice, Italy. E-mail: ccarraro@unive.it

Abstract

International negotiations on climate change control are moving away from a global cooperative agreement (at least from the ambition to achieve it) to adopt a bottom-up framework composed of unilateral pledges of domestic measures and policies. This shift from cooperative to voluntary actions to control GHG emissions already started in Copenhagen at COP 15 in 2007 and became a platform formally adopted by a large number of countries in Paris at COP 21. The new architecture calls for a mechanism to review the nationally determined contributions (NDCs) of the various signatories and assess their adequacy. Most importantly, countries' voluntary pledges need to be compared to assess the fairness, and not only the effectiveness, of the resulting outcome. This assessment is crucial to support future, more ambitious, commitments to reduce GHG emissions. It is therefore important to identify criteria and quantitative indicators to assess and compare the NDCs.

Keywords: climate change negotiations, Paris agreement, GHG emissions, mitigation

1. Introduction

Last December, the long-awaited Paris Conference on Climate Change (COP 21) approved a new, comprehensive deal that will guide international action to control climate change from 2020. The Paris agreement is just a first step: countries need to find common ambitions not only on mitigation objectives, but also on adaptation measures, financing to support developing countries plans, as well as technology transfers. Nevertheless, the Paris agreement is an important step: for the first time, all the most important GHG emitters are committed to keep their own GHG emissions under control.

A key pillar of the Paris agreement are the so called NDCs (Nationally Determined Contributions), that are a new type of instrument under the UNFCCC, through which both developed and developing countries declare the actions they intend to undertake to tackle climate changes at the national level.

Going beyond the historical dichotomy between Annex I and Non-Annex I countries, the Paris agreement asks indeed for the participation of all countries, which agreed to communicate their targets and plans to reduce GHG emissions well in advance the two-week negotiations in Paris. Table 1 summarizes the main Nationally Determined Contributions adopted at COP 21.

Table 1. *Nationally Determined Contributions* by the top 10 GHG emitting countries (2011 GHG emissions, source CAIT Data Explorer)

Country	Emission reduction target	Reference year	Period of implementation
China	Peak in 2030; 60-65% CO2 per unit of GDP	2005	by 2030
United States	26-28%	2005	2020-2025
European Union	≥40%	1990	2021-2030
India	33-35% CO2 per unit of GDP	2005	by 2030
Russia	25-30%	1990	2020-2030
Japan	26%	2013	1 April 2021 –31 March 2031
Brasil	37%	2005	by 2025
	43%		by 2030
Indonesia	29-41%	BAU	2030
Mexico	22-36%	BAU(from 2013)	2020-2030
Canada	30%	2005	by 2030

Despite serving the same purposes, the submitted NDCs show many substantial differences. From one side, most advanced economies, including the US and the EU, proposed economy-wide emissions reduction targets from a base year. On the other side, it is not uncommon to find intensity targets among developing nations, as in the case of China, Singapore, and Tunisia, which chose a reduction of GHG emissions per unit GDP, or more frequently, a percentage deviation from a Business as Usual (BaU) scenario.

As for developing countries, usually a lower "unconditional" bound and an upper "conditional" bound were proposed, the latter to be implemented only with financial and technological support from the international community. Moreover, developing countries' contributions usually put more emphasis on adaptation measures than developed counterparts, which conversely continue to focus mainly on mitigation actions.

Against this background, attempts to evaluate and compare such a fragmented picture recently start to emerge. Looking, for example, at the emission targets pledged by four among the major emitters, namely EU, US, China and Russia, it can be seen that, if absolute levels of emissions are compared, the EU will support a higher effort compared to the other countries. On the contrary, when changes in the GHG/GDP ratio are taken into account, China and Russia will bear a larger burden of the climate action (see Table 2).

Table 2. Comparison of US, EU, China and Russia's Targets

Comparison among NDCs targets	Country			
	US	EU	Russia	China (Emissions to peak by 2030)
GHG emissions change (%)				
wrt 1990	-16 a -14	-40	-30 a -25	+265 a +291
wrt 2005	-28 a -26	-35	+10 a +18	+76 a +89
Changes in GHG/GDP ratio (kgCO$_2$eq/US$)				
wrt 1990 (%/year)	-3.0 a -2.9	-2.8	-3.7 a -3.5	-4.7 a -4.5
wrt 2005 (%/year)	-3.6 a -3.5	-2.9	-4.5 a -4.2	-5.0 a -4.7

This kind of analysis, even though often proposed by governments and NGOs, is however quite superficial. A proper analysis and comparison of NDCs should rather focus on a more precise effectiveness metric, for example the distance of each NDC from the domestic optimal emission pathway to achieve the 2C target. And it should consider that effectiveness is not the only, and possibly not the most important, metrics when comparing different countries' efforts to reduce GHG emissions. Fairness, and therefore relative costs, is also very important. Cost of reducing GHG emissions can for example be computed with respect to the business as usual emission pathway in each country. The cost in one country can then be compared with costs in the other ones to provide information of the fairness of the proposed NDCs. Efficiency, namely the distribution of marginal costs, is also crucial. Marginal costs should indeed be equalized for the NDCs to be fully efficient. Is this the case for the NDCs approved in Paris?

The objective of this paper is to carry out an analysis of the NDCs for the three major world economies - US, EU and China – covering almost 60% of total GHG emissions. Before proceeding with the analysis, let us note two important features of the NDCs and the Paris agreement.

First, for the first time, more than 180 countries agreed to control their own GHG emissions, including many emerging and developing economies. This is an important step to move beyond the traditional dichotomy between developed and developing countries, the latter claiming that GHG emission control should pertain to developed nations to avoid restrictions to their own economic development.

Second, the Paris agreement reflects the move from a "top down and cooperative" to a "bottom –up and non-cooperative" approach to climate change control. This move can be explained by the failures of many previous negotiations over a global cooperative agreement (the Kyoto protocol, for example, covered only about 14% of total GHG emissions) and by the consequent attempt to achieve a broad agreement with a large number of signatories (the Paris Agreement covers about 95% of total GHG emissions). A crucial question is therefore whether this non-cooperative approach, accompanied by measures to support developing nations' efforts to reduce GHG emissions, and by a strong reputation effect inducing large nations to compete for ambitious emission reduction effort, is sufficient to keep global GHG emissions on track to achieve the 2°C target.

2. Effectiveness and Efficiency of NDCs: How Far from Achieving the 2°C target?

Let us look at the basic facts. China has committed to peaking their emissions by 2030, if not before, with an intensity target of - 60-65% with respect to 2005 (see Table 1). The US says it is shooting for emissions reductions of - 26-28% by 2025 (from 2005 levels). The EU target is -40% of total EU27 GHG emissions in 2030 with respect to 1990.

The US and China also signed a bilateral deal in which they commit themselves not only to reduce emissions but also to develop a joint Research and Development program focused on renewables to increase the share of renewables in their own domestic energy mix. China is committed to achieve a share of renewables equal to 20% by 2030.

A first initial comparison of NDCs requires harmonizing the reference year. By achieving its target of reducing emissions by 26-28% by 2025 (from 2005 levels), the US will have achieved a 16.3% reduction in GHG emissions compared with 1990 levels. Though notable, this target is decidedly less than the about 30% reduction decided by the EU for 2025 (recall that the EU committed to reduce its GHG emissions by 40% from 1990 levels by 2030).

Even so, broadly speaking, both the US and the EU are on track to achieve the 2°C target. To prove this statement let us compare the EU and US targets with the emission levels that would be consistent with the 2°C pathway.

Let us consider three sets of scenarios for US future emissions, all consistent with the achievement of the 2°C target by the end of the century. The first set of scenarios (EMF) is produced by the Energy Modeling Forum (Cf. EMF, 2014). The second set of scenarios (LIMITS) comes from LIMITS, an important project funded by the European Commission (Cf. LIMITS, 2013). The third set of scenarios (SSP) is produced by the IPCC (Cf. IPCC, 2010). The three sets of scenarios identify cost efficient US GHG emissions within the socio-economic pathway leading to a 2°C temperature increase by the end of the century.

As shown in Figure 1, in all scenarios the emission reduction target adopted by the US administration is consistent with the 2°C objective. An important additional effort will be necessary beyond 2025, but the target for 2025 seems to be ambitious enough.

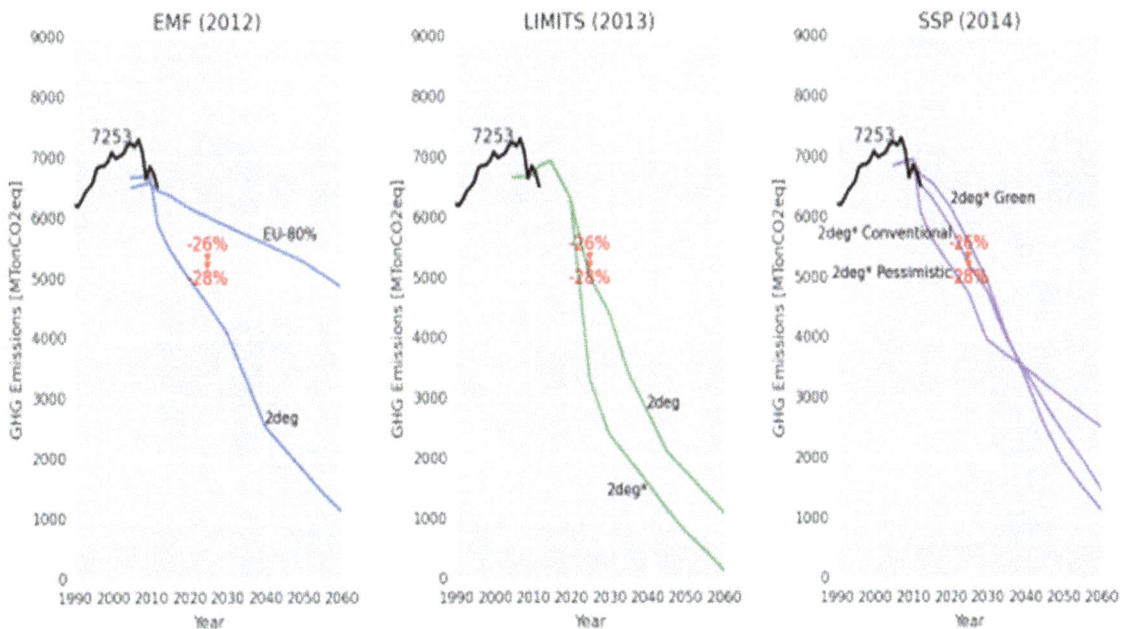

Figure 1. Scenarios of US GHG emissions from 1990 to 2100 consistent with achieving the global 2C target. In red, the target adopted at COP 21 in Paris

A different question is whether the US emission reduction objective is feasible. This target will mean doubling the pace of emissions reduction set for 2005-2020 thereafter. Doubling the US effort to mitigate GHG emissions is likely to be technically and economically feasible. Both US per capita emissions and US emissions per unit of GDP are larger than the EU ones. Hence, the marginal mitigation effort is smaller in the US than in the EU. However, in the US, the most significant barrier to climate change action will be a political one. The US lawmakers in Congress, currently a Republican-dominated body, may oppose any action to effectively reduce GHG emissions. In response to these political obstacles, President Obama may develop a climate action framework through his regulatory power that does not need to be passed through Congress. The most notable of these regulatory mechanisms are the Clean Power Plan, energy efficiency standards, heavy-duty engines and vehicles standards.

Even with this practical approach to avoiding Congress, it is however unlikely that these policies will be able to achieve all of the 26-28% emissions cut needed unless new clean energy technologies are developed. With the Congress obstacle still standing, this could mean looking to private investment to facilitate the development of such clean energy technologies. With lowering costs and an opening market for some renewables in the US, attracting this finance may be viable. This may be the case particularly for solar power, the production of which has grown by 139,000 percent in the US in the past decade.

Let us consider the European Union. The EU target is -20% in 2020 (already adopted a few years ago) and -40% in 2030 (pledged at COP 21 in Paris). The 2014 EU climate and energy framework also contains the indication to reduce GHG emissions by 80% to 95%, compared to 1990 levels, in 2050.[1]

---

[1] In 2014, the European Council agreed on the 2030 climate and energy policy framework for the EU and endorsed new targets on greenhouse gas emissions, renewable energy and energy efficiency for 2030 (EC, 2014). In 2015, the EU adopted an Energy Union Strategy to ensure that Europe has secure, affordable and climate-friendly energy and achieve its climate and energy goals for 2030 (EC, 2015).

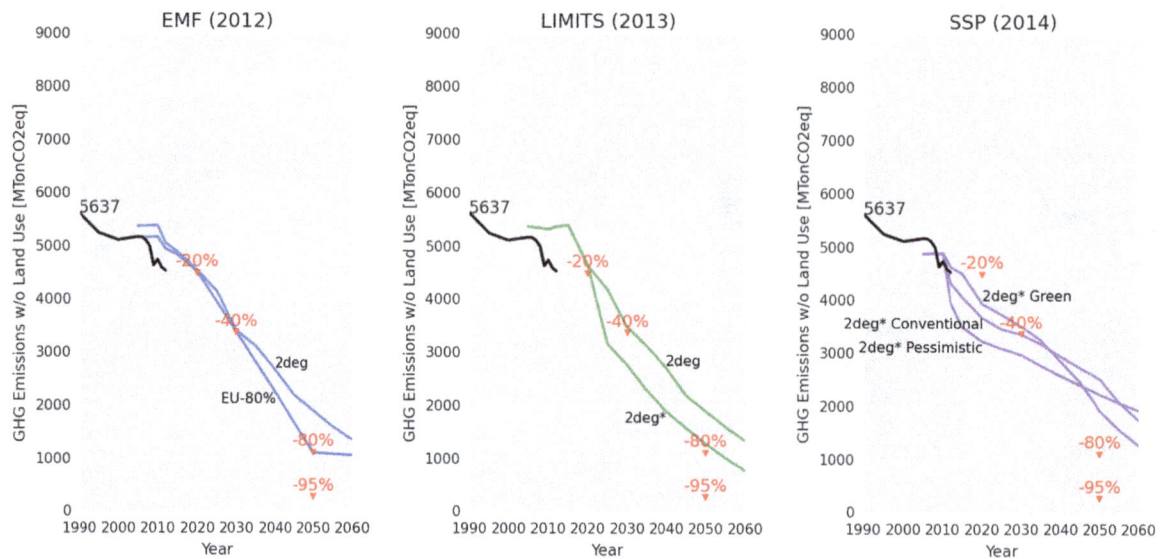

Figure 2. Scenarios of EU GHG emissions from 1990 to 2100 consistent with achieving the global 2C target. In red, the target adopted at COP 21 in Paris

Again, the EU COP 21 pledge is consistent with the three scenarios that different research institutes suggest as consistent with achieving the global 2°C target. The effort required to EU member countries is considerable. The EU 2030 target will mean increasing the pace of emissions reduction set for 2005-2020 from 1% per year to 1.4% per year in the period 2020-2030. The reduction in GHG emissions needed between the 2030 target level (– 40% below 1990) and the 2050 EU objective (at least 80% below 1990) will have to be two to three times steeper than the necessary reduction between current levels and the 2030 target. Nevertheless, the EU is likely to achieve the 2020 target and shows enough political commitment to achieve the 2030 target. Additional efficiency gains in the EU energy system remain feasible and the diffusion of renewable is proceeding rapidly.[2] EU Member States are now developing low carbon development strategies outlining concrete steps to turn EU-wide, long-term ambitions into national and local actions.

The situation is less positive in China. Prior to the recent commitment to peaking emissions by 2030, China had only a carbon intensity target of reducing emissions per dollar of economic output by 40-45% in 2020 (from 2005 levels). The intensity target has been raised to 60-65% in 2030. Modeling projections from the IEA and EIA suggest that a 45% carbon intensity target would result in overall emissions that are a slightly more ambitious than (IEA) or the same as (EIA) China's business-as-usual trajectory (IEA, 2015). The new intensity target is therefore a commitment to concretely reduce GHG emissions wrt BAU. Most importantly, peaking emission in 2030 is an absolute commitment that constitutes an important step into the right direction. However, can we also conclude that China is on track for the 2°C by 2100 warming limit?

Let us consider again the three scenarios previously outlined and their implications for China. As shown in Figure 3, a peak of emissions in 2030 does not seem to be consistent with the 2°C target in any of the three scenarios. Chinese emissions should peak in 2020-2025 for the 2°C target to be achieved by the end of the century. Nevertheless, the enhanced effort by China is worth being positively considered. Under the old commitment (2005-2020), China was supposed to reduce energy intensity by about 3% per year (a target that China is likely to achieve). Under the new commitment (peaking emissions by 2030) the implicit pace of emission reduction is about 4% per year. It is not a doubling of the emission reduction effort, but it's a significant and costly one.

As for implementation, China seems to rely on two major mechanisms: a nation-wide emission trading scheme from 2017, as announced by President Xi in New York last September, and an effort to increase the share of

---

[2] In October 2014, EU heads of state or government agreed to a target of raising the EU-wide RES share to at least 27% by 2030. Assuming that the current pace of renewable energy deployment in the EU (+ 0.8 percentage points per year on average) is sustained until 2030, the 2030 RES share in the EU would be above the 2030 target (EEA, 2015).

non-fossil fuels (renewables and nuclear) in the domestic energy mix up to 20% in 2030.

It is important to stress how important the non-fossil fuel energy target is for China. Solar energy in China is developing at unprecedented rates. Nuclear energy is also growing fast. Unfortunately, China's coal use and overall growth are also developing at unprecedented rates. With the number of coal plants being built, China is already locked into a high level of carbon emissions no matter what actions they take now.

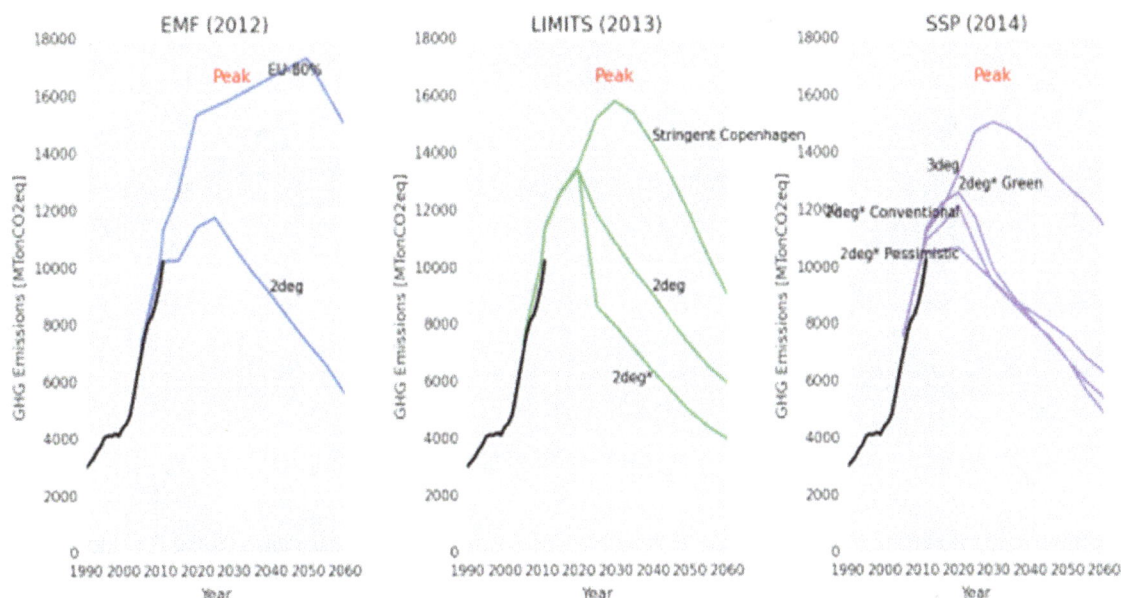

Figure 3. Scenarios of China GHG emissions from 1990 to 2100 consistent with achieving the global 2C target. In red, the target adopted at COP 21 in Paris

With this in mind, hope comes anyway from China's concrete aim of increasing the total share of non-fossil fuel energy to 20% by 2030 at the latest. This commitment is certainly demanding. At the moment, only 10% of China's energy mix comes from non-fossil fuel energy sources. The 20% "clean" energy target would require China to deploy an additional 800-1,000 gigawatts of wind, solar, nuclear and other carbon neutral technologies by 2030. This is greater than the capacity of all coal-fired power plants currently operating in China.

Given this analysis of the pledges coming from the three major players in the global GHG emission game, let us focus on the global effectiveness of the Paris agreement. A the end of October 2015, the UNFCCC published a report synthesizing the aggregate effect of all the NDCs submitted, in order to assess the effectiveness of the proposed actions towards the objective of limiting the global temperature increase to 2°C (UNFCCC, 2015). Results can be summarized as follows. The implementation of the communicated NDCs is estimated to result in aggregate global emission levels of 55.2 Gt CO2 eq in 2025 and 56.7 Gt CO2 eq in 2030. Compared with global emissions in 1990, 2000 and 2010, global aggregate emission levels resulting from the NDCs are expected to be higher 8–18 per cent in 2025 and 11–22 per cent in 2030 in relation to the global emission level in 2010. While these figures show that global emissions considering the NDCs are expected to continue to grow until 2025 and 2030, the growth is expected to slow down substantially. The relative rate of growth in emissions in the 2010–2030 period is expected to be 10–57 per cent lower than that over the period 1990–2010, reflecting the impact of the NDCs (UNFCCC, 2015).

In addition, global average per capita emissions considering the NDCs will decline by 8 and 4 per cent by 2025 and by 9 and 5 per cent by 2030 compared with the levels in 1990 and 2010, respectively. And the implementation of the NDCs would lead to lower aggregate global emission levels than in pre-NDC trajectories (UNFCCC, 2015).

Our calculations using the WITCH Model (Aleluja, Carraro and Tavoni, 2015) are shown in Figure 4 and 5. We consider the second Shared Socioeconomic Pathway (SSP2) as business as usual scenario (BAU). SSP2 is a Socioeconomic Pathway with almost no greening of world economic development (IPCC, 2010), unless some effective policy decisions are implemented. In addition to the BAU (the red line), we consider the sum of the

NDCs in two cases: when all unconditional pledges are aggregated (the blue line), and when the conditional pledges are also taken into account (the green line). Figure 4 shows that COP 21 pledges reduce emissions by 19-23% with respect to the SSP2 emission pathway.

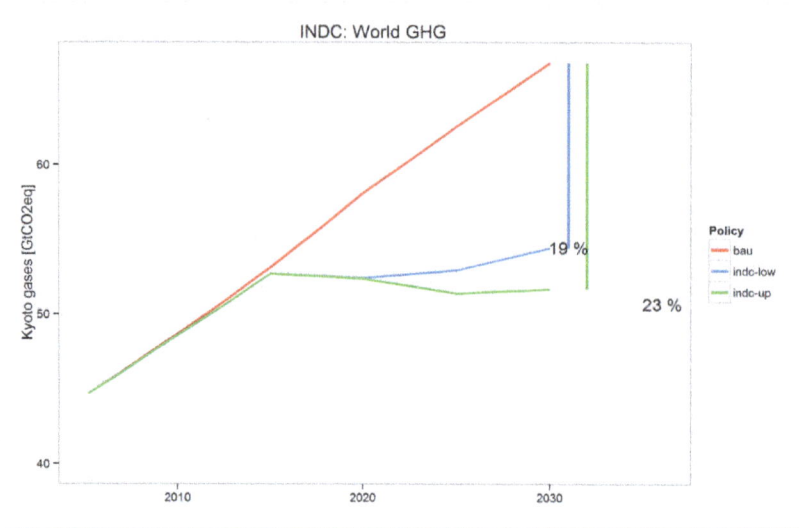

Figure 4. The effectiveness of aggregate NDCs

If the comparison is carried out with respect to the global emission pathway consistent with the 2°C target, results are less encouraging. As shown in Figure 5, COP 21 pledges are slightly above the trajectory that would be required to keep global temperature below 2°C by the end of the century. Additional contributions along with improved negotiating efforts are therefore needed in future climate negotiations. This is not surprising, because the non-cooperative approach adopted in Paris is likely to lead countries to pledge emission reductions close to the ones they would have implemented on the basis of their own domestic interests, without any cooperative effort to control climate global externality.

Is this enough to conclude that Paris COP21 failed to achieve its objectives? Certainly not and for multiple reasons. First, even though not "deep" enough, the Paris agreement is very "broad". For the first time, a large group of countries, notably US and China, committed to reduce their own GHG emissions with the obvious consequence that, for the first time, a cap on total emissions is likely to be achieved. Second, emission targets are just one of the components of the Paris agreement. Many countries are implementing multilateral and bilateral investments into R&D, which aim to drive the technology innovations and price reductions required to catalyze a clean energy future. Third, the big issue behind climate negotiations of the last years, and Paris COP 21 was no exception, is finance. Many developing and emerging economies are not going to make any effort to achieve their own NDC unless adequate financing support is received by developed countries. The Green Climate Fund, albeit insufficient, is certainly a step forward into the right direction. Fourth, the Paris agreement must be considered the first mile of a long journey. More ambitious emission reduction commitments will be adopted in the coming years. What we do need now is a sound monitoring and verification system to guarantee that all countries actually implement, through domestic policies, what they promised to do in Paris.

There is another element of Paris COP 21 to be underlined. As argued above, the likelihood of keeping the increase of global temperature below the 2°C "security threshold" is far from being at hand. All IPCC scenarios show that the 2°C target can only be achieved not only by progressively reducing the current flow of emissions, but also by removing, at least partially, the stock of emissions already in the atmosphere. As a consequence, the opportunities and constraints in deploying large-scale carbon capture and storage (CCS) systems are thus of the utmost actuality, as the technology promises to get rid of the most common greenhouse gases produced in industrial and energy plants before they reach the atmosphere (or even to achieve "negative" emissions, if combined with biomass).

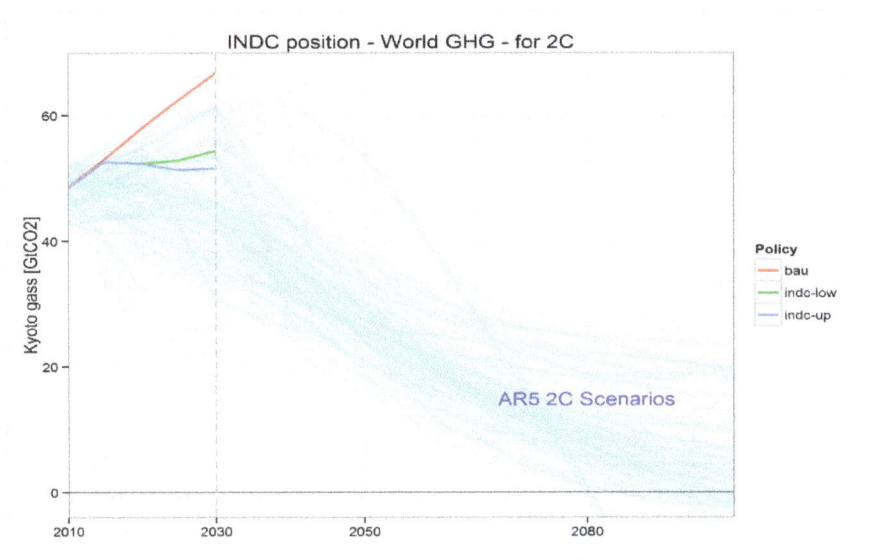

Figure 5. Consistency of COP 21 pledges with the 2°C emission pathways

The potential of CCS is widely recognized: many global climate models cannot reach concentrations of about 450 ppm CO_2eq by 2100 (corresponding to the 2°C target) without CCS. Moreover, in the Fifth IPCC Assessment, scientists observed that mitigation costs become consistently higher if CCS is excluded from the mitigation scenarios. Nevertheless, the challenges that CCS deployment is facing can raise doubts about the role it can play in future climate strategies and plans.

The IEA Greenhouse Gas R&D Program (IEAGHG) recently published a special issue on CCS, with the aim of marking the 10th year anniversary of IPCC's Special Report on CO_2 Capture and Storage (SRCCS), issued in 2005, and outlining the progress made in the field in the last 10 years. According to the report (IEAGHG, 2015), substantial progress has been made in the last decade concerning CO_2 capture, storage efficiency, and methods to assess leakage impacts and risks of induced seismicity. However, the high costs and high-energy penalties of CCS remain a concern and are among the highest barriers to the wide deployment of CCS in the energy sector, where the majority of GHGs are produced.

Over the past 14 years, governments have committed around USD 24 billion to fund CCS projects, and companies have spent at least USD 9.5 billion since 2005 (14). While only one CCS system on a commercial power plant is currently in operation several other projects have been dismissed or are facing investment shortage, such as the FutureGen project of a CCS-equipped coal plant in Illinois, from which the US government pulled out earlier this year. The financial viability of CCS in the power sector is likely to remain a constraint without clear actions leading to credible carbon prices, technology requirements or emissions standards, ideally at the global level.

A recent report by Citigroup noted that CCS represents "a potentially enormous game-changer for energy markets" but its application has been slow and its future deployment may prove to be "too little, too late" with respect to other more cost-competitive, low-carbon technologies (Citigroup, 2015). "Despite progress on the technical front, the industry believes there is a need for government policy to support the business case for broad scale implementation. While the fossil fuel industry, particularly coal, has tended to resist carbon pricing developments, ironically the lack of carbon pricing means there has been no business case for large scale CCS deployment" (Citigroup, 2015). The public acceptability of CCS is a consequent issue of the need for government support, related to the potential competition for public funding between CCS and other low carbon options, and to the real and perceived risks of deploying CCS at the local level.

Summing up, the attempt to move quickly towards a development path consistent with the 2°C target depends more on technology development (for CCS in particular) and financial transfers (the full funding of the Green Climate Fund at least) than on the quantitative emission reduction commitment that will be adopted at Paris COP 21. The NDCs are an important decision, but without financial support to developing countries and without rapid technological improvements in CCS technologies, humanity will have to adapt to a temperature increase larger than 2°C.

Let us conclude this first part of our analysis with a brief comment on the efficiency of the NDCs approved at COP 21. According to our assessment (see Aleluja, Carraro and Tavoni, 2015), high marginal abatement costs are likely to characterize the abatement efforts of the EU, Korea, Australia and the US. The abatement cost will be much smaller in developing and emerging economies, namely Latin America, East Asia, South Asia, China and Transition Economies. Marginal costs are likely to be smaller in these regions both because targets are less ambitious and because abatement opportunities are often much cheaper.

This opens the way to the introduction of measures to increase efficiency. For example, by pricing carbon worldwide, or by adding to the Paris agreement a set of measures to enable the exchange of credits from emissions reductions implemented by a given country in another country, or by linking emission trading schemes implemented in various countries. This kind of mechanisms would progressively move towards the equalization of marginal abatement costs in different world regions.

3. Fairness of the NDCs: Do They Provide an Equitable Distribution of Mitigation Costs?

As stated in the Introduction, fairness is another important ingredient of the Paris agreement. We have already seen how the EU, the US and China have different objectives. China, in particular, is committed to emission reduction less ambitious than the EU and the US, and less consistent with the optimal 2°C trajectory. The reason is likely not to be a lower consciousness of climate change threats in China. It is probably the other way around: China's commitments is enhanced by the high levels of pollution in China's cities and by the high local benefits, in addition to the global ones, generated by China's GHG emission reductions.

The main reason for China slower progress towards the 2°C trajectory is likely to be fairness, namely the total cost of reducing GHG emissions in China with respect to the EU and the US. Consider indeed the cost of implementing emissions reductions consistent with the achievement of the 2°C target in the EU under the three sets of scenarios previously utilized to assess the effectiveness of NDCs. Let us recall that the first set of scenarios (EMF) is produced by the Energy Modeling Forum. The second set of scenarios (LIMITS) comes from LIMITS, an important project funded by the European Commission. The third set of scenarios (SSP) is produced by the IPCC. For the three studies, three different 2°C consistent trajectories are considered. For example, in the case of SSP, figures below show the costs (wrt BAU) of achieving the 2°C objective of three different pathways: SSP1 (green) SSP2 (conventional) and SSP3 (pessimistic).

The results are shown in Figures 6, 7 and 8. Costs (wrt BAU) to achieve the 2°C target seem to be low in the EU in all scenarios. The cost in 2050 is estimated to be between 3% and 6% of EU GDP.

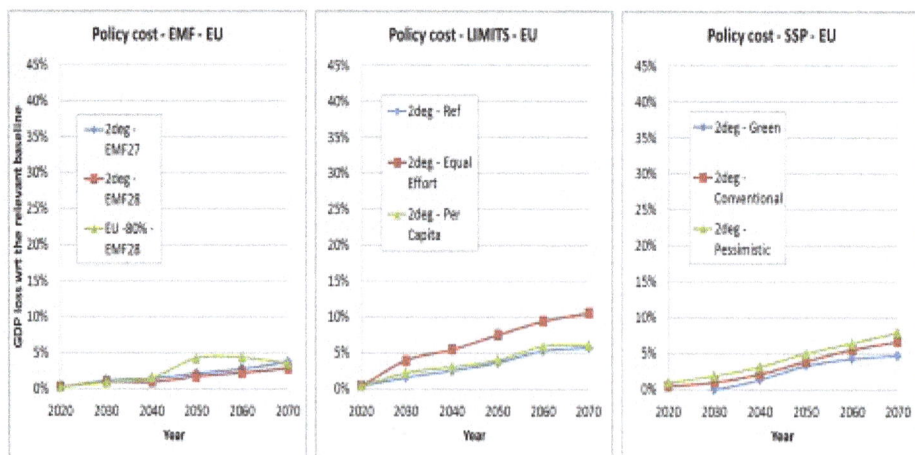

Figure 6. The cost of achieving the 2°C emission reduction trajectory in the EU

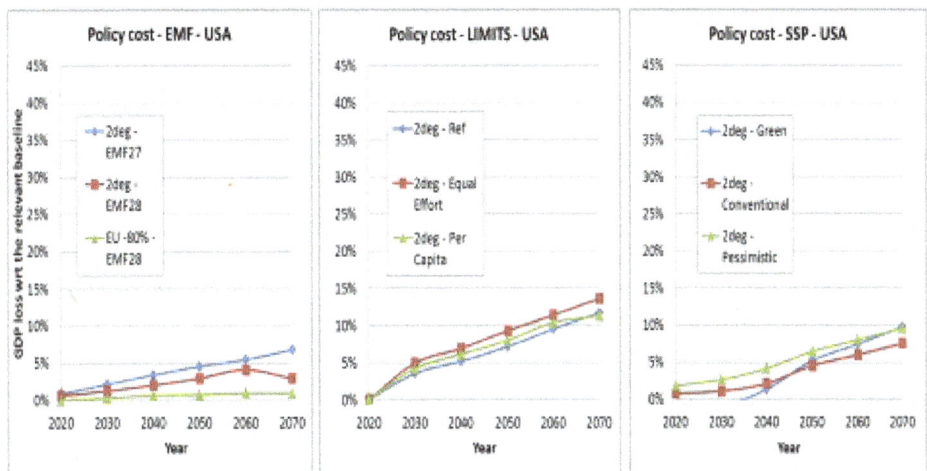

Figure 7. The cost of achieving the 2°C emission reduction trajectory in the US

As shown in Figure 7, costs are slightly higher in the US. In 2050, the range is 3-10%. However, costs are higher in China. Figure 8 shows that costs of reducing GHG emissions consistently with cost-effective 2°C scenarios range between 5 and 20% of China GDP in 2050.

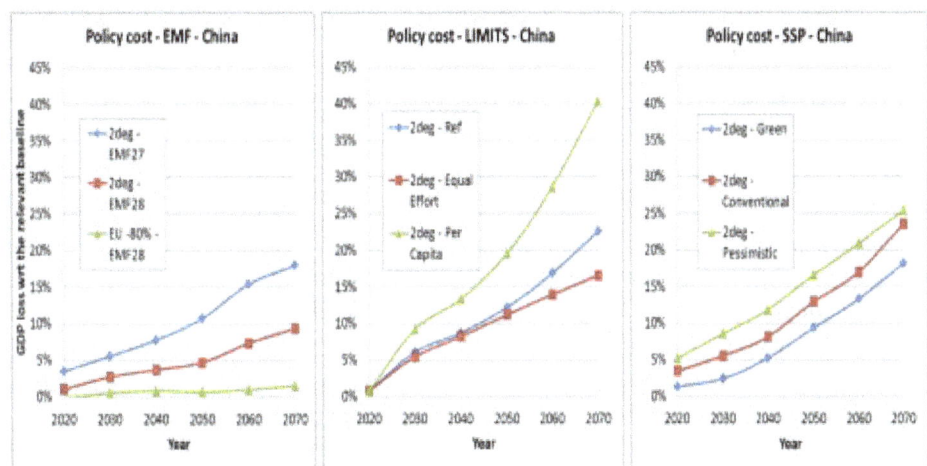

Figure 8. The cost of achieving the 2°C emission reduction trajectory in China

As a consequence, one of the reasons behind China's less ambitious pledge at COP 21is likely to be the higher relative cost for China to reduce emissions consistently with the 2°C target. For the EU and the US is seems to be easier, at least in terms on GDP losses, to attain the 2°C target. Therefore, their pledges are more consistent with this objective than China's one.

4. Conclusions

The non-cooperative bottom-up approach adopted to prepare the Paris Agreement has been unable to deliver a highly effective deal at COP 21. Nevertheless, COP 21 represents a crucial and innovative step towards GHG emission control. After four decades of increasing emissions (and the last decade is the one with the highest emission growth rate), for the first time GHG emissions will be stabilized: emissions in 2030 will almost equal to emissions in 2015. For the first time, almost all countries have submitted concrete and operational pledges to reduce or control their GHG emissions: the divide between developed and developing countries has been largely removed.

These important results have been achieved because a bottom-up non-cooperative approach has been adopted. A broad, even though not deep, agreement has been signed in Paris. This is likely to create consensus for further future actions, when more ambitious efforts need to be implemented in the coming decades.

The bottom-up non-cooperative Paris agreement obviously lacks effectiveness. Given the absence of coordinated economic instruments (a global carbon price or linked emission trading schemes) it also lacks efficiency. However, it contains a degree of fairness, which explains the large consensus on the Paris agreement emerged at COP 21.

Most importantly, the commitments and pledges adopted in Paris are just the first step in a long journey. Additional emission reduction efforts will need to be implemented in the coming years, and more effective policy measures will need to be adopted, both domestically and internationally. The adoption of robust systems for measurement, reporting, and verification facilitates compliance, but without supporting enforcement measures, countries are unlikely to achieve large emissions reductions. What is missing is both enforcement and vision. It is not enough to agree on a temperature target. It is now urgent to agree on a societal transformation path, which, in market economies at least, can be driven only by a change in relative prices.

For this process to move quickly towards the objective, a set of metrics to measure and assess the effectiveness, efficiency and fairness of the abatement efforts implemented in various countries is necessary. This paper is a first attempt to provide this crucial assessment. Extensions to a larger number of countries and the use of a larger number of models would certainly improve the robustness and usefulness of the results.

Acknowledgments

The author is grateful to Lara Aleluja and Massimo Tavoni for the WITCH modeling runs that provided the results described in this paper. He is also grateful to participants at the Second Protection and Sustainability Forum held in Bath, UK, 9-11 April 2015, for helpful comments and suggestions.

References

Aleluja, L., Carraro, C., & Tavoni, M. (2015). *The Paris Agreement: Assessment and Comparison of National Emission Reduction Efforts.* FEEM Nota di Lavoro, Milan.

Barrett, S., Carraro, C., & de Melo, J. (Eds.). (2015). *Towards a Workable and Effective Climate Regime.* VoxEU.org eBook, CEPR, London.

Citigroup. (2015). *ENERGY DARWINISM II: Why a Low Carbon Future Doesn't Have to Cost the Earth.* Citi GPS: Global Perspectives & Solutions, August 2015, London.

EC. (2014). *Communication from the Commission to the European Parliament, the Council, the European Economic and Social Committee and the Committee of the Regions: A policy framework for climate and energy in the period from 2020 up to 2030.* COM(2014) 15 final Brussels. Retrieved from http://eur-lex.europa.eu/legal-content/EN/TXT/PDF/?uri= CELEX:52014DC0015&from=EN

EC. (2015). *Energy Union Package. A framework strategy for a resilient Energy Union with a forward-looking climate change policy.* Communication from the Commission to the European Parliament, the Council, the European Economic and Social Committee, the Committee of the Regions and the European Investment Bank COM(2015) 80 final Brussels. Retrieved from http://eur-lex.europa.eu/resource.html?uri=cellar:1bd46c90-bdd4-11e4-bbe1-01aa75ed71a1.0001.03/DOC_1&format=PDF

EEA. (2015). *Trends and projections in Europe 2015. Tracking progress towards Europe's climate and energy targets.* Report No 4/2015, Copenhagen.

EMF. (2014, April). The EMF27 Study on Global Technology and Climate Policy Strategies. *Climatic Change, 123*(3-4).

IEA. (2015, June). *World Energy Outlook 2015 Special Report.* International Energy Agency, Paris.

IEAGHG. (2015). *Assessment of emerging CO2 capture technologies and their potential to reduce costs.* International Energy Agency, Paris, December 2014.

IPCC. (2010). Workshop Report. In Edenhofer, O., Pichs-Madruga, R., Sokona, Y., Barros, V., Field, C. B., Zwickel, T., ... C. von Stechow (Eds.), *IPCC Workshop on Socio-Economic Scenarios.* 1-3 November 2010, Berlin, Germany.

LIMITS. (2013, November). LIMITS Special Issue on Durban Platform scenarios. *Climate Change Economics, 4*(4).

UNFCCC. (2015). *Synthesis report on the aggregate effect of the intended nationally determined contributions,* UNFCCC, Paris, Oct 2015.

16

Conservation Agriculture Promotion and Uptake in Mufulira, Zambia-A Political Agronomy Approach

Bridget Bwalya Umar[1]

[1] University of Zambia, Zambia

Correspondence: Bridget Bwalya Umar, University of Zambia, Zambia. E-mail: bridget.bwalya@gmail.com

Abstract

This study utilized 120 semi-structured interviews with smallholder farming households and two focus group discussions; as well as several key informant interviews with experts to explore the promotion and uptake of conservation agriculture (CA) in Mufulira, Zambia. Results reveal that ridges and flat culture continued to be the preferred tillage systems (97 per cent and 55 per cent respectively) despite the farmers having been trained in the use of a minimum tillage technique. None of the interviewed farmers perceived CA as a solution to any of their agricultural related problems. The NGO promoting CA in the district had framed it as suited for and claimed to target labour constrained HIV/AIDS affected households. Conversely, farmers complained that CA was challenging for them due to its high labour demands (23 per cent); poor harvests (18 per cent) and was unsuited to the rainfall patterns of the area (10 per cent). Local agricultural experts contested the promotion of basins in Mufulira. The framing of CA as a solution to labour constraints did not seem to hold in the study area. This effectively limited the contestation spaces available to the public officials with dissenting views on the suitability of basin CA in the district.

Keywords: contested agronomy, smallholder farmers, minimum tillage, crop rotations, exotic goats, Heifer International

1. Introduction

1.1 Agricultural Technology Promotions and Framing

The low agricultural productivity of smallholder farming households in Sub Saharan Africa (SSA) has spurred the development and dissemination of agricultural productivity enhancing technologies from a range of actors. Adoption rates of these agricultural technologies have generally been low and were for a long time blamed on the conservatism of the targeted beneficiaries, the smallholder farming households. Increasingly there has been a recognition that the lack of uptake of agricultural technologies occurs because farmers are constrained in resources such that investment in a new technology not only influences what must be done in one field, but involves trade-offs with other activities from which the farmers generate their livelihoods (Giller et al., 2006). Some scholars have noted that the low adoption rates result from the promotion by agricultural experts of 'blueprint technologies' that are inappropriate and unsuited to smallholders for various economic or socio-cultural reasons. Actors engaged in agricultural development interventions choose and privilege specific agricultural technologies over others and promote these to smallholder households as scientifically proven to be the best technologies. The rhetoric around such technologies is framed in a way that conveys the message that the decisions to develop and disseminate particular technologies are made in apolitical environments without contestations. To frame, in Entman's words is to select some aspects of a perceived reality and to make them salient in a communicating text in such a way as to promote a particular problem definition, causal interpretation, moral evaluation, and/or treatment recommendation for the item described (1993: 52). Sumberg et al. (2013:77) assert that framing determines to a significant degree how much attention the problem receives and the approach taken to address it, and thus prefigures the eventual solution. In a similar vein, Tisenkopfs et al. (2014) explain that in framing, agents emphasize some aspects and downplay others, employing particular definitions and defining the scope of the issue.

Closely related to framing is the question of how agronomic research priorities are determined. While a large literature assumes that prioritization is (or should be) a rational, technical process, an alternative view sees it as a process in which power and politics are of utmost importance. Sumberg et al. (2013: 76) observed that there was 'a tendency, supported by professional, institutional, business and political pressures, for powerful actors and institutions to attempt to "close down" or limit discussion in favour of particular research agendas and development pathways'. A result of this dynamic is the promotion of universal approaches to policy that obscure alternative framings and development pathways (Sumberg et al., 2012:11). This has been reported in cases of Integrated Pest Management (IPM), and Conservation Agriculture (CA).

Orr (2003) and Orr and Ritchie (2004) documented an IPM project in southern Malawi which was promoted on the basis of what turned out to be a wrong assumption that crop losses from pests and diseases were a critical constraint on smallholder crop production. Lessons learnt during the project were not implemented as this would have challenged the vested interests of the national research system and the donor agency (Orr & Ritchie, 2004:48). Andersson and Giller (2012:1) outlined how CA became a policy success sanctioned by religion, despite earlier agronomic research suggesting the value of other options, evidence of disadoption and contestation over the suitability of particular CA technologies in Zimbabwe. This was a follow up on earlier critical analysis of CA in SSA by Giller et al. (2009) which had questioned whether CA was the best approach or whether its suitability for smallholder farmers in SSA had been established given their diverse settings. Both publications show a huge policy drive to promote CA in southern Africa by an array of organizations including the FAO, DFID, the EU, NORAD; international research and development organizations (CIMMYT, CIRAD, ICRAF and ICRISAT), and numerous NGOs that came about due to increased attention to CA in international policy discourse in the late 1990s and early 2000s and a shift in donor support from government-linked agricultural research to the NGO sector (Andersson & Giller, 2012:2). This policy drive continued despite existence of scientific reports contradicting the claims of CA proponents. Why has the push for CA continued, and even intensified despite the persistent low adoption rates by its targeted beneficiaries, the smallholder farmers of SSA?

1.2 Political Agronomy

Sumberg and his colleagues (2012, 2013) are advocating for what they have termed 'political agronomy' when it comes to analysing agronomic issues. They have defined political agronomy as "the study of relationships and processes which link political, economic, and social forces and factors to the creation and use of agronomic knowledge and technology (2014:978). Political agronomy takes account of the contestation that can arise around the generation and promotion of new agronomic knowledge and technologies. This perspective is useful in analysing why some challenging ideas and innovations are successfully integrated into the agronomic research agenda while others are not (Sumberg et al., 2013: 77).

In their book *contested agronomy - agricultural research in a changing world* Sumberg and others discuss that new spaces of contestation have resulted from the arrival of new research actors and funders such as NGOs and other civil society organisations, and private agribusinesses. With the phenomenal development of the internet and other ICTs, websites have been set up through which information and experiences from a broad range of sources (peer and non-peer reviewed) are shared reflecting the interests of diverse communities, which include public relations units of research organisations, funders, development organisations and private firms promoting their scientific achievements and innovations. This development, they argue, means unsupported evidence and dubious conclusions can be widely propagated, and claims and counter claims about the impacts, outcomes and potentials of agronomic research technologies and practices may be partial and ill-informed. A political agronomy approach includes a concern for how legitimacy of research is determined, and how the presentation and interpretation of results supports or counters particular narratives and policy framings, or promotes particular political projects and agendas (2012: 9-10).

Political agronomy analysis can bring some important perspectives to the analysis of CA in southern Africa which has been the focus of much agronomic research and policy, and dissemination of technologies backed by a conglomeration of actors with diverse ideological backgrounds. This study uses political agronomy approach to analyse the implementation of CA in Zambia using a case study from a mining district. It differs from other studies of CA in Zambia, which are mostly apolitical, by explicitly focusing on the contestations surrounding the use of agronomic knowledge and technologies to promote an agenda that did not serve the interests of the targeted community and glossed over the contestations of stakeholders with alternative views.

1.3 Conservation Agriculture in Zambia

Conservation Agriculture has many definitions but there is a general consensus that its practice must encompass the simultaneous application of the three principles of (i) minimum or no mechanical soil disturbance (ii) permanent organic soil cover consisting of a growing crop or a dead mulch of crop residues, and (iii) diversified crop rotations or associations (Kassam et al., 2009). These principles are operationalized in many numerous ways by different actors which results in disparate technologies and practices being promoted under the umbrella term of CA. Thus conservation agriculture is not an actual technology but rather refers to a wide array of specific technologies and practices that are based on the simultaneous application of the three principles. A publication that was an outcome of a joint workshop organized by the Food and Agricultural Organization of the United Nations (FAO) and three agricultural organizations defined CA as a toolkit of agricultural practices that combines, in a locally adapted sequence, the simultaneous principles of reduced tillage or no-till; soil surface cover and crop rotations and/or associations, where farmers choose what is best for them (FAO, 2009:9). They elaborated that CA is essentially an approach that advocates the concept of sustainable intensification of production by picking the best possible options that farmers can apply at their own conditions.

CA as promoted in Zambia is packaged differently for the two distinct categories of target farmers; smallholder or commercial farmers. Smallholder farmers predominantly utilize manual or animal powered farming implements (hand held and ox-drawn); produce rain-fed crops for own subsistence and sale; and disproportionately use family labour in all farming operations. Traditional smallholder farming practices include complete soil inversion, shifting cultivation, burning of crop residues, and minimal use of external inputs such as hybrid seeds, herbicides and mineral fertilizers. Commercial farming is characterized by the use of motorized farm machinery, hybrid seeds, herbicides, mineral fertilizers, irrigation, hired labour, and modern management (Siegel & Alwang, 2005). Smallholder farming households cultivate between less than one and up to 20 hectares of land, and constitute 92 per cent of the estimated 1,305,783 farming households in the country while the rest (8 per cent) are commercial farmers who till from 20 to 1000 hectares of land (Central Statistical Office [CSO], 2003).

The CA toolkit for smallholder farmers in Zambia contains several key practices and technologies. The key practices are dry-season land preparation using minimum tillage systems, retention of crop residues in the field, input application (seeds, mineral fertilizers, manure, and lime)in fixed planting stations or along ripped furrows, early and continuous weeding, leguminous crop rotations, and agro-forestry (CFU, 2006; 2009a, 2009b). For hand hoe farmers, use is made of a specially designed hoe-called a *Chaka* hoe- which is made of a one meter wooden handle and a relatively heavier blade. The *Chaka* hoe is used to make permanent planting basins which are supposed to be accurately measured. The recommended dimensions are 30cm depth, 15cm width, and 20cm breadth, set out in a 90cm×70cm matrix (CFU, 2009a). For farmers using animal draught power, the oxen has attached to it a plough-like implement called a ripper. This is pulled behind the oxen to make furrows which are 15-20cm deep and spaced 90-100cm apart (CFU, 2009b).

Promotion of CA among smallholder farmers in Zambia has been on-going since the early 1990s. Development actors of various ideological persuasions have been involved and have differed in their agronomic practices preferred, in the incentives offered to adopters and pedagogically. Most organizations have engaged in CA promotion as part of broader programmes aimed at rural livelihood improvements and poverty reduction. These include the Conservation Society of Zambia and its COMACO (Community Markets for Conservation) Programme, DAPP (Development Aid from People to People), GART (Golden Valley Agricultural Research Trust), and the Ministry of Agriculture and Livestock.

In contrast the Conservation Farming Unit (CFU) was created in 1997 to exclusively promote CA in Zambia and has become the dominant player nationally and has made inroads regionally as well. CFU and most other actors in the promotion of CA in Zambia have restricted their activities to the low to medium rainfall regions. Concomitantly, CFU oversaw the development of the basin technology specifically tailored for low to medium rainfall areas, for capturing moisture and reducing the susceptibility of crops to intra-seasonal water stresses. Empirical evidence has shown that differences in yields from planting basins and traditional tillage methods are highest during seasons with below normal rainfall with planting basins showing a clear superiority (Langmead, 2004; Umar, 2011). Crops grown under CA are more drought tolerant (Grabowski and Kerr 2014). Conversely, during periods of above normal rainfall, the planting basins were waterlogged, which lead to reduced yields when remedial measures were not undertaken. Planting basins are thus unsuitable for high rainfall areas.

In 2013, The European Union (EU), FAO, and the Government of the Republic of Zambia (GRZ) launched a four year programme to increase crop production and productivity by 315 000 smallholder farmers by promoting CA in nine out of Zambia's ten provinces. During the programme launch, the FAO representative contended that conservation agriculture had become a preferred means of promoting agricultural development for the 40per cent of rural population in Zambia who depend on agriculture for their survival; provides better resilience against droughts and prevents soil erosion (www.fao.org).This programme covers Zambia's high rainfall areas where the use of basin technology is likely to result in water logging and associated yield losses. One of the high rainfall areas in which basin technology based CA is being promoted is Mufulira town in the Copperbelt province of

Zambia. In the rest of this paper I examine the promotion of CA in Mufulira using a political agronomy framework. Four research questions guided my study (i) What CA technologies/practices have been introduced and promoted? (ii) How was CA introduced (the process and the main actors) in Mufulira? (iii) What are the experiences of smallholder farmers and other actors concerning CA in Mufulira? The rest of the paper begins with a brief description of the study area, followed by the methods used for collecting and analyzing the data. This is followed by presentation of the results and discussion, and finally the conclusion.

2. Method

2.1 Description of Study Area

Mufulira town is located between latitudes 12° 30' South and 12° 40' South, and longitudes 28°10' South and 28°20' East (Figure 1). It experiences tropical savannah climate characterised by three typical seasons; hot and wet season from November to March; cool and dry season from April to July and; hot and dry season from August to October (Kapungwe 2013).

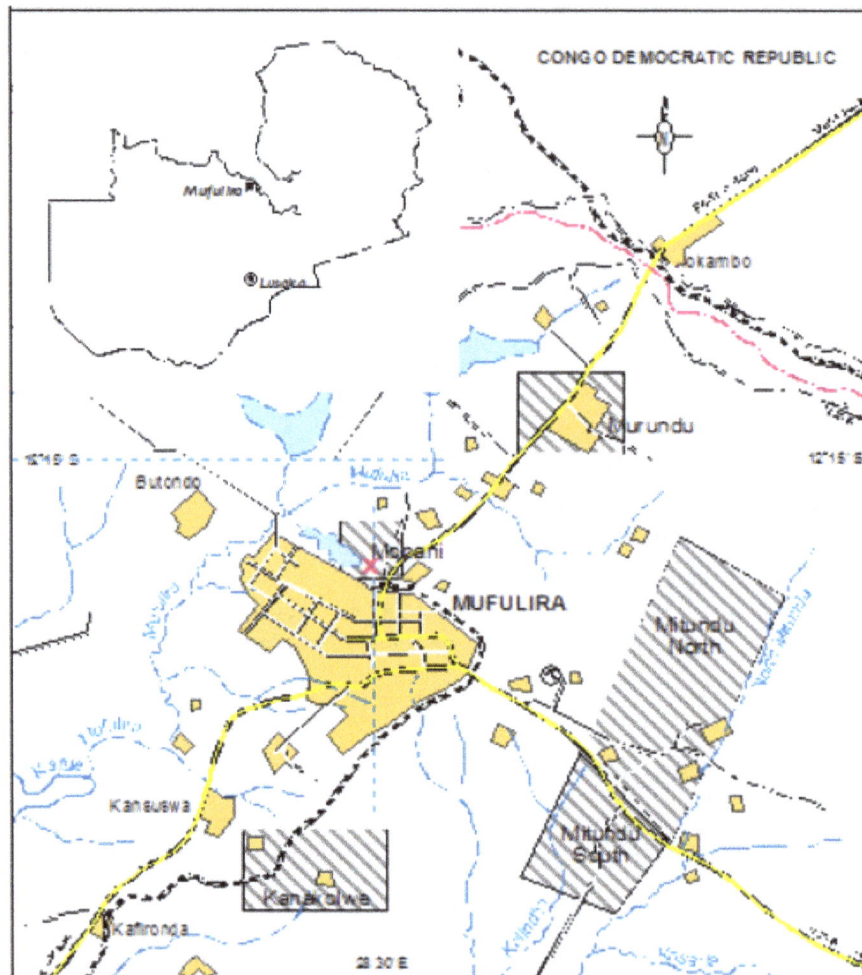

Figure 1. Map of MufuliraTown showing the study sites

The town is located in the high rainfall agro-ecological region which is part of the Central African Plateau. It received average annual rainfall of 1263.8mm during the period 1969 to 2013. The minimum and maximum temperatures during the same period were 12 and 29 °C respectively (Zambia Meteorological Department, 2015). The high rainfall has resulted in highly leached, and consequently highly acidic sandy soils (Mutamba, 2007).The district is characterized by a crop growing period of 190 days (Kapungwe, 2013).

Mufulira's economy is dominated by copper mining. The town owes its genesis and growth to mining activities which started in 1932 and today include mining, smelting and refinery of copper; as well as other mining related industries and services. Agricultural activities are common in both its urban and peri-urban environs and are predominantly smallholder crop production. Smallholder crop production is mostly rain fed and is focused on crops such as maize (*Zea mays*), cassava (*Manihot esculenta*), groundnuts (*Arachis hypogeae*), sweet potatoes (*Ipomoea spp*), pumpkins and other cucurbits (*Cucurbita spp*), and African eggplant *(Solanum macrocarpon)*. Some households own or lease agricultural plots near perennial streams and engage in irrigated production of vegetables during the dry season. Smallholder farmers use simple farming implements such as hand hoes and sickles. Livestock husbandry is minimal and mainly restricted to goats, pigs and rarely cattle. Collection and utilization of non-timber forest products is important among the peri-urban and rural households who appropriate firewood, mushrooms, tubers, fruits, bark rope, thatch grass, medicinal plants and bamboos. Farmers have relatively good linkages to stable urban markets due to the high population densities in the town and province at large, and proximity to other several highly urbanized mining towns.

The town shares borders with the Democratic Republic of Congo and is well connected by road to other mining towns in the province. The national census of 2010 projected the total population for the town to reach 161 601 by 2014 and population density at 98.7 persons per Km^2 (CSO, 2013). This is much higher than the national average population density of 17.3 and is due to the lure of the copper mine which has pulled people from other regions of the country into the district for over 80 years. There are no traditional structures (such as chiefs or head persons) and state structures control all facets of local governance. Much of the land in Mufulira is state land held under leasehold tenure (Note 1). Its residents are of mixed tribal groupings due to the historical migrations into the area but most of them speak *ichibemba.*

2.2 Data Collection and Analysis

Fieldwork for this study was conducted from June to August 2014. Semi structured interviews were conducted with 120 smallholder farmers from four communities, 30 from each community. The four communities were selected because CA had been promoted there since 2010 by the non-governmental organization, Heifer International Zambia under its 'Action on HIV and AIDS Nutrition and Food Security Project'. This project aimed at improving availability and access to sources of food and income for poor and vulnerable HIV and AIDS affected smallholder farmers (reference). This was to be achieved through the donation of exotic goats and training in CA to a selected sample of farming households in the study area. The households to be interviewed within the four communities were identified with the help of the local agricultural extension officers in conjunction with gatekeepers who had knowledge of the households that had participated in the Heifer International's CA project. A political analysis framework for conservation agriculture (Figure 2) was used for setting the questions to be asked to all the respondents and for analysis of responses. The framework was composed of a series of questions that drew attention to the framing of CA promotion, and possible sources of contestations among different actors in the agricultural sector. Respondents were asked the same series of questions and given considerable latitude in how they answered them, with follow up questions to clarify answers or get more information on issues that arose from their answers. The semi structured interviews were carried out by the author with the help of three research assistants in the local language.

Key informant interviews were conducted with agricultural and development experts and political leaders based in the town. The questions put to the experts included why and how CA was introduced into the town, who the main actors were, and the processes involved. Two focus groups discussions (FGDs) were conducted in two of the four study sites. The focus group discussants included both men and women farmers who had been trained in the use of CA technologies and those who had not received any training. The FGDs addressed issues related to planting basins, the use and availability of *Chaka* hoes, input use, weed management, and maize yields under CA. The responses from the semi structured interviews was typed into Microsoft Excel spreadsheet then copied to the qualitative data analysis software QDA Miner 3.2 (Provalis Research, 2010). Thematic analysis was used to systematically examine the answers to each question for themes and categories created to include the whole range of answers given to each question by all the respondents. Each response was then examined and placed in the relevant category. Frequencies and percentages were then calculated for each category.

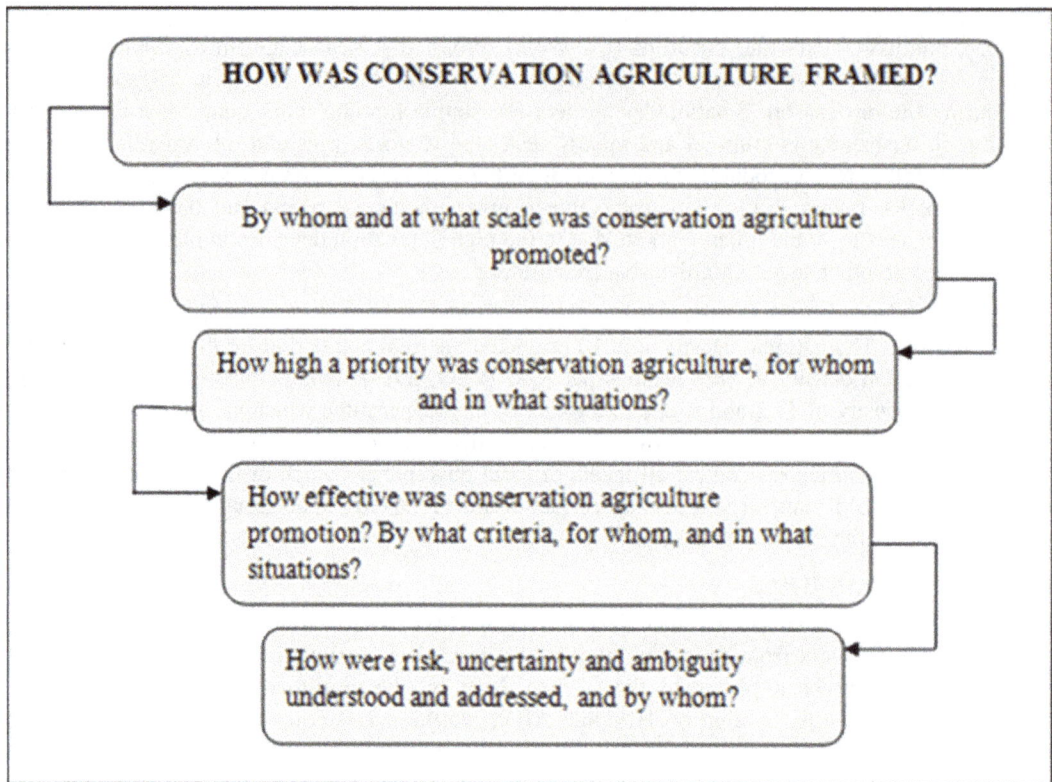

Figure 2. A political agronomy framework for conservation agriculture (*after Contested Agronomy Conference 2016*)

3. Results and Discussion

The mean household size for the sample was 7 with a mean of 3.5 members taking part in agricultural activities. The mean age of household heads was 54 years, and 70 per cent of the households in the sample were headed by men while the rest were women headed. The sampled households owned 6.2 hectares of land on average but only 41.4 per cent of this was under cultivation (mean cultivated size was 1.5 hectares).

3.1 Experiences with Goat Project

Livestock ownership was quite low with less than 1% of the sampled households owning cattle, 5 per cent owned pigs, 16 per cent owned goats and 71 per cent owned chickens. Several challenges were cited to explain the relatively low importance of livestock in the study area (Figure 2). Almost half of the respondents alluded to poultry diseases that have wreaked havoc on the free ranging chickens in the area. The poultry disease epidemics are common in the hot and dry season and have resulted in many of the sampled households losing all their chickens.

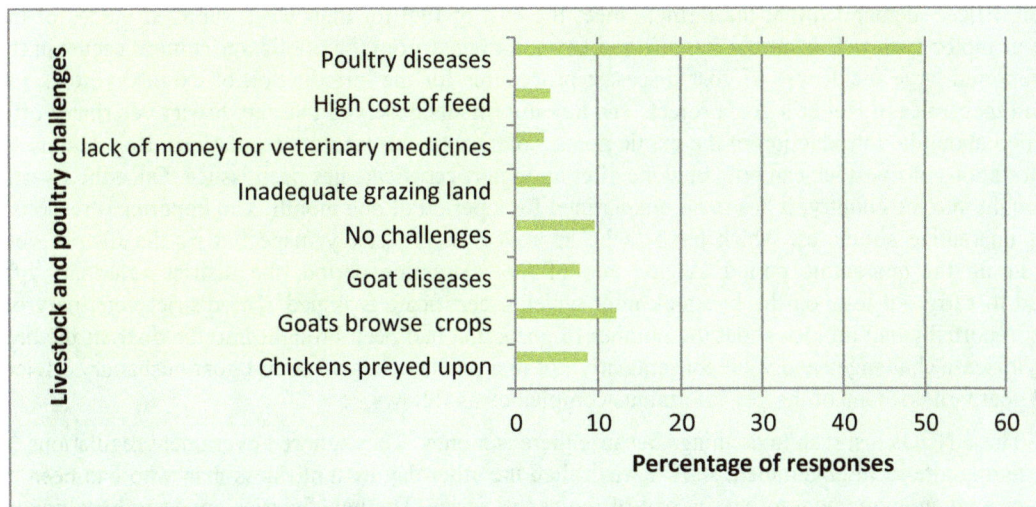

Figure 3. Livestock and poultry challenges faced by farmers in Mufulira, Zambia

For those that own goats, lack of grazing land which necessitates increased stall feeding has proved challenging. The lack of communal grazing areas means that once goats are let out of their shelters, they invariably end up browsing somebody's crops. Incidences of goat owners being fined were reported by both goat owners, and agricultural extension officers who are usually called upon to assess the economic value of the crop damage. The sampled households that owned goats had received them as gifts from Heifer International Zambia with a proviso to pass the gift on to others. Heifer International operates on the concept of 'passing on the gift'; a revolving fund in which one group of beneficiaries receives gifts of livestock from the NGO. The offspring of these animals are passed on as gifts to another group on the project. Time is given for the first generation to reproduce whose offspring are then passed on to the next group and so on. On average, passing on the gift lasts for at least nine generations, regardless of the project (Heifer International, 2015). The first beneficiaries reported that they have experienced many problems with their exotic goats as many of them were affected by several diseases, died and had to be replaced by the project. One respondent explained her experience as follows;

> I received five goats from Heifer's Pass-on project. All the five goats from the first batch died due to diseases. The goats were not used to the local climate. They came from abroad so they easily succumbed to disease here. The project restocked the goats and now I have six [46 year old female farmer, goat project beneficiary, Murundu Area, July 2014]

Another respondent expressed the following views

> These goats were sleeping on carpets where they came from. Now they are confused because the air has changed, that is why they are getting sick here.....they are Caucasians and thus susceptible to the diseases here [51 year old male, Local political leader and executive member of the goats projects committee].

The goat beneficiaries observed that they found it challenging to rear the goats due to the land tenure system in the area which does not have provision for communal grazing land. When they release their goats to range freely, they invariably end up in other people's fields and their owners have to compensate the crop owners. The alternative, which is full time stall feeding, was said to be too costly. Goat owners therefore have to always be on the lookout monitoring their goats' movements and ensure goat shelters are clean and the goats are given sufficient food. One respondent exclaimed the following in relation to this dilemma:

> I am not interested in Heifer's goat rearing project. It is too demanding. Their exotic goats need constant monitoring which effectively reduces one to being a guard. Cases of goats browsing on crops in peoples' fields are common. This has discouraged me from taking part in the pass-on project [54 year old male Murundu farmer, July 2014].

All the goat owners referred to the challenge of diseases and upon being probed on how they addressed this challenge, they claimed that they had difficulties getting help from the public veterinary officers as they were met with "excuses" that there was no transport to enable them conduct field visits. It was alleged that veterinary extension officers demanded that the farmers meet the cost of fuel for their work supplied modes of transport (vehicle or motor cycle). This author interviewed key informants from the public agricultural sector in the town on the reported large incidences of goat diseases; procedures for the introduction of exotic livestock into new areas, and their roles in Heifer's goat project. The key informants observed that the district veterinary office was not notified about the introduction of the exotic goats, contrary to current state regulations. Regulations stipulate that importations of livestock can only be done after an import certificate has been issued. Once the livestock has been brought into the country, it has to be quarantined for a period of one month. The importer is responsible for securing quarantine structures, which have to be approved and routinely inspected by the district veterinary officer during the quarantine period. At the end of the quarantine period, the district veterinary officer is mandated to carry out tests on the livestock after which a certificate is issued. The district veterinary office in Mufulira reportedly had no idea about the number of goats that had been brought into the district; the breed and its likely disease challenges and were consequently not in a position to give sound goat husbandry advice to the affected goat farmers. One of the key informants complained as follows;

> These NGOs just rush to do things because there is money. They ignore government regulations and ignore local government staff. I was called the other day by a business man who had been offered an exotic goat to buy by one of the beneficiaries. The beneficiaries are secretly selling off the goats and claiming they died. Yet before any livestock is sold and transported, a veterinary official has to be called in to certify it; same when there is death from disease. This is why livestock disease outbreaks are never ending in this country, because regulations are flouted [Interview with Key informant, 29th January, 2015].

3.2 CA Practices and Technologies –Local Experiences

Ridges were the most common tillage system used, followed by flat culture (Table 1). Ridges were of two variants; the traditional large version and smaller ones locally known as 'agriculture' because they were introduced by the state's Department of Agriculture in the recent past. The 'agriculture ridges' are smaller than the traditional ones and were being used by 10 % of the respondents.

Table 1. Tillage systems and reasons for their use in the study area

Tillage System used	Reasons for using system	Percentage of respondents using tillage system (n= 120)
Ridges	Easy to weed (32%) Give good yields (27%) Prevent crops getting waterlogged as high rainfall in area (6%) Easier to make ridges than flat culture (17%) Suitable for certain crops (12%) Retain nutrients as crop residues are buried (3%) Easy to plant in ridges (3%) Following local tradition (7%) Good for groundnuts (63%)	96.7
Flat culture	Easier to weed (5%) Yields higher than from ridges (23%) Suitable for their soils (3%) Facilitates herbicide use (2%) On trial basis (29%)	54.5
Planting basins	Prevents fertilizer run-off (6%) Keeps soil fertile (12%)	14

| Weeding retains soil near crop roots (6%) |
| Reduces labour needs (6%) |
| Higher crop density (6%) |
| What we have been taught to use (6%) |

The figures in parentheses represent the percentages among those who used the tillage system that expressed the view.

Flat culture which involves complete turning of soil using a hand held hoe was used by about 55 per cent of the sample and mostly for groundnuts; and on virgin land. The respondents reported that this is the only suitable method for cultivating groundnuts in the area and is used despite its higher labour demands compared to making ridges. Basins were used by 14 per cent of the sampled households. For the entire sample, 27 per cent of the households had tried and disadopted CA while 60 per cent had never tried it on their individual plots despite having been trained in the use of CA technologies and practices. Most of the respondents had tried CA by testing it on group or demonstration plots. It is common practice for farmers in the study area to belong to a farmers group and attend training sessions in groups. These are the groups that were targeted for trainings in CA. The groups then tried out the CA practices on a quarter hectare plot for a complete season after which members were expected to move to individual plots. After the group trials, almost 60% made up their minds not to try CA on individual plots while 27% tried it but discontinued. They explained that several bottlenecks hindered their fully fledged adoption of CA (Table 2). This seems to suggest that the CA promotion had not been very effective as the majority of the farmers trained in the use of CA technologies and practices did not adopt it at household level.

Table 2. Bottlenecks to adoption of conservation agriculture in study area

Bottlenecks to CA adoption	Percentage of responses
Basin digging too labour intensive	24.3
Difficulties in accessing CA equipment and inputs	23.4
Poor harvests	17.8
Basins unsuited to high rainfall	10.3
Weed burden/pressure too high	9.3
Insufficient knowledge on CA	5.6
Drudgery of precise measurements in CA	4.7
Increased termite attacks	2.8
Chaka hoe is too heavy for women	1.9

The making of planting basins was reported as being very hard work. The need for precisely measuring the dimensions of basins using ropes, and bottle tops; and application of accurately measured out lime, mineral fertilizer, manure and seed added to the drudgery. CA promoters had encouraged the farmers to spread out the work of making basins by commencing in the dry season rather than the conventional practice of waiting for the commencement of seasonal rains before starting land preparation. The farmers noted that this practice was unsuited in a high rainfall area like theirs because after the onset of the rains, the already made basins were overrun by weeds by the time they were ready to sow their crops. They further explained that with the high rainfall, the basins were easily waterlogged resulting in reduced yields and poor harvests. Thus although the CA promoter had framed CA as cardinal for labour constrained farming households, this view was not shared by the respondents.

Most complained about the stunted maize that gave poor harvests. Thus unlike in the low rainfall regions of Zambia, in the study area, CA is a system associated with lower yields. One farmer put this view across as follows:

> I stopped conservation agriculture due to excessive labour demands and bad harvests. I could start if weed killer will be ready, given to me for free as its expensive to buy. No conservation agriculture without weed killer [40 year old female farmer, Murundu July 2014].

Another farmer expressed the following views on why he disadopted CA

> I disadopted conservation agriculture due to poor harvests after heavy rainfall. I have no intentions of re-starting conservation agriculture because the rainfall is just too much for the practice of conservation agriculture [70 year old male farmer, Murundu July 2014]

Hiring labour for making basins was more expensive for a given area compared to ridges, and was reportedly less profitable given that "maize yield levels from the waterlogged basins were very low". Key informants from the district agricultural office were interviewed about why the practice of making basins was being promoted in the town, when it seemed that it was agronomically unsuited. This author was informed that the district office encouraged other CA practices such as crop rotations, improved efficiency in input application but not the making of basins. Actors involved in the promotion of basins had not formally consulted the district agricultural department. Thus for the public sector experts, CA was not a high priority in the agricultural development agenda for the district.

The key informant from Heifer International explained that the organization had promoted CA and basins in particular in the study area based on recommendations from a consultant that basins would ease the work load of HIV and AIDS affected households. The organization trained 137 farmers and provided them with seeds, *Chaka* hoes, and teren ropes for demonstration purposes. The seeds provided were for the leguminous crops sun hemp (*Crotalaria juncea*), velvet beans (*Mucuna pruriens*) and pigeon peas (*Cajanus cajan*). The organization reportedly found it difficult to implement other CA technologies and subsequently evaluate which were the most appropriate for Mufulira due to the short duration of the project. When asked about the organization's experience with basins, the key informant said "the basins only did well in 2010/2011 season when we had moderate rains. In 2011/2012 season they did not do very well and farmers were advised to cover them a bit". In this way, attempts were made to address the hazard of water logged basins by making small ridges around the standing crops, which is what is conventional done in the area.

The farmers alluded to the unavailability of CA implements and inputs. Several explained that they could not continue with CA because they did not have all the required implements and inputs. There was a perception that the practice of CA necessitated too many inputs which had to be available at the right time and in right quantities. The *Chaka* hoe was cited as especially difficult to access as it is not available for sale locally but only accessible through the project. The need to have all the inputs mandatory for CA practice in order to start is in sharp contrast to what farmers do with conventional farming practices. In addition, the farmers seemed reluctant to dedicate their own inputs to the practice of CA but seemed to prefer that somebody else, preferably CA promoters meet these costs. Sentiments such as " if I am provided with inputs, I will start conservation agriculture " or " they [CA promoters] did not bring the inputs as promised so we did not start conservation agriculture". They seemed unwilling to use own inputs in testing the new agricultural system. Thus for most, adoption of CA was premised on receiving free inputs as incentives. In a way, the risks associated with CA adoption were expected to be borne by the CA promoter. In response, the CA promoter supplied free agricultural inputs and implements needed for correct practice of CA to the targeted farmer groups. It was incumbent upon the CA promoter to sort out the ambiguities associated with CA, if farmers were to adopt it.

The provision of incentives to spur adoption or participation is now a pervasive feature of development interventions in Zambia. Project beneficiaries have become accustomed to being given material incentives for them to partake in development interventions. Project implementers work with local gatekeepers to identify and organize beneficiaries. For agricultural projects, the farmer cooperatives are used as they are already organized and registered entities. Farmer cooperatives have been formed to fulfil the Ministry of Agriculture pre-condition for accessing subsidized farming inputs from the state. Farmer cooperatives are selected or agree to take part in projects depending on their perceived usefulness to their members. When it is thought that benefits could be obtained for members that in some way improve their overall welfare, cooperatives take part in development interventions whose goals they may not necessarily agree with. In this study, I found that most cooperative members did not think basins could ever work well in the area but they went for trainings in the hope that they could access hybrid seed, lime and fertilizer which they could use to sow larger areas of their agricultural plots.

Focus group discussants agreed that they did not think CA and the goat project addressed their pressing challenges as farmers and did not provide the kind of assistance that they were most in need of. Results from the household semi-structured interviews show that access to mineral fertilizers, and hybrid seed was the most frequently reported form of assistance the respondents would like to receive (Figure 2). Over three-quarters want access to (more) subsidized hybrid seed and mineral fertilizer. Those already receiving seed and fertilizer subsidies called for larger quantities and to be delivered on time. In the recent past the subsidized inputs have been delivered late. This resulted in late sowing and concomitantly reduced yields. Subsidy recipients also complained of receiving expired hybrid seeds which failed to geminate, and receiving top dressing fertilizer (which is to be applied when the maize stalks are at knee height) at time of planting and only receiving the basal fertilizer (which requires pre-emergence application) when it is too late to use it as it is delivered when the maize has already germinated.

Farmers also requested for assistance with tractors and other mechanized machinery to help lighten labour demands, especially during land preparation and weeding.

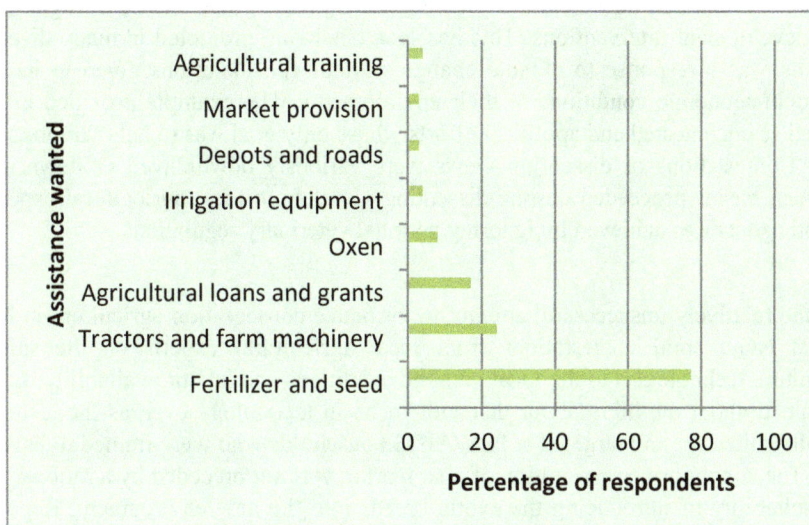

Figure 4. Assistance wanted by farmers in study area

The major bottlenecks to enhancing agricultural productivity seems to be labour shortages during land preparation and weeding given the short period in which these activities have to be performed. The use of oxen, tractors and herbicides significantly reduces manual labour demands which are met by household members and occasionally hired labourers. Most farmers are engaged in irrigated vegetable farming and would like to expand their operations. For this they require pumps and other equipment. Farmers in two of the four study sites are renowned for their vegetable gardens and supply the bulk of the vegetables sold in all the markets in the town and beyond. Their proximity to highly urbanized areas with high population densities, good road connections and existence of several perennial streams has provided them with a comparative advantage not available to their more rural based counterparts.

Mufulira town is a high rainfall area and receives some of the highest annual amounts of rainfall nationally (Zambia Meteorological Department 2015). About 11 per cent of the interviewed farmers observed that their area received too much rainfall, which caused their fields to become waterlogged. Five per cent noted that the rainfall distribution has been variable in the last several years with late onset and early end to season. The rest of the respondents (74 per cent) had not experienced any rainfall challenges and thought the area received more than adequate rainfall for proper crop development. This further suggests the inappropriateness of basins which have been developed essentially as a water harvest technology for low rainfall areas. Development interventions that aim at addressing such farmer identified challenges would probably spur more spontaneous adoption.

By working with already established groups, project implementers failed to reach non-group members. The Action on HIV and AIDS Nutrition and Income Project is said to have targeted HIV/AIDS affected vulnerable households yet in its selection of beneficiaries to be trained in the use of CA technologies, the criterion used was membership to a farmer cooperative. It is difficult to see how targeting farmer cooperatives would result in helping labour constrained HIV/AIDS affected households improve their nutrition and incomes, as advanced by the project rhetoric. In addition, promoting basins among labour constrained households *as a labour saving technology* seems inimical to project goals given that basin making demands high labour inputs. During interviews with women farmers and FGDs, sentiments were expressed on how women found it especially challenging to use *Chaka* hoes, and how basin making exacerbated their labour constraints. Recent research has shown that basin making does not reduce labour inputs compared to conventional manual based tillage systems (Andersson 2011; Umar 2012; Rusinamhodzi 2015). This seems to suggest that the privileging of basins over other tillage systems as a labour saving technology did not result from empirically based evaluations.

This leads to the question of why the promotion of CA in this project was privileged over other agricultural systems. Over the past decade CA has arguably become a hegemonic paradigm in scientific and policy thinking about sustainable agricultural development (Andersson & D'souza, 2014:116). In the last few years, it evolved from being framed only as a sustainable agricultural intensification option to climate smart agriculture (CSA). As CSA, it is advocated as an important option for smallholder farmers in SSA in responding to climate variability and climate change. With the increased visibility of climate change in discourses or 'climate washing' among international development agencies, many NGOs have jumped onto the bandwagon and added CA to their repertoire of development interventions. This has seen CA being promoted in many diverse environments to smallholder farmers as a response to climate change without considerations given to its suitability to the bio-physical and socio-economic conditions of their environments. The example provided in this study shows interventions framed as uncontested and apolitical efforts whose only goal was to help various selected groups of vulnerable people. Contestations or dissenting views were variously downplayed or downright ignored. The decision to implement basins proceeded despite dissenting views from local agricultural experts and while the introduction of exotic goats was achieved by ignoring national veterinary regulations.

4. Conclusion

This study shows the relatively unsuccessful attempt to introduce conservation agriculture in Mufulira, Zambia by an international NGO amid contestations from local agricultural experts on the suitability of some conservation agriculture technologies to the local climatic conditions and labour availability. Labour constrained households complained about the high labour demands of basin technology even as the technology promoters hailed it as especially suited for and targeted at HIV/AIDS households who were framed to have the most severe labour constraints. The supply driven promotion of goat rearing was not preceded by a rational technical process of evaluating the suitability of introducing the exotic breeds into the new environment. Reports of introduced goats succumbing to diseases were common among the goat project beneficiaries while the local veterinary officers were handicapped to provide services as they had not been made privy to the goat project as required by national veterinary regulations. The short duration of the conservation agriculture and goat project did not provide sufficient time for learning and incorporation of lessons learnt so as to better tailor the interventions to local context and aspirations of project beneficiaries.

The effect of this experience on the community is increased negative perceptions towards development interventions which could limit their participation in such endeavours in future. The use of the political agronomy framework has enabled the consideration of contestations among the various development actors involved in agricultural development initiatives in the district and shown the important role that vested interests play in privileging one set of factors other others. It is important to pay attention to such contestations as they influence the success of proposed agricultural initiatives at present and in the future, and their explicit consideration could provide for space for alternative interventions and higher chances of their success.

Acknowledgments

This publication is part of the author's research work at Umeå University, thanks to a Swedish Institute Post-Doctoral Guest Researcher Scholarship.

References

Andersson, J. A., & D'Souza, S. (2014). From adoption claims to understanding farmers and contexts: A literature review of Conservation Agriculture (CA) adoption among smallholder farmers in southern Africa. *Agriculture, Ecosystems & Environment, 187*(0), 116-132. http://dx.doi.org/10.1016/j.agee.2013.08.008

Andersson, J. A., & Giller, K. E. (2012). On heretics and God's blanket salesmen: Contested claims for conservation agriculture and the politics of its promotion in African smallholder farming. *Contested Agronomy: Agricultural Research in a Changing World.* J. Sumberg, Thompson, J. London, Earthscan.

Andersson, J. A., Giller, K. E., Mafongoya, P., & Mapfumo, P. (2011). Diga-Udye or Diga-ufe? (Dig-and-eat or Dig and die): Is Conservation Agriculture Contributing to Agricultural Involution in Zimbabwe? Regional

Conservation Agriculture Symposium for Southern Africa. Johannesburg South Africa, FAO Regional Emergency Office for Southern Africa.

CFU. (2006). *Reversing Food Insecurity and Environmental Degradation in Zambia through Conservation Agriculture*. Lusaka, Conservation Farming Unit.

CFU. (2009a). *Conservation Farming and Conservation Agriculture Handbook for Hoe Farmers in Agro-Ecological Regions I and IIa-Flat Culture*. 2009 Edition. Lusaka, Conservation Farming Unit.

CFU. (2009b). *Conservation Farming and Conservation Agriculture Handbook for Ox Farmers in Agro-Ecological Regions I and IIa*. 2009 Edition. Lusaka, Conservation Farming Unit.

CSO. (2003). Zambia *2000 Census of Population and Housing. Agriculture Analytical Report*. Lusaka, Central Statistical Office.

CSO. (2013). *2010 Censu of Population and Housing. Population and Demographic Projections 2011-2035*. Lusaka, Central Statistical Office.

Entman, R. M. (1993). Framing: Toward Clarification of a Fractured Paradigm. *Journal of Communication, 43*(4), 51-58. http://dx.doi.org/10.1111/j.1460-2466.1993.tb01304

Giller, K. E., Rowe, E. C., de Ridder, N., & van Keulen, H. (2006). Resource use dynamics and interactions in the tropics: Scaling up in space and time. *Agricultural Systems, 88*(1), 8-27. http://dx.doi.org/10.1016/j.agsy.2005.06.016

Grabowski, P. P., & Kerr, J. M. (2014). Resource constraints and partial adoption of conservation agriculture by hand-hoe farmers in Mozambique. *International Journal of Agricultural Sustainability, 12*(1), 37-53. http://dx.doi.org/10.1080/14735903.2013.782703

Heifer International. (2015). The Heifer Way. Big change starts small. Retrieved February 26, 2015, from http://www.heifer.org/ending-hunger/the-heifer-way/index.html

Kapungwe, E. M. (2013). Heavy metal contaminated water, soils and crops in peri-urban wastewater irrigation farming in Mufulira and Kafue towns in Zambia. *Journal of Geography and Geology, 5*(2), 55-72. http://dx.doi.org/10.5539/jgg.v5n2p55

Kassam, A., Friedrich, T., Shaxson, F., & Pretty, J. (2009). The spread of Conservation Agriculture: justification, sustainability and uptake. *International Journal of Agricultural Sustainability, 7*(4), 292-320. http://dx.doi.org/doi:10.3763/ijas.2009.0477

Langmead, P. (2004). *Hoe Conservation Farming of Maize in Zambia*. Lusaka.

Mutamba, M. (2007). *Farming or Foraging? Aspects of Rural Livelihoods in Mufulira and Kabompo Districts of Zambia*. Bogor, Center for International Forestry Research.

Orr, A. (2003). Integrated Pest Management for Resource-Poor African Farmers: Is the Emperor Naked? *World Development, 31*(5), 831-845. http://dx.doi.org/10.1016/S0305-750X(03)00015-9

Orr, A., & Ritchie, J. M. (2004). Learning from failure: smallholder farming systems and IPM in Malawi. *Agricultural Systems, 79*(1), 31-54. http://dx.doi.org/10.1016/S0308-521X(03)00044-1

Provalis Research. (2009). QDA Miner Version 3.2. Montreal, Provalis Research.

Rusinamhodzi, L. (2015). Tinkering on the periphery: Labour burden not crop productivity increased under no-till planting basins on smallholder farms in Murehwa district, Zimbabwe. *Field Crops Research, 170*(0), 66-75. http://dx.doi.org/10.1016/j.fcr.2014.10.006

Siegal, P. B., & Alwang. J. (2005). *Poverty reducing potential of smallholder agriculture in Zambia: Opportunities and Constraints*. Africa Region Working paper Series No. 85. Washington, D.C, The World Bank.

Sumberg, J., Thompson, J., & Woodhouse, P. (2012). *Contested agronomy: Agricultural research in a changing world. Contested agronomy: Agricultural research in a changing world*. J. Sumberg, Thompson, J. London, Earthscan.

Sumberg, J., Thompson, J., & Woodhouse, P. (2013). Why agronomy in the developing world has become contentious. *Agriculture and Human Values, 30*(1), 71-83. http://dx.doi.org/10.1007/s10460-012-9376-8

Tisenkopfs, T., Kunda, I., & Sumane, S. (2014). Learning as Issue Framing in Agricultural Innovation Networks. *The Journal of Agricultural Education and Extension, 20*(3), 309-326.

http://dx.doi.org/10.1080/1389224X.2014.887759

Umar, B. B., Aune, J. B., Johnsen, F. H., & Lungu, I. O. (2012). Are Smallholder Zambian Farmers Economists? A Dual-Analysis of Farmers' Expenditure in Conservation and Conventional Agriculture Systems. *Journal of Sustainable Agriculture, 36*(8), 908-929. http://dx.doi.org/10.1080/10440046.2012.661700

Umar, B. B., Aune, J. B., Johnsen, F. H., & Lungu. O. I. (2011). Options for Improving Smallholder Conservation Agriculture in Zambia. *Journal of Agricultural Sciences, 3*(3), 50-62. http://dx.doi.org/10.5539/jas.v3n3p50

Zambia Meteorological Department. (2015). *Monthly Rainfall Records.* Lusaka, Zambia Meteorological Department.

Notes

Note 1. Zambia has a dual land tenure system; customary and statutory. Customary land is controlled by the chiefs and their village heads but act with the consent of their people and govern according to local customary practices. Statutory land is held under leasehold tenure.

Permissions

List of Contributors

Angula Nahas Enkono
Department of Geography, History and Environmental Studies, University of Namibia, Namibia

Alfons W Mosimane
Multidisciplinary Research Centre, University of Namibia, Namibia

Babacar Faye
West African Science Service Center on Climate Change and Adapted Land Use (WASCAL), School of Agriculture, University of Cape Coast, Ghana
Institut Sénégalais de Recherches Agricoles (ISRA), Bambey BP: 211, Senegal

Heidi Webber and Thomas Gaiser
Crop Science Group, Institute of Crop Science and Resource Conservation (INRES), University of Bonn, Katzenburgweg 5, D-53115 Bonn, Germany

Mbaye Diop
Institut Sénégalais de Recherches Agricoles (ISRA), Bambey BP: 211, Senegal

Joshua D. Owusu-Sekyere
Department of Agricultural Engineering, School of Agriculture, University of Cape Coast, Cape Coast, Ghana

Jesse B. Naab
West African Science Service Center on Climate Change and Adapted Land Use (WASCAL) Competence Center, Ouagadougou, Burkina Faso
Doris Fuchs and Berenike Feldhoff
University of Münster, Germany

Mohammadehsan Zarringol and Mohammadreza Zarringol
Geotechnical Engineering Department, University of Guilan, Rasht, Iran

Balgah Sounders Nguh
University of Buea, Buea, Cameroon

Jude Ndzifon Kimengsi
Catholic University of Cameroon (CATUC), Bamenda, Cameroon

Rada Khoy
Graduate School of Bioresource and Bioenvironmental Sciences, Kyushu University, Fukuoka, Japan

Teruaki Nanseki and Yosuke Chomei
Faculty of Agriculture, Kyushu University, Fukuoka, Japan

Ingy M. El Barmelgy
Architecture Department, Faculty of Engineering, Cairo University, Egypt

Motaz S. Aly
General Organizations for Physical Planning, Ministry of Housing, Cairo, Egypt

Nurliza and Eva Dolorosa
Agribusiness Department, Faculty of Agriculture, University of Tanjungpura, Pontianak, Indonesia

Briana N. Dobbs
Graduate Student, College of Design, Construction, and Planning, University of Florida, USA

Michael I. Volk
Assistant Research Professor, School of Landscape Architecture and Planning, College of Design, Construction, and Planning, University of Florida, USA

Nawari O. Nawari
Associate Professor, School of Architecture, College of Design, Construction, and Planning, University of Florida, USA

Rowan Kushinga Machaka and Lakshmanan Ganesh
Centre for Research, Christ University, Bangalore, India

James Mapfumo
Christ College, Harare, Zimbabwe

Maiko Ebisudani and Akihiro Tokai
Laboratory of Environmental Management, Division of Sustainable Energy and Environmental Engineering, Graduate School of Engineering, Osaka University, Osaka, Japan

Haoyu Yin
China Energy Storage Alliance, Beijing, China

Fei Mo
Beijing, China

Derek Wang
Business School, China University of Political Science and Law, Beijing, China

Jena Trolio, Molly Eckman and Khanjan Mehta
Humanitarian Engineering & Social Entrepreneurship
(HESE) Program, Pennsylvania State University,
University Park, PA, USA

Everson James Peters
University of the West Indies, Trinidad and Tobago

Carlo Carraro
Ca' Foscari University of Venice and FEEM, Venice,
Italy
Vice-Chair, IPCC WG III

Bridget Bwalya Umar
University of Zambia, Zambia

Index

www.ingramcontent.com/pod-product-compliance
Lightning Source LLC
Chambersburg PA
CBHW080657200326

41458CB00013B/4888